NMR SPECTROSCOPY
IN ORGANIC CHEMISTRY

PHYSICAL METHODS IN ORGANIC CHEMISTRY

B. I. Ionin and B. A. Ershov
  NMR Spectroscopy in Organic Chemistry, 1970

V. I. Minkin, O. A. Osipov, and Yu. A. Zhdanov
  Dipole Moments in Organic Chemistry, 1970

NMR SPECTROSCOPY
IN ORGANIC CHEMISTRY

B. I. Ionin and B. A. Ershov
Lensovet Institute of Technology
Leningrad, USSR

Translated from Russian by
C. Nigel Turton and Tatiana I. Turton

⨎ SPRINGER SCIENCE+BUSINESS MEDIA, LLC 1970

Library of Congress Catalog Card Number 78-80753

ISBN 978-1-4684-1787-6   ISBN 978-1-4684-1785-2 (eBook)
DOI 10.1007/978-1-4684-1785-2

Additional material to this book can be downloaded from http://extras.springer.com.

The original Russian text was first published by Khimiya Press in Leningrad in 1967. The present translation is published under an agreement with Mezhdunarodnaya Kniga, the Soviet book export agency.

Б. И. ИОНИН, Б. А. ЕРШОВ

ЯМР-СПЕКТРОСКОПИЯ В ОРГАНИЧЕСКОЙ ХИМИИ

YAMR–SPEKTROSKOPIYA V ORGANICHESKOI KHIMII

Preface

In recent years high-resolution nuclear magnetic resonance spectroscopy has found very wide application in organic chemistry in structural and physicochemical investigations and also in the study of the characteristics of organic compounds which are related to the distribution of the electron cloud in the molecules. The vigorous development of this method, which may really be regarded as an independent branch of science, is the result of extensive progress in NMR technology, the refinement of its theory, and the accumulation of large amounts of experimental material, which has been correlated by empirical laws and principles. The literature directly concerned with the NMR method and its application has now grown to such an extent that a complete review of it is practically impossible. Therefore the authors have limited themselves to an examination of only the most important, fundamental, and general investigations.

The book consists of six chapters. In the first chapter we have attempted to present the fundamentals of the NMR method in such a way that the reader with little knowledge of the subject will be able to use the method in practical work for investigating simple compounds and solving simple problems. The three subsequent chapters give a deeper analysis of the method, while the last two chapters and the appendix illustrate the various applications of NMR spectroscopy in organic chemistry. Thus, Chapters V and VI are more in the nature of reviews and include the material of many investigators working in various fields. In this connection we would like to thank all the scientists who have kindly offered us their work on nuclear resonance.

The authors are very grateful to Professor A. A. Petrov, who directed this work, and to Professor T. I. Temnikova for interest and help in the preparation of this book. The authors are also grateful to V. B. Lebedev and A. I. Kol'tsov for valuable consultations on the theory and equipment of nuclear resonance and for help in the preparation of individual spectra.

The section on the double nuclear resonance method (Chapter IVB) was written at the request of the authors by É. T. Lippmaa (Cybernetics Institute of the Academy of Sciences of the Estonian SSR, Tallin). The translation was edited by Dr. Roy. H. Bible, Jr., G. D. Searle & Co., Chicago.

B. I. Ionin
B. A. Ershov

Contents

CHAPTER II. Chemical Shift

Chapter I

The Fundamentals of NMR Spectroscopy

Nuclear magnetic resonance (NMR) spectroscopy is one of the youngest physical methods of investigating organic compounds. The phenomenon of NMR was first observed experimentally in 1945 although it had been predicted theoretically considerably earlier [1]. The practical application of NMR spectroscopy to the study of the structure of complex organic compounds became possible only after the discovery in 1951 that the spectrum of ethyl alcohol consists of three individual signals corresponding to the resonance of the methyl, methylene, and hydroxyl protons [2] and that the signals of different groups of magnetic nuclei in liquid molecules give rise to further fine splitting which depends on the number and character of the nuclei* contained in the molecule [5]. The nuclear resonance of liquid substances or of solutions, which makes it possible to investigate the number, position, and intensity of the lines in a spectrum, has been called high-resolution NMR spectroscopy in contrast to the resonance of solids which is called broad-line NMR spectroscopy. At the present time high-resolution NMR spectra include largely those spectra in which the width of individual lines does not exceed a few hertz. This definition is undoubtedly not the final one and in the near future the requirements for high-resolution spectra will become still more rigid.

The widespread use of NMR spectroscopy of organic compounds dates from the middle of the fifties, when the industrial

---

*Proctor and Yu were the first to discover these characteristics of NMR spectra in 1950 [3, 4].

production of high-resolution spectrometers began. At the same time the foundations were laid for the application of the method in the investigation of organic compounds and these are reflected in the fundamental monograph of Pople, Schneider, and Bernstein (see §12). Since then the method has been developed very vigorously, covering an increasing range of organic compounds. The experimental and calculation methods and apparatus were rapidly refined. Although the tempo of this development increases continuously, now we can already see the completion of a definite stage, which is characterized by the development in principle of the theory of chemical shifts and spin—spin splitting, methods of analyzing spectra, and the accumulation of a large amount of experimental material on the correlation of these parameters with the structure of molecules. Essentially the method constitutes an independent field of physical organic chemistry. The further development of NMR spectroscopy of organic compounds will involve a more thorough application of the basic principles for the investigation of compounds of complex structure and also substances containing various magnetic isotopes (in addition to hydrogen and fluorine).

1. MAGNETIC PROPERTIES OF MATTER.
NUCLEAR MOMENTS. NUCLEAR RESONANCE

NMR spectroscopy is based on the absorption of the energy of radio-frequency radiation by a substance placed in a strong uniform magnetic field.

The interaction of a substance with a uniform magnetic field leads to a change in this field in accordance with the well-known equation of magnetostatics

$$H = H_0 (1 + \chi_v) \tag{I-1}$$

where H and H_0 are the strengths of the resulting and initial magnetic fields and χ_v is the bulk magnetic susceptibility of the substance.

The magnetic susceptibility (the bulk or molar χ_m if it relates to 1 mole of the substance) may be positive or negative, i.e., the substance may either increase or decrease the magnetic field showing either paramagnetic or diamagnetic properties. The first

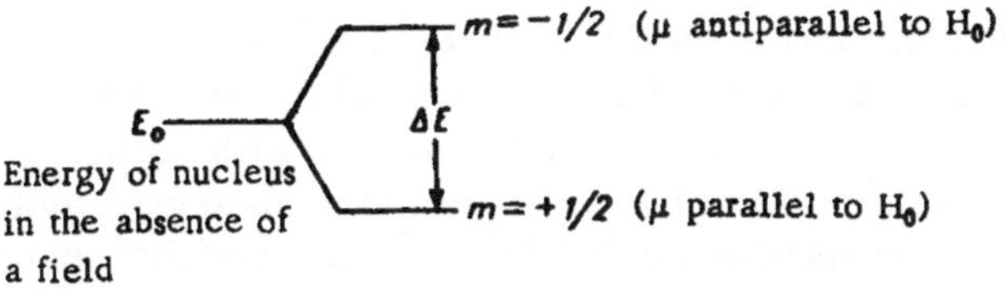

Fig. I-1. Splitting of the energy levels of a nucleus in
a magnetic field.

case is usually realized when molecules of the substance contain
unpaired electrons and is the special subject of the investigation
of electron paramagnetic resonance (EPR). The NMR method is
concerned almost exclusively with diamagnetic substances.

Diamagnetism is a macroscopic property of a substance
and is studied by classical magnetostatics. The effect of the diamagnetic susceptibility on the magnitude of the external magnetic
field is small (on the order of 10^{-5} of the magnitude of this field)
and though it must be taken into account in high-resolution experiments, this phenomenon is still of secondary importance in
NMR spectroscopy.

The macroscopic magnetic properties of a substance are
made up from the magnetic properties of its elementary component particles (nuclei and electrons). The nuclei of many isotopes
have a positive magnetic susceptibility. Nuclear paramagnetism
is considerably smaller than electronic paramagnetism. At normal temperatures the contribution to the bulk magnetic susceptibility of a substance due to nuclear paramagnetism is 10^{-10} or less
of the static magnetic susceptibility so that essentially it has no
effect on the macroscopic magnetic properties of the substance as
a whole. Therefore magnetostatic methods are unsuitable for
studying it.

Nuclear paramagnetism arises because the nuclei of some
isotopes have a magnetic moment. Quantum mechanics relates
the magnetic moments of nuclei to their angular momentum. The
value I, which is called the s p i n q u a n t u m n u m b e r , characterizes the maximum measured value of the angular momentum
of the nucleus and, in accordance with the principle of discreteness,
it may have the values 0, $\frac{1}{2}$, 1, $1\frac{1}{2}$, 2, ..., etc., and a nucleus with
a spin I may be in $2I + 1$ states. When I $= 0$ the nucleus has no magnetic moment and consequently no paramagnetism. Such nuclei

(this includes the isotopes C^{12}, O^{16}, Si^{28}, S^{32}, etc.) do not appear directly in NMR spectra. Of greatest value for NMR spectroscopy of organic compounds are magnetic nuclei with a spin of $\frac{1}{2}$ and these are protons, which have the highest magnetic moment of the stable nuclei, and also the nuclei F^{19}, P^{31}, C^{13}, and Si^{29}, each of which may be in $(2 \cdot \frac{1}{2} + 1) = 2$ states. In the absence of a magnetic field (a case which essentially is purely speculative) both possible states are equivalent in energy (in quantum mechanical terminology, a doubly degenerate state); when an external magnetic field is present the degeneracy is eliminated and this effect is greater, the higher the strength of the applied field. In other words, when a substance containing magnetic nuclei with a spin of $\frac{1}{2}$ is introduced into a uniform magnetic field of sufficiently high strength the two states become energetically nonequivalent (Fig. I-1). The value ΔE, which characterizes the difference in the energy of the two states, is proportional to the strength of the external field H_0 and depends on the magnetic properties of the nucleus examined.

$$\Delta E = 2\mu H_0 = \gamma \frac{h}{2\pi} H_0 \qquad \text{(I-2)}$$

The latter are determined by the magnetic moment of the nucleus μ or a more convenient value, the gyromagnetic ratio of the nucleus γ, which includes both the magnetic moment and the spin of the nucleus (h is Planck's constant). Each of the states is associated with a definite value of the magnetic quantum number m, which in the case of nuclei with a spin of $\frac{1}{2}$ has the values $+\frac{1}{2}$ (μ parallel to H_0) and $-\frac{1}{2}$ (μ antiparallel to H_0).

With a transition of a magnetic nucleus from a lower to an upper energy level there is the absorption or with the reverse transition, the liberation of the energy $\Delta E = h\nu_0$, which equals the difference in the energies of the two states

$$\Delta E = h\nu_0 = \gamma \frac{h}{2\pi} H \qquad \text{(I-3)}$$

Hence

$$\nu_0 = \frac{\gamma}{2\pi} H = \frac{\gamma}{2\pi} H_0 (1 + \chi_0) \qquad \text{(I-4)}$$

Within these transitions lies the basic meaning of the phenomenon of nuclear magnetic resonance and equation (I-4) is its basic equation.

The possibility of the experimental observation of nuclear resonance is connected with the fact that the number of magnetic nuclei occupying each of the levels is different. The system normally tends to go to the state with the lower energy ($m = +\frac{1}{2}$) and this leads to an increase in the population of the lower level. This tendency is limited by random thermal motion, which tends to equalize the populations of the levels. As a result an equilibrium is established in which the difference in the populations of the levels in accordance with Boltzmann's principle equals $2\mu H/kT$, where k is Boltzmann's constant and T is the absolute temperature. At normal temperatures and fields the difference in population does not exceed 10^{-5} of the total number of magnetic nuclei, i.e., it is very low. Nonetheless, if in addition to the uniform magnetic field we apply to a system of spins a high-frequency field with a frequency ν_0, which satisfies equation (I-4), there will be absorption of the energy of the rf field. An NMR experiment consists of recording this absorption by electronic means.

Nuclei with a spin greater than $\frac{1}{2}$ have in addition to the dipole magnetic moment an electrical quadrupole moment, which must be taken into account in investigating nuclear resonance. Important nuclei of this type are D^2 and N^{14}. The nuclei of many other isotopes which are of importance in organic chemistry, primarily halogens (apart from fluorine), also have magnetic and quadrupole electric moments, but their effect on the spectra is only insignificant so that they may be regarded as equivalent to nonmagnetic nuclei in practical work.

2. OBSERVATION OF NUCLEAR MAGNETIC RESONANCE

The information given above essentially determines the principles of construction of nuclear magnetic resonance spectrometers. An instrument consists of three main parts: a magnet which produces a uniform field of high strength H_0, a stable oscillator for creating a high-frequency alternating magnetic field with a frequency ν_0 [see equation (I-4)], and an electronic device with which it is possible to detect the absorption by the sample of

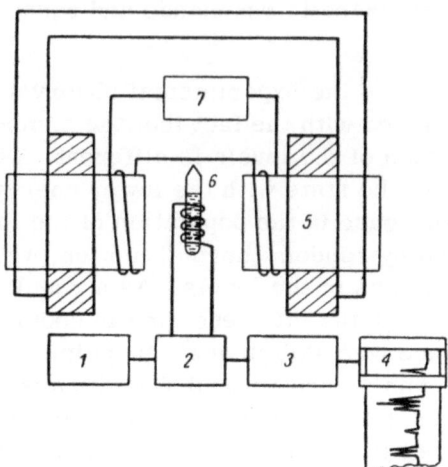

Fig. I-2. Schematic diagram of a nuclear magnetic resonance spectrometer: 1) high frequency oscillator; 2) bridge; 3) amplifier and detector; 4) recorder; 5) electromagnet; 6) sample investigated; 7) field sweep oscillator.

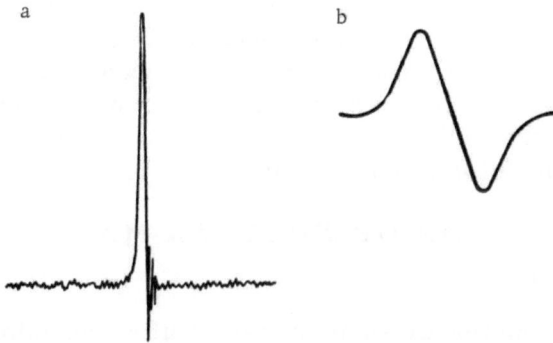

Fig. I-3. NMR spectrum of a substance containing only equivalent magnetic nuclei: a) absorption signal with an average sweep rate; b) dispersion signal with slow passage.

energy of rf radiation (Fig. I-2). Either the magnetic field of the
instrument or the frequency of the oscillator is swept to obtain a
nuclear magnetic resonance spectrum. In practice the sweep of
the field is carried out in such a way that with the oscillator at
constant frequency, the magnetic field is established with a
strength somewhat lower than that required for resonance;
then it begins to increase smoothly and at the same time the chart
of the recording potentiometer, which measures the potential at
the detector output, begins to move. At the moment of passage
through resonance, i.e., when the strength of the external field H_0
satisfies equation (I-4) there is a deviation of the recorder pen
from the zero position and then it again returns to the initial position so that the trace on the chart is in the form of a more or less
narrow peak of definite amplitude. The spectrum is recorded
analogously in the case of frequency sweep and in both cases it is
possible to sweep in either direction. Figure I-3a shows a typical
spectrum of a substance in which all the protons are equivalent.

The scale of the spectrum is usually marked in parts per
million (ppm) of the applied field or frequency or in frequency
units (Hz).

In the plotting of NMR spectra there arise important problems connected with the selection of the strength of the magnetic
field and the rate of passage through resonance and also other prob·
lems, for the solution of which it is necessary to have an idea of
the mechanism of the interaction of the substance with the field.
Below we will examine some conclusions from the theory of the
absorption of energy by nuclear magnetic resonance.

3. SPIN-LATTICE RELAXATION

When the conditions for nuclear resonance are fulfilled, some
of the nuclei absorb the energy of the rf radiation and pass from
the lower to the upper energy level. The equilibrium Boltzmann
distribution of the nuclear spins is then upset. The system must
in some way return to the equilibrium state as otherwise the populations of the levels are rapidly equalized with the result that no
further resonance absorption is observed, i.e., so-called saturation occurs.

The possibility of the spontaneous transition of a nuclear spin
to a lower level with the emission of a quantum is very low. If the

system were to return to equilibrium only through spontaneous transitions, then the time to establish equilibrium would be measured in millions of years. In practice, equilibrium is established as a result of resonance phenomena in the substance. The presence of electrical charges in the molecules and their rapid movement produces local fluctuations in the magnetic fields and among them there will always be fields whose fluctuation frequency corresponds to the nuclear resonance frequency in the given field. These fields initiate transitions between levels and thus promote the establishment of equilibrium in the spin system. The energy thus liberated is converted into t h e r m a l   e n e r g y   o f   t h e l a t t i c e .   The phenomenon described is called spin-lattice (or longitudinal) relaxation. The efficiency of this process is characterized by the time T_1 in which the system returns to the equilibrium Boltzmann distribution. Thus, spin-lattice relaxation provides the conditions for the continuous absorption of rf energy by the substance. Moreover, the value of T_1 is one of the factors which determines the form of the NMR signal.

In practice the spin–lattice relaxation time depends on many factors of which the most important are the viscosity of the substance and the presence of paramagnetic impurities. An increase in the viscosity, for example due to a decrease in temperature of the sample, and the introduction of paramagnetic additives leads to a decrease in the relaxation time. A characteristic example, which demonstrates the effect of various factors, is the relaxation time (in sec) of benzene protons at 25°C [6].

	T_1
Pure benzene	19.3
Benzene in the presence of air.	2.7
Solution of benzene in carbon disulfide (11%) .	60

The decrease in relaxation time in the presence of air is evidently connected with the paramagnetic properties of oxygen; the increase in the relaxation time in carbon disulfide solution indicates the importance of intermolecular interactions and may be connected with a decrease in the viscosity. The relaxation time of most organic liquids has a value on the order of a few seconds.

The presence of a quadrupole moment, for example, that of

the nucleus N^{14}, leads to a considerable decrease in the relaxation
time of nitrogen compounds. Since this effect depends on the gradient of the electrical field, in ammonia compounds, which have a
spherically symmetrical distribution of the electron cloud, the
effect of the quadrupole moment of nitrogen on the relaxation time
is small.

4. LINE WIDTH IN NMR SPECTRA

As the characteristic of the line width in NMR spectroscopy
it is customary to use the width at half the height from the base
line.

A quantum mechanical examination of nuclear resonance
leads to the conclusion that transitions occur to frequencies which
correspond exactly to nuclear resonance and also to frequencies
close to this. This leads to broadening of the lines of the spectrum, though the total area remains constant.

The line width observed experimentally $\Delta \nu$ consists of the
so-called natural width of the line $\Delta \nu_n$, which depends on the
structure and mobility of the molecules, and the broadening caused
by the apparatus $\Delta \nu_a$ (largely the nonuniformity of the external
magnetic field):

$$\Delta \nu = \Delta \nu_n + \Delta \nu_a \qquad \text{(I-5)}$$

The natural width of the line $\Delta \nu_n$ (in Hz) is determined by the
relaxation time T_2, which takes into account the effect of all the
relaxation mechanisms

$$\Delta \nu_n = \frac{1}{\pi T_2} \qquad \text{(I-6)}$$

In solids $T_2 \ll T_1$, and the spectra of solids consists of broad
lines, whose width reaches 10^4 Hz. Therefore NMR spectroscopy
of solids is unsuitable for the investigation of organic molecules.
In mobile liquids and gases $T_2 \approx T_1$ (on the order of a few seconds,
which corresponds to a natural line width of fractions of a hertz).

The broadening of the line connected with the nonuniformity
of the external magnetic field $\Delta \nu_a$ likewise does not usually exceed fractions of a hertz in modern spectrometers.

If the relaxation time T_2 is short (fractions of a second and less), then it determines the width of the line. This is observed in solutions of paramagnetic substances and also viscous liquids. Therefore high resolution may be achieved only in the investigation of mobile liquids (and gases) which do not contain paramagnetic substances.

If the relaxation times are quite long (on the order of several seconds), the quality of the magnetic fields produced by the spectrometer becomes decisive since the width of the spectral lines depends essentially only on the resolving power of the instrument.

The usual line width in proton resonance spectra is 0.3-0.5 Hz, but it may often be increased because of the overlapping of adjacent transitions, which do not exactly coincide, but still cannot be resolved into separate lines. This broadening may be used for correlation with the structure of the molecule. A source of line broadening which is connected with the properties of the substance investigated may be the presence of nuclei with a spin greater than $\frac{1}{2}$ (for example, nitrogen) and chemical exchange. Figure A-25, -34, and -38* show the degree of broadening of the line of protons attached to nitrogen of an amide group.

5. SATURATION

According to Bloch's theory† with slow passage through resonance the area of the absorption signal is determined by the relation

$$S \equiv \frac{\chi_0 H_1}{(1 + \gamma^2 H_1^2 T_1 T_2)^{1/2}} \tag{I-7}$$

and its amplitude by the relation

$$A \equiv \frac{\chi_0 H_1 T_2}{1 + \gamma^2 H_1^2 T_1 T_2} \tag{I-8}$$

where χ_0 is the magnetic susceptibility of the sample, H_1 is the

---

*The letter "A" in front of the number indicates that these spectra are in the Appendix.
†For a detailed account of Bloch's theory see the general literature and the monographs of Pople et al., Abragam, and Leshe.

strength of the rf field, γ the gyromagnetic ratio, and T_1 and T_2 are the spin-lattice and spin-spin relaxation times. With small values of H_1 the numerator in both formulas is close to unity and consequently the area and the amplitude of the signal are proportional to the strength of the rf field. With an increase in H_1 the intensity of the signal first increases, reaches a certain maximum, and then falls and with a considerable increase in H_1 it may become zero. This phenomenon is called s a t u r a t i o n .

Saturation usually appears even with a slight increase in the rf field as distortion of the form of the signal. Therefore, to obtain a high-quality spectrum it is necessary to work with the minimum potential at the output of the rf oscillator. It is particularly important to take into account saturation in quantitative measurements, based on the comparison of signal areas or amplitudes. Since saturation depends on the relaxation time (the latter is not essentially the same for different nuclei in the system), the areas of signals under conditions close to saturation may not be proportional to the number of nuclei. In individual cases to reduce saturation it is possible to resort to the artificial reduction of the relaxation time by adding paramagnetic impurities. However, then there is substantial broadening of the lines.

6. RATE OF SWEEP OF SPECTRUM

By slow passage we mean a rate of sweep of the field H_0 such that the time that the sample is under resonance conditions substantially exceeds the relaxation time, i.e., at each moment the system of spins is in an equilibrium state. To reveal the details of the spectrum one usually tries to plot the spectrum with the slowest possible sweep. In practice it is only possible to achieve slow-passage conditions for samples with a very short relaxation time, which are essentially unsuitable for high-resolution spectroscopy. For normal substances with a relaxation time of the order of seconds, slow passage is not feasible due to the instability of the magnetic field of the spectrometer with time so that a very slow sweep leads to deterioration of the resolution. The normal rate for sweeping the spectrum in modern spectrometers is 0.5-2 Hz/sec and in high quality instruments it may be reduced to 0.02 Hz/sec. If the relaxation time is 1 sec and the line width 0.5 Hz, then the sweep rate and the relaxation time will be of the same order. Therefore in normal spectra there appear details which are characteristic of

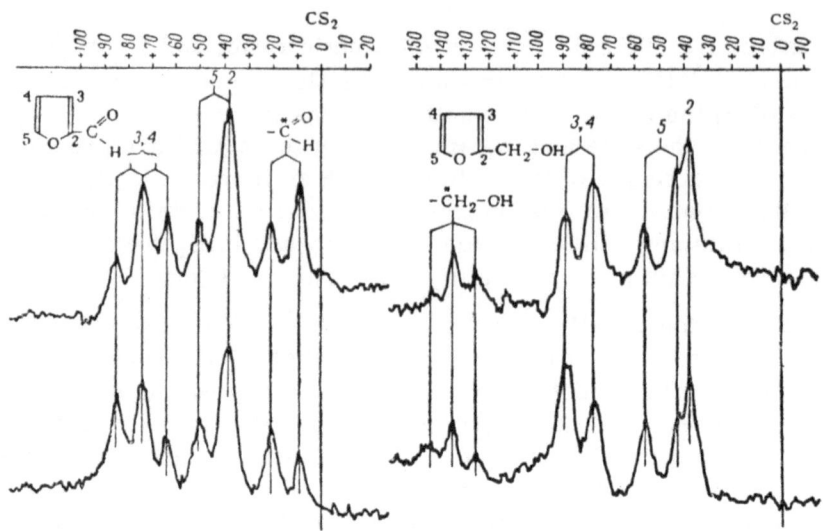

Fig. I-4. C^{13} NMR spectrum of furfural and furfuryl alcohol with the natural content of the isotope; the spectra were plotted at a frequency of 15 MHz with a rate of sweep of the magnetic field of 1.3 G/min (the sweep was carried out in two directions), rf field 15 mG.

fast passage. These are primarily the so-called wiggles, which are characteristic kicks below the base line after passing through the signal, which have the form of damped oscillations ("ringing") as in Fig. I-3a. With low uniformity of the magnetic field the ringing is reduced and therefore its presence usually indicates a high-quality spectrum.

In individual cases it is useful to use the other extreme of the conditions for plotting the spectrum, namely, adiabatically fast passage through resonance. With a very high rate of sweep of the magnetic field (or frequency) the sample is under resonance conditions for a short time, which is much less than the relaxation time. As a result the spin system is unable to reach equilibrium. This makes it possible to use an rf field H_1 of high strength since saturation does not occur under these conditions. Thus, fast passage makes it possible to increase the intensity of the signal considerably. Adiabatic fast passage was used to observe the C^{13} signal with the natural content of these nuclei in the sample. The apparatus was tuned so that it recorded the so-called d i s p e r s i o n   s i g n a l   and not the usual absorption signal. With slow-passage conditions the dispersion signal has the form given in Fig. I-3b, but with adiabatic passage it is distorted and assumes the form of a normal absorption signal. Figure I-4 gives a C^{13} NMR spectrum plotted with adiabatically fast passage [7]. If we consider that the amplitude of the C^{13} signal is 1.6% of the amplitude of protons in the same field, while the natural content of the isotope C^{13} is 1.1%, the value of this method becomes obvious.

7. MAIN PARAMETERS OF NMR SPECTRA

Of primary interest in NMR spectroscopy is the study of the effect of the electron shells of the molecule, mainly the valence electrons, on the nuclear resonance. This effect leads to the appearance of the most important NMR parameters, namely, the chemical shift and spin—spin coupling between nuclei. The main aim of plotting a spectrum is to determine these parameters so that by comparing them with available data and using various theoretical considerations they may be correlated both with the structure and with the distribution of electron density in the molecules. A detailed analysis of the chemical shift and spin—spin coupling is given in the next two chapters; below we examine only some fundamental principles and also practical procedures connected with the determination of these parameters.

Chemical Shift

When an isolated atom is placed in a magnetic field of strength H_0 the electrons precess about the direction of the field in accordance with the laws of electromagnetic induction. These currents induce a magnetic field H' in the opposite direction to H_0 and as a result there acts upon the nucleus an external field which is reduced by the value H' in comparison with that which would have acted in the absence of the electron shell. Therefore, to produce resonance of diamagnetic screened nuclei it is necessary to apply a somewhat greater external field and this increase will be different for nuclei with different electronic environments. The shift in the nuclear resonance signal as a result of the electronic environment is called a chemical shift. If the chemical shifts exceed the width of the lines, then the spectrum of the substance or mixture consists of separate bands. It is possible to obtain such spectra by using magnets with a high magnetic field strength, usually 10-25 kG, which are stable and uniform through the volume of the sample to 10^{-8}-10^{-9} of the nominal value.

The electron currents and the secondary field H' associated
with them and hence the absolute magnitude of the chemical shift
are proportional to the field H_0, i.e., the working field of the spectrometer. In order to compare the magnitudes of chemical shifts
obtained on different spectrometers it would be most convenient
to use the absolute value of them, determined as the ratio H'/H_0.
However, under normal conditions it is very difficult to determine
experimentally the a b s o l u t e chemical shift, i.e., the shift relative to an unscreened nucleus. This makes it necessary to use a
standard substance whose signal provides a reference point. The
use of standards is connected with the fact that the accuracy of ex-
isting methods of determining the absolute frequency of oscillations
and the strength of a magnetic field is very much less than the accuracy of determining the d i f f e r e n c e s of chemical shifts.
Therefore, in practice the chemical shift δ is measured in relative
units, namely, parts per million (ppm) of the applied field (or frequency) relative to the standard reference substance

$$\delta = \frac{H_{sam} - H_{st}}{H_0} \cdot 10^6 = (\sigma_{st} - \sigma_{sam}) \cdot 10^6 \qquad (I-9)$$

where σ, which equals H'/H_0, is the so-called s c r e e n i n g con-
s t a n t and the subscripts "st" and "sam" denote that the parameter refers to the standard substance and the sample investigated,
respectively. The references used and the scales of chemical
shifts connected with them are examined below.

In contrast to atoms or certain ions which have a spherical
symmetrical electron shell, in molecules the electron currents induce a secondary field which is not essentially in the opposite direction to the field H_0, but may have a complex configuration.
Moreover, the character of this field depends on the position of
the molecule relative to the poles of the magnet of the instrument.
As a result of the random motion of the molecules the nuclei are
subject to the effect of the secondary field averaged in all directions and in the general case this effect may be both screening, i.e.,
shifting the signal to a stronger magnetic field (with a constant
working frequency) and descreening. Therefore, theoretical calculations of chemical shifts in molecules encounter great difficulties. In practice it is usual to use empirical data on the chemical shifts of protons in characteristic structures (see the correla-

tion diagram on pp. 15-16). As the figure shows, the range of chemical shifts of protons is ~16-20 ppm. In the general case the magnitude of the chemical shift in proton magnetic resonance (PMR) spectra is determined both by the electron density at the nucleus (i.e., the electronegativity of the substituents) and the effect of secondary magnetic fields produced by the circulation of electrons in neighboring atoms and by interatomic currents (i.e., by the magnetic anisotropy of neighboring atoms and bonds).

The chemical shift may be affected substantially by external factors, primarily the nature of the solvent and the concentration of the solution, in addition to the structure of the molecule. The chemical shift of protons attached to heteroatoms (oxygen and nitrogen) is particularly sensitive to the effect of solvents. Strictly speaking, true chemical shifts are shifts in rarefied gases or at least in solutions in inert nonpolar solvents, extrapolated to infinite dilution. However, in practice PMR spectra are usually plotted in solutions at concentrations of 5-10% (by volume). The most convenient solvent used in PMR spectroscopy is carbon tetrachloride. Since it contains no protons, it can be used over the whole range of the spectrum. Other solvents used widely are chloroform and deuterochloroform, cyclohexane, and acetone and for water-soluble compounds, heavy water. Hydrocarbons and compounds which contain no polar groups may be investigated in a pure form since their chemical shifts depend little on the dilution.

The presence of magnetically anisotropic aromatic rings in a sample often leads to a deviation in the chemical shifts of protons beyond the limits of typical values. Therefore, aromatic compounds cannot be used as solvents in the determination of chemical shifts and aromatic hydrocarbons should be investigated in solutions.

The temperature of the sample investigated often may also have a substantial effect on the chemical shift. Unfortunately the stabilization of the temperature of the sample is quite a complex problem technically and many spectrometers are not fitted with appropriate equipment. Most measurements are carried out at 30-35°C due to the great evolution of heat by the electromagnets. A change in the temperature may also have an indirect effect on the form of the spectrum, for example, leading to a change in the ratio between conformations in molecules with hindered internal rotation.

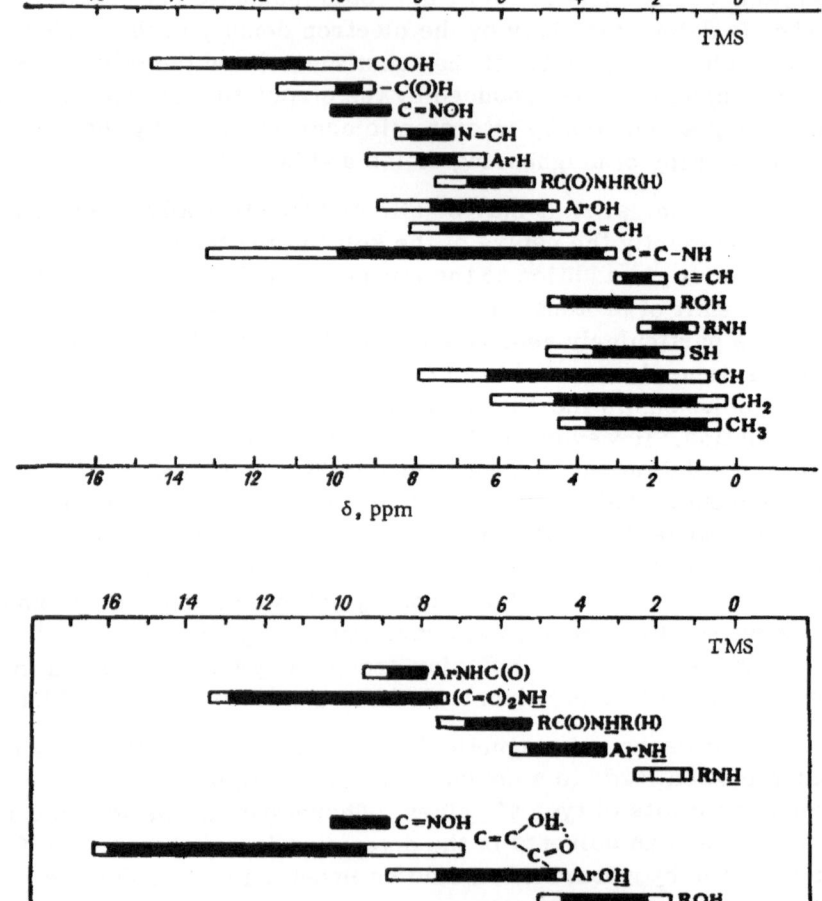

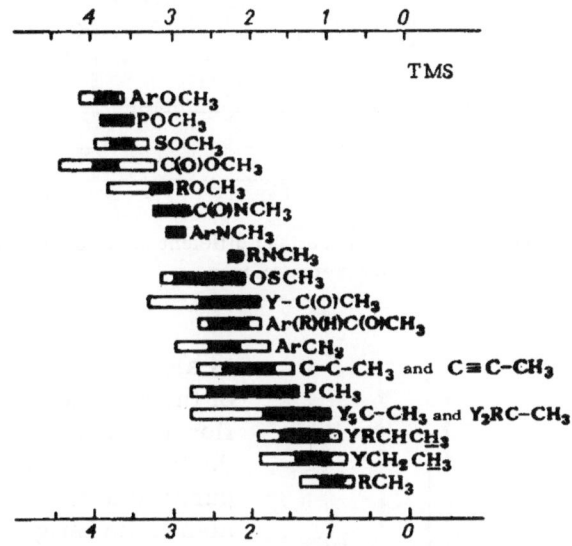

Correlation diagram of chemical shifts in organic
compounds [8]

$$R = -CH_3; \ -CH_2; \ -\underset{\diagdown C}{\overset{\diagup C}{CH}}; \ -\underset{\diagdown C}{\overset{\diagup C}{C}}-C$$

Y) usual structures of organic groups apart from R.

Scale of Chemical Shifts of Protons.

References

The scale of the chemical shifts depends essentially on the
choice of the reference substance and the method of using it. There
are two possible methods of introducing the reference substance
into the sample investigated, namely, direct mixing (internal reference) and placing the substance and the reference in separate am-
poules, one of which is placed inside the other (external reference).
The advantage of the first method is that there is no need to take
into account changes in the magnetic field strength inside the sample due to its magnetic susceptibility since this field is the same
for both the sample and the reference. When an internal reference is used the effect of solvents on the chemical shift is eliminated to a considerable extent. It is necessary to use an external
standard when investigating interactions in solutions. This method

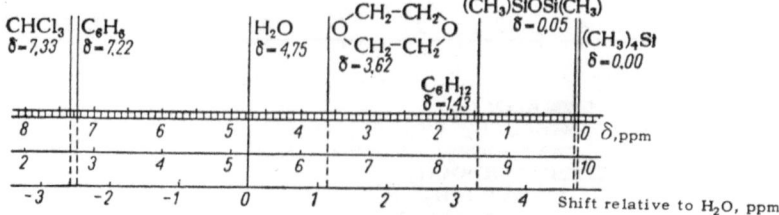

Fig. I-5. Comparison of references and different scales of chemical shifts for proton resonance.

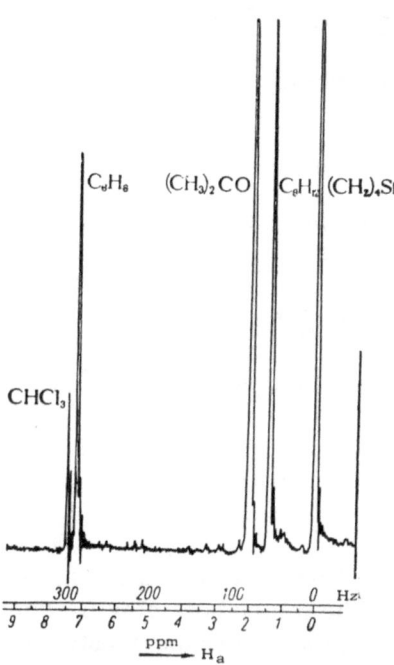

Fig. I-6. Proton resonance spectrum of a mixture of substances, each of which contains only equivalent protons.

is also convenient in the use of the NMR spectrum for flow control, when it is undesirable to introduce an impurity into the sample, and also in investigating active compounds which are capable of interacting with the reference. The external reference is usually introduced into the tube with the sample in a sealed glass capillary. The use of an external reference in the determination of chemical shifts inevitably necessitates the introduction of a correction for the diamagnetic susceptibility of the sample (see Ch. II).

There is a series of requirements for substances used as internal and external references in NMR spectroscopy: they must give a single narrow intense signal, which is as remote as possible from the signals of other nuclei; the chemical shift of the reference should not depend markedly on the external conditions; substances used as internal references should be chemically inert and readily soluble. Unfortunately there are no substances which satisfy all these requirements.

The reference used most widely in PMR spectroscopy is tetramethylsilane (TMS). This is quite an inert compound, which gives a signal at higher fields than the signals of most protons. Therefore, tetramethylsilane is a recognized international reference for the measurement of chemical shifts of protons. There are two scales of chemical shifts based on tetramethylsilane, namely, the scale of δ in which the signal of tetramethylsilane is taken as the reference point and the chemical shifts increase toward lower fields, and the scale of τ in which a chemical shift of $+10$ ppm is taken arbitrarily for tetramethylsilane and the signals at lower fields have lower values and sometimes they may even be negative. Another system, which is essentially most strict, is also used and this is analogous to the scale of δ, but with negative shifts in the direction of lower fields. We will subsequently use the scale of δ with shifts to lower fields taken as positive values.

In addition to tetramethylsilane, the standard references used include cyclohexane, benzene, chloroform, and hexamethyldisiloxane. The convenience of the latter lies in the fact that its signal is close to the signal of tetramethylsilane (δ 0.05 ppm), but it is less volatile and more accessible. Cyclohexane gives an intense signal and this compound is one of the most inert, but unfortunately its signal lies in the most densely populated region of the absorption of methyl and methylene groups. Cyclohexane is convenient for the investigation of aromatic compounds. Benzene and chloroform are less suitable standards since their chemical shifts depend appreciably on the concentration. In old work water in a capillary tube was used as an external reference. However, this reference cannot be recommended for accurate investigations since the chemical shift of water depends appreciably on the temperature and it is very sensitive to the presence of small amounts of impurities. Internal standards proposed for aqueous (and D_2O) media include 2,2-dimethyl-2-silapropanol, sodium 4,4-dimethyl-4-silapentane-1 sulfonate and also tetramethylammonia bromide [9, 10]. These compounds provide the basis for independent scales of chemical shifts.

Figure I-5 gives the chemical shifts of various standard substances relative to tetramethylsilane (TMS), by means of which it is possible to convert from shifts relative to these standards to the generally accepted scale.

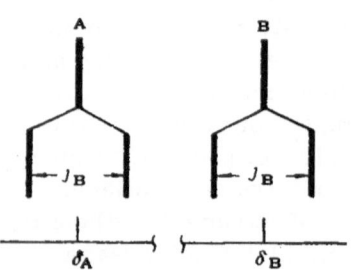

Fig. I-7. Origin of splitting in the spectrum of a system of two spins.

The nuclei of one molecule, which are equivalent from the chemical point of view and give NMR signals with the same chemical shift, are said to be c h e m i c a l l y   e q u i v a l e n t. The six protons of benzene and the six protons of ethane are chemically equivalent; in pyridine there are three groups of chemically equivalent protons, namely, the two α-protons, the β-protons, and the one protons in the γ-position. In cases where the molecule contains one group of chemically equivalent protons, its spectrum consists of a single peak. Figure I-6 shows the spectrum of a mixture of substances, the molecules of each of which contain one group of chemically equivalent protons. This spectrum clearly shows the value of the information which may be obtained by NMR spectroscopy. This information is greatly increased if the molecule contains nonequivalent nuclei with the result that in the spectra of such molecules there appears another important characteristic of NMR spectra, namely, spin—spin coupling between nuclei.

Spin — Spin Coupling.   Simple First-Order

Spectra

The origin of spin—spin coupling is examined most simply by using nuclei with spin of $\frac{1}{2}$, such as protons, as example. As has already been pointed out, in a magnetic field such nuclei may be in two energy states. Each of these states may be associated with a definite direction of the magnetic moment vector, namely, with the field (with the lower energy) and against the field (with the higher energy). From the quantum mechanical point of view, the nuclear spin states must be described by the wave function ψ. The wave function of an isolated nucleus with a spin $\frac{1}{2}$ placed in a magnetic field may have two values, which are denoted by α and β, respectively, so that for the state with the higher energy $\psi = \alpha$ and for the state with the lower energy, $\psi = \beta$. If there is spin—spin coupling between the magnetic nuclei of the molecule, then the spin state of one nucleus affects the resonance of another and vice versa. Let us examine a substance, each of whose molecules contains two magnetic nuclei (A and B)* with spin—spin coupling. In molecules containing the nucleus A in the state α, i.e., with the vector of the magnetic moment in the opposite direction to the external field, the strength of the local field acting on the nucleus B is somewhat lower than the strength when there is no magnetic moment at the nucleus A. As a result, the resonance of the nucleus B is observed at a higher strength of the external field H_0. In molecules containing the nucleus A in the state β, the resonance of the nucleus B is shifted toward lower fields by the same value. Since the states α and β are equally probable (the slight difference in population of the two levels may be neglected), the resonance of the nuclei B appears in the spectrum in the form of two lines of equal intensity. The difference between these lines characterizes the energy of spin—spin coupling between the nuclei and is called the s p i n — s p i n   c o u p l i n g   c o n s t a n t J. By an analogous argument we arrive at the conclusion that the signal of the nucleus A should also appear in the form of two peaks with the same spin—spin coupling constant. Thus, the spectrum of a substance whose molecules contain two nuclei with spin—spin coupling consists of four lines of equal intensity, whose formation may be illustrated diagramatically (Fig. I-7).

Analysis of this spectrum makes it possible to obtain three parameters, namely, the chemical shifts of the nuclei A and B (δ_A and δ_B), which are determined from the distance from the signal of the reference to the centers of the doublets and which are related to the resonance of these nuclei, and the spin—spin coupling constant. It is important to note that since the spin—spin coupling constant characterizes the energy of an intramolecular interaction it is independent of the strength of the applied magnetic field H_0; as a result of this, in contrast to the chemical shift, it should be expressed in frequency units.

A more complex picture arises if the molecule contains a considerable number of magnetic nuclei and they are not all chem-

---

* The elementary theory of spin—spin coupling presented here, which has a limited application, must be distinguished from the stricter theory examined in later chapters. In the Russian edition, the authors used Russian letters here to denote the magnetic nuclei. The editors of the English edition feel that the use of Russian letters would cause needless confusion, therefore, magnetic nuclei are denoted by A, B, and C.

ically equivalent. Let us examine the analysis of such a spectrum using the example of a substance containing three magnetic nuclei of one type (A) and two of another type (B) with the spin−spin coupling constants J_{AB} for any pair of nuclei A and B equal to each other. The latter condition indicates that nuclei in each of the groups (A_3 and B_2) are not only chemically, but also magnetically equivalent. Ethyl bromide (CH_3CH_2Br) is an example of a compound in which these conditions hold well as a result of unhindered rotation about the C−C bond.

Let us examine the resonance of the nuclei A. In different molecules these nuclei see the nuclei B in different spin states and the following variants are possible:

1. α,α: the spin vectors of both nuclei B are in the opposite direction to the field;
2. α,β: the spin vector of one of the nuclei B is against the field and the other with the field;
3. β,α: analogous to the previous case but with the opposite numeration of the nuclei B;
4. β,β: the spin vectors of both nuclei B are in the same direction as the field.

It is clear that each of the variants is equally probably so that the number of molecules in which a particular variant is realized is 25% of the total number of molecules in the sample. In cases 2 and 3 the spin vectors of the nuclei B are in opposite directions and balance each other so that in 50% of the molecules the resonance of the nucleus A is not shifted relative to the position in which it would be without spin−spin coupling with the nuclei B (for example, had the latter been replaced by their nonmagnetic isotopes). In the other molecules the resonance of the nuclei A is shifted toward lower (1) or higher (4) fields. As a result the signal of the nuclei A consists of three peaks (a triplet) with a ratio of intensities of $1:2:1$ and with the same distances between the components, equal to J_{AB}.

As regards the resonance of the nuclei B, the signal is connected with the fact that they see the equally probable states: 1) $\alpha\alpha\alpha$, 2) $\alpha\alpha\beta$, 3) $\beta\alpha$ $\alpha\beta\alpha$, 6) $\beta\alpha\beta$, 7) $\beta\beta\alpha$, and 8) $\beta\beta\beta$. As a result, the signal of the group B consists of a quartet with a ratio of intensities of $1:3:3:1$ with distances between the components equal to J_{AB}. At the same time

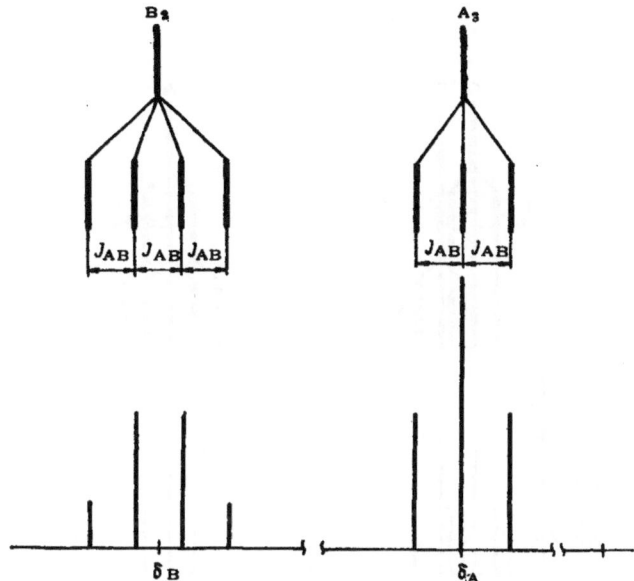

Fig. I-8. Theoretical spectrum of a system of five spins A_3B_2.

the ratio of the total intensities of the signals of A and B is 3 : 2 in accordance with the number of nuclei in these groups. Figure I-8 shows the theoretical spectrum of the system of nuclei A_3B_2 with the chemical shifts δ_A and δ_B and the spin−spin coupling constant J_{AB} and a scheme illustrating its formation. In the construction of the theoretical spectrum the relative intensity (more accurately, the area) of each of the peaks is represented by the height of the corresponding line. Like the previous one, this spectrum makes it possible to obtain three parameters, namely, the two chemical shifts and the spin−spin coupling constant.

A spectrum of this type (A_3B_2) is given by protons of the ethyl group of ethanol (Fig. A-10). Since there is no observable spin−spin coupling between the hydroxyl proton and the other protons of the molecule (for reasons which will be considered later), the signals of the ethyl and hydroxyl protons are independent of each other.

By developing the principles examined above it is possible to formulate the basic rules for the multiplicity of signals and the distribution of line intensities in the multiplets.

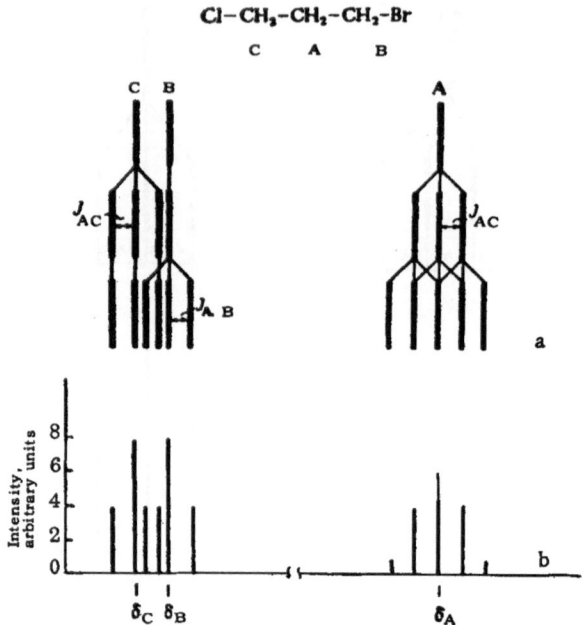

Fig. I-9. Theoretical spectrum of 1-bromo-3-chloropropane
at 60 MHz.

The number of lines in a multiplet produced by spin–spin coupling with a group containing n magnetic nuclei with a spin of $\frac{1}{2}$ equals $n+1$. Below we give the distribution of line intensities in the multiplet, which is determined by the binomial coefficients in the expression $(a+1)^n$.

Form of signal	Number of lines $(n + 1)$	Distribution of intensities
Singlet	1	1
Doublet	2	1 : 1
Triplet	3	1 : 2 : 1
Quartet (quadruplet)	4	1 : 3 : 3 : 1
Quintet	5	1 : 4 : 6 : 4 : 1
Sextet	6	1 : 5 : 10 : 10 : 5 : 1
Septet	7	1 : 6 : 15 : 20 : 15 : 6 : 1

In practice it is rarely possible to observe multiplicity above five lines (quintet) since the outer peaks of low intensity are usually lost in noise.

It should be noted that in cases where the splitting of the signal is produced by spin—spin coupling with two or more non-equivalent groups of equivalent nuclei the multiplicity of the signal is determined in the general case by the product of the multiplicities due to each of the groups separately. Thus, if a group of protons is subject to the effect of spin—spin coupling with two other groups, each of which contains two equivalent protons, the signal may contain 9 lines. However, the spectrum is simplified if the two spin—spin coupling constants are the same. As an example, let us examine the spectrum of 1-bromo-3-chloropropane. In this compound (Fig. I-9, A-13) the central methylene group (A) has spin—spin coupling with the protons of the two other methylene groups (B and C) and the spin—spin constants are the same $(J_{AB} = J_{AC})$. At the same time, there is hardly any coupling between the outer methylene groups (B and C), i.e., $J_{BC} = 0$). As a result, the signal of the groups B and C consists of triplets and that of the group A, a quintet, formed as a result of the superposition of three triplets. The spectrum is complicated by the fact that the chemical shifts of the groups B and C (δ_B and δ_C) are similar to each other with the result that their signals overlap. The formation of the spectrum is illustrated by the scheme in Fig. I-9a.

Let us now derive the relation of the peak intensities in this spectrum. As is already known, the intensities of the components of the triplet are defined by the ratio $1 : 2 : 1$. The signal of the group A consists of three triplets (the middle triplet having twice the intensity as the other two) and those of the groups B and C consist of one triplet each. Since there is the same number of protons in each of the three groups, the components of the triplets of B and C are four times as intense as the corresponding components of the triplets which form the signal of the group A. Therefore, in order to have whole numbers, we must take for the first two ratios of intensities $4 : 8 : 4$. The ratio of the intensities of the quintet of A may be determined now by summing the intensities of the corresponding components of the triplets forming it:

$$\begin{array}{r} 1:2:1 \\ +\quad 2:4:2 \\ 1:2:1 \\ \hline 1:4:6:4:1 \end{array}$$

The theoretical spectrum obtained from this calculation is given in Fig. I-9b. With the arbitrary units we adopted the total intensity of all the lines of this spectrum equals 48; which corresponds to 8 intensity units to each of the protons. It can be said that the theoretical spectrum has been normalized so that the total intensity of the lines due to the resonance of each of the nuclei equals 8 units on the scale adopted. The equality of the total intensities of the lines connected with the resonance of each of the nuclei in the molecule (we are considering nuclei of the same magnetic isotope) is one of the most common fundamental rules in NMR spectra and is called the r u l e   o f   t h e   s u m s   o f   i n t e n s i t i e s .

The experimental spectrum of 1-bromo-3-chloropropane plotted at a frequency of 60 MHz is given in Fig. A-3. In this spectrum the ratio of peak intensities derived above is not maintained exactly; the reasons for the deviations, which are quite general, will be examined later. The intensity summation rule holds and deviations from it may be caused only by instrument error.

Let us now examine an example of a spectrum of three groups of nuclei with three different spin—spin coupling constants. Figure I-10a and b gives diagramatic spectra of the diacid fluoride of methylphosphinic acid (CH_3POF_2) at a frequency of 40 MHz in two different fields, corresponding to resonance of H^1 and F^{19} nuclei. The problem consists of deriving all possible parameters from these spectra and also constructing the theoretical spectrum of P^{31}.

The spectrum of the protons in the molecule (Fig. I-10a) consists of 6 lines, which form two triplets. It is obvious that the triplet splitting is due to the presence of the two fluorine atoms so that the constant J_{HF} may be determined by measuring the distance between neighboring components in the triplets and taking the mean value. The presence of one phosphorus atom produces doublet splitting of the proton and the constant J_{HP} obviously may be obtained by measuring the distances between the corresponding components belonging to the two different triplets, for example, between the central peaks of these triplets. To obtain a more accurate result it is useful to take the mean of the three possible measurements. Moreover, from the proton resonance spectrum it is possible to determine the chemical shift of the protons by measuring the distance from the center of the signal of the methyl group to the reference signal. All the measurements are carried out at the tops of the peaks.

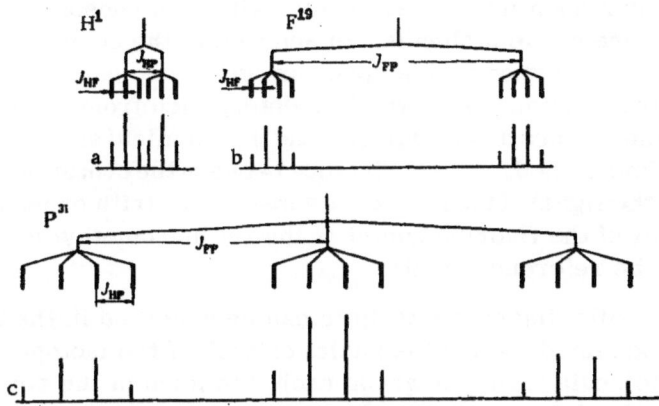

Fig. I-10. Theoretical spectra of the diacid fluoride of methylphosphinic acid for resonance of H^1 (a), F^{19} (b), and P^{31} (c) nuclei.

TABLE I-1. Spin—Spin Coupling Constants of Protons [11]

Grouping	Coupling constant, Hz	Grouping	Coupling constant, Hz
>C< with H, H	12—15	>C=C< with CH, H	4—10
>CH—CH<	2—9	>C=C< with CH, H	0.5—2.0
>CH—(C)ₙ—CH<	0	>C=CH—CH=C<	10—13
>C=C< with H, H	0—3.5	>CH—CH=O	1—3
H >C=C< H	6—14	>CH—C≡CH	2—3
H >C=C< H	11—18	⟨benzene⟩—H ortho / meta / para	7—10 / 2—3 / Up to 1

The F^{19} NMR spectrum consists of two quartets, produced by spin–spin coupling of these nuclei with both phosphorus and the three protons of the molecule. In addition to the constant J_{HF}, which is already known, this spectrum gives two parameters, namely, the constant J_{FP}, which is determined from the distance between the centers of the quartets, and the chemical shift of fluorine, which is determined from the distance between the center of this spectrum and the signal of the reference substances, trifluoroacetic acid (the signal of the fluorine nuclei of the trifluoromethyl group is taken as the reference point).

The following characteristic can be observed in the two spectra examined: since the nuclei of each of the isotopes present in the molecule form one group each, the form of the spectrum is determined solely by the spin–spin coupling constants and consequently it is independent of the working frequency of the spectrometer. The same may be said of the spectrum of the phosphorus nuclei. The theoretical P^{31} resonance spectrum of this molecule consists of three quartets with the distances between the lines determined by the two spin–spin coupling constants. The relative intensities of the peaks are given by the ratios $1:3:3:1$, $2:6:6:2$, and $1:3:3:1$. The theoretical spectrum and the scheme for its formation are given in Fig. I-10c. Since the spectrum is due to the resonance of one nucleus, there is no point in talking of its normalization.

Here we have examined the main principles of the analysis of so-called first order spectra, which are characterized by the fact that 1) the differences in the chemical shifts of the signals of groups of equivalent nuclei substantially exceed the constants of spin–spin coupling between nuclei of these groups, and 2) nuclei which belong to any group participating in spin–spin coupling are both chemically and magnetically equivalent. These rules cannot be used for the accurate analysis of more complex NMR spectra in which the spin–spin coupling constants are comparable with or greater than the chemical shifts (expressed in the same units). It was first assumed essentially that the state of the nuclear spin is independent of whether the other spins coupled with it are under resonance conditions or not. At the same time, this assumption is only valid when the spin–spin coupling constants are considerably less than the differences in the chemical shifts for the same nuclei. Otherwise the distances between the components in the multiplet are not exactly equal to the coupling constants, the chemical shifts cannot be determined from the center of the multiplet, and the ratios of the intensities differ from those derived above. Methods for complete analysis of complex spectra are examined in Ch. IV; the same chapter gives a more strict classification of spectra. In many cases complex spectra may be analyzed approximately by means of the rules for first order spectra. In practice, if the chemical shift (expressed in Hz) is 5-6 times the constant of spin−spin coupling between the nuclei, the differences in the results of analysis of the spectrum by means of the rules for a first order spectrum and complete analysis do not exceed the instrument errors. The first order rules are assuming increasing importance because of the appearance of spectrometers with magnets having higher working field strengths. As has already been stated, an increase in the field strength leads to an increase in the chemical shift with the spin−spin coupling constants remaining unchanged; thus, with an increase in the field strength complex spectra become increasingly closer to first order spectra.

The main rules for the analysis of first order spectra are formulated below.

1. The number of lines in a multiplet produced by spin−spin coupling with a group containing n magnetically equivalent nuclei with a spin of $\frac{1}{2}$ equals $n+1$. In the general case, if there is a group of n nuclei with a spin of I, the number of lines in the multiplet equals $2In+1$.

2. The distance between two adjacent lines in the multiplet (in Hz) equals the spin−spin coupling constant J.

3. The chemical shift of a group of chemically equivalent nuclei δ is determined from the distance (in ppm) from the center of the multiplet to the signal of the reference.

4. The intensities of the lines of a multiplet produced by spin−spin coupling with one group of nuclei are in the same ratio as the coefficients in the expansion $(a+1)^n$, where n is the number of nuclei in the neighboring group.

5. The total intensities of the lines due to resonance of each of the nuclei of one isotope are equal to each other.

6. The multiplicity of a signal due to unequal spin−spin coupling

with several groups of magnetically equivalent nuclei equals the products of the multiplicities produced by each of these groups. In this case rules 2 and 4 may be used only after the corresponding analysis.

7. If the magnetic nuclei of one molecule may be divided into several groups, between which there is no spin–spin coupling, the spectrum of the whole molecule may be represented as the superposition of the spectra of these groups.

8. If the sample contains a mixture of substances which do not interact chemically, then its spectrum may be represented as the superposition of the spectra of the separate components at the same concentration.

The spin–spin coupling constant of protons in typical groupings are given in Table I-1.

8. SOME FEATURES OF HIGH-RESOLUTION NMR SPECTRA

In the previous section we examined the basic characteristics of NMR spectra. Below we give the features of the spectra connected with the definite structure of the molecules and the properties of individual groups, which must be taken into account in practical work.

Chemical Exchange

In many organic compounds individual nuclei are often capable of more or less rapid exchange both between different positions inside the molecule and between different molecules. The form of the NMR spectrum depends on the rate of this exchange.

To characterize the exchange rate it is useful to introduce the time constant τ (usually measured in seconds), which characterizes the mean life of a nucleus in a definite state. Thus, for example, for the keto–enol equilibrium

$$-\underset{\underset{O}{\|}}{C}-CH\big\langle \;\rightleftarrows\; -\underset{\underset{OH}{|}}{C}=C\big\langle$$

it is possible to introduce two such constants τ_k and τ_e, which characterize the mean lifetimes of the ketonic and enolic forms.

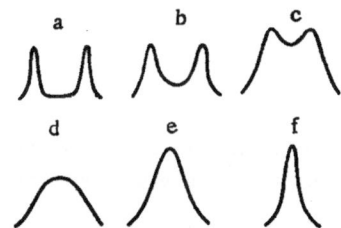

Fig. I-11. NMR spectra with different rates of exchange of the magnetic nucleus between two positions
in the molecule.

Under equilibrium conditions the number of transitions from the ketonic to the enolic form and back remains constant and it follows from this that the lifetime is proportional to the molar content of the corresponding form in the mixture.

The protons in the ketone and the enol are in different electronic environments and consequently they give signals with different chemical shifts. If the difference in the chemical shifts is expressed in frequency units (Hz), then a comparison of the lifetime with the reciprocal of this difference characterizes the exchange rate from the point of view of nuclear resonance.

Let us denote the chemical shifts of the protons in the two states by ν_1 and ν_2 (in Hz) and the lifetimes corresponding to these states by τ_1 and τ_2 (in sec). We should note that the value $1/(\nu_1 - \nu_2)$ also has the dimensions of seconds.

If τ_1 and τ_2 are much greater than the absolute value of $1/(\nu_1 - \nu_2)$, then from the point of view of NMR there is slow exchange. In this case the spectrum shows two separate signals, which are connected with the resonance of the nuclei in the two chemical states (Fig. I-11a). Slow exchange leads only to broadening of the signals (Fig. I-11b) and the broadening is greater, the smaller the value of τ for the given state. It is important that since the value of ν depends on the field strength of the spectrometer the width of the line may differ, depending on the conditions under which resonance is achieved.

With smaller values of τ the signals may overlap (Fig. 11c) and then merge into one broad signal (Fig. 11d). In individual cases, the broadening may be so great that the signal is practically impossible to observe and it apparently disappears from the spectrum. In systems with still shorter lifetimes of the different states the signal, on the contrary, becomes narrower (Fig. 11e) so that with rapid exchange one common narrow signal is observed (Fig. 11f) and its position is determined by the contributions of the separate states:

$$v_{av} = v_1 p_1 + v_2 p_2 \tag{I-10}$$

where p is the probability of finding the proton in the given chemical state, in other words, the mole fraction of the given structure.

The same rules govern the relation between the exchange rate and the splitting of the signal due to spin–spin coupling. Slow exchange leads to broadening of the spin multiplets and with sufficiently high rates there may be merging into a single signal. In the case of rapid exchange (the lifetimes are considerably smaller than the reciprocal of the spin–spin coupling constant) the spin–spin coupling of the exchanging nuclei with other nuclei in the molecule does not appear in the spectrum. This is shown particularly clearly by alcohols. A trace of water or acids leads to acceleration of the exchange of the hydroxyl proton of the alcohol, leading to the breakdown of spin–spin coupling of the hydroxyl and other protons in the molecule. Careful removal of impurities and also the addition of substances which form a strong hydrogen bond with a hydroxyl proton reduce the rate of exchange and make it possible to observe spin–spin coupling.

Internal Rotation

The effect of internal rotation in the molecule on the NMR spectrum is essentially analogous to chemical exchange. In the case of slow rotation the spectrum consists of slightly broadened signals, produced by the resonance of each conformation. An increase in the rotation rate (for example, by raising the temperature of the sample) leads to broadening of the lines and at a definite temperature only one broad signal is formed. If the rotation rate increases so much that the lifetime of each conformation becomes considerably less than the reciprocal of the difference in the resonance frequency of the signals corresponding to the individual conformations, the lines become narrower and a single narrow signal appears. In contrast to rapid exchange, with rapid internal rotation the spin–spin coupling between nuclei does not disappear from the spectrum, but is only averaged out.

Saturated hydrocarbons and their derivatives are typical examples of compounds in which the rotation about C–C bonds pro-

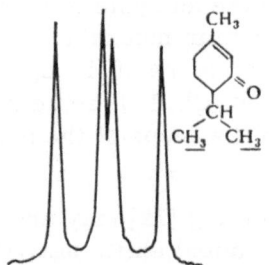

Fig. I-12. Spectrum of the
methyl protons of the isopropyl group of the piperitone
molecule.

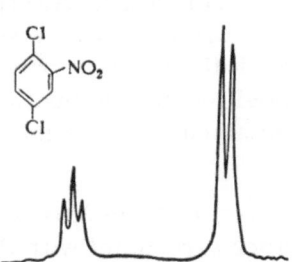

Fig. I-13. Spectrum of 2,5-
dichloronitrobenzene with
masked spin–spin coupling.

ceeds so rapidly at normal temperatures that the protons attached
to a single carbon atom (for example, in a CH_3 group) become magnetically equivalent.

If the nuclei are nonequivalent in each of the stable conformations, then free rotation does not lead to magnetic equivalence of
the nuclei. This situation, which is called m o l e c u l a r   a s y m -
m e t r y , is realized, for example, in molecules of the type
ACH_2-CDEF, where A, D, E, and F are different substituents.

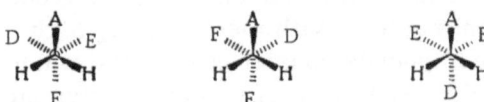

Figure I-12 gives the spectrum of the methyl groups of piperitone. In this case the presence of the three different substituents
on the carbon atom bearing the isopropyl group introduces asym-
metry which causes the protons in the two methyl groups to be nonequivalent. Their signal consists of two doublets with chemical
shifts corresponding to the two different positions. The doublet
splitting is produced by spin–spin coupling with the proton at the
third carbon atom [12].

Spectra with Masked and Virtual Spin– Spin

Coupling

These terms denote spectra which appear to be simple at
first glance, but in actual fact cannot be analyzed in accordance

with first-order rules. Although a complete analysis of these
spectra requires, strictly speaking, a quantum mechanical examination of the relation between the parameters of the spectrum
and the position and intensities of the lines (Ch. IV), some data
may be obtained by using methods within the scope of the present
chapter.

Spectra with masked spin−spin coupling [12] may arise in
cases where two nuclei with the same chemical shifts have different spin−spin coupling constants with a third nucleus in the molecule. A typical example of a spectrum of this type is given by
2,5-dichloronitrobenzene (Fig. I-13). The spectrum consists of a
characteristic doublet and triplet with approximately equal distances between the components (1.6 Hz). It is easy to assume mis-
takenly that this distance is the spin−spin coupling constant of the
proton in the position 6 with each of the protons in positions 3 and
4. At the same time, data from the investigation of benzene derivatives indicate that the spin−spin coupling constants of meta
and para protons differ substantially (~3 and 0-0.3 Hz, respectively). In this case the splitting is close to the expected mean
value.

Masking of spin−spin coupling is often observed in the spectra of para-disubstituted derivatives of benzene, containing two
pairs of equivalent nuclei. With insufficiently high resolution the
spectra of such compounds consist of 4 peaks, which may be readily taken as a typical spectrum of two interacting nuclei (two
doublets). In actual fact, the distance between the components of
each of the doublets equals the sum of the constants $J_{AB\,ortho}$
$+J_{AB\,para}$. A typical spectrum of this type is given by 4-methyl-
acetophenone (Fig. A-29). It should be noted that in the particular
case of benzene derivatives, because of the very small value of
J_{para}, the distances between the components in these doublets
make it possible to estimate reliably the change in the constant
J_{ortho} for a series of compounds without resorting to a complete
analysis.

The existence of spectra with masked spin−spin coupling
emphasizes the importance of differences between chemical and
magnetic equivalence of the nuclei.

Virtual (effective) spin−spin coupling [13] has a meaning
which is largely opposite to masked coupling. Virtual coupling

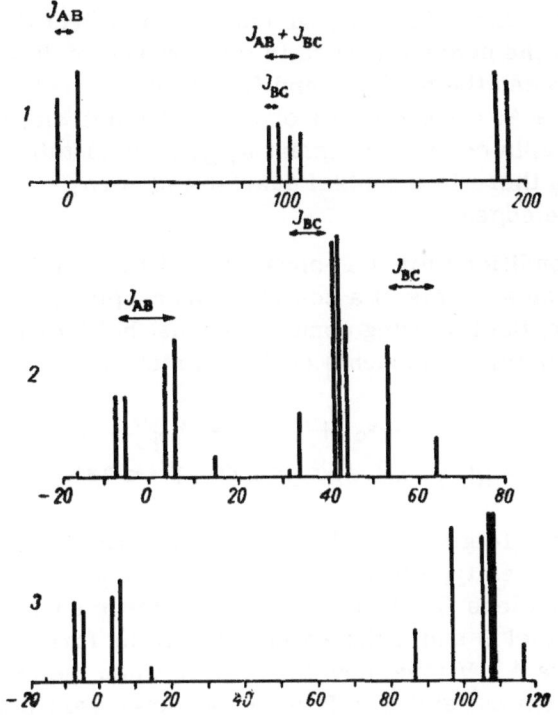

Fig. I-14. Theoretical spectra of a system of three spins.

causes complication of the spectra of compounds whose structure should lead at first glance to simple spectra of the first order.

Let us examine a fragment of a molecule with the linear structure

$$-\underset{\underset{A}{\overset{|}{H}}}{\overset{|}{C}}-\underset{\underset{B}{\overset{|}{H}}}{\overset{|}{C}}-\underset{\underset{C}{\overset{|}{H}}}{\overset{|}{C}}-$$

If the chemical shift of the nucleus A differs substantially from the shifts of the nuclei B and C, the constant $J_{AC} = 0$ and $|\nu_A - \nu_B| \gg J_{AB}$, so that from the point of view of the first order rules, the signal of the nucleus A may be regarded as a simple spectrum independently of the relation between the chemical shift and the coupling of the nuclei B and C. In actual fact, under certain conditions the spin of the nucleus C may affect the spin state of A through the nucleus B even in the absence of direct spin–spin coupling between the nuclei A and C. In the general case the signal of the nucleus A, instead of the expected doublet with splitting equal to the spin–spin coupling constant J_{AB}, may contain up to 6 lines, among which there is a central quartet and w i n g s of low intensity at the edges.

The conditions for the appearance of the quartet and wings follow from an analysis of a complex three-spin system (Ch. IV). In particular, the following conditions must hold for the splitting between the outer components of the quartet to equal 0.1 J_{AB}:

$$
[(|\,\nu_B-\nu_C\,|\,+^{1}/_{2}J_{AB})^2+J^2_{BC}]^{1/2}
$$
$$
-[(|\,\nu_B-\nu_C\,|\,-^{1}/_{2}J_{AB})^2+J^2\,_{BC}]^{1/2} \leqslant 0.8 J_{AB} \tag{I-11}
$$

Figure I-14 gives the theoretical spectra of a system of three spins of the given type in which we are interested only in the resonance of the nucleus A. With a reasonably large chemical shift between the nuclei B and C the spectrum has the form of normal first order spectra A. Virtual coupling arises when the difference between the chemical shifts of B and C becomes very small and the signal of A obeys the following rule:

1. The constant J_{AB} may be obtained directly from the spectrum from the distance between the outer lines of the quartet.

2. The quartet changes into a doublet (the splitting of the outer bands $> 0.1 J_{AB}$) if

a) $|\,\nu_B-\nu_C\,| \geqslant 2J_{AB} \simeq J_{BC}$;

b) $|\,\nu_B-\nu_C\,| \geqslant J_{AB} \simeq {}^4/_3 J_{BC}$

or, this means roughly that

$$
|\,\nu_B-\nu_C\,|+^{1}/_{2}J_{AB} > 2J_{BC}
$$

3. The splitting of the doublet into the quartet increases with an increase in the ratio $J_{AB}/|\,\nu_B-\nu_C\,|$ or $J_{AC}/|\,\nu_B-\nu_C\,|$, and the signal gradually changes to a triplet.

4. The form of the signal is independent of the sign of $\nu_B - \nu_C$, while the value of the splitting is independent of the ab-

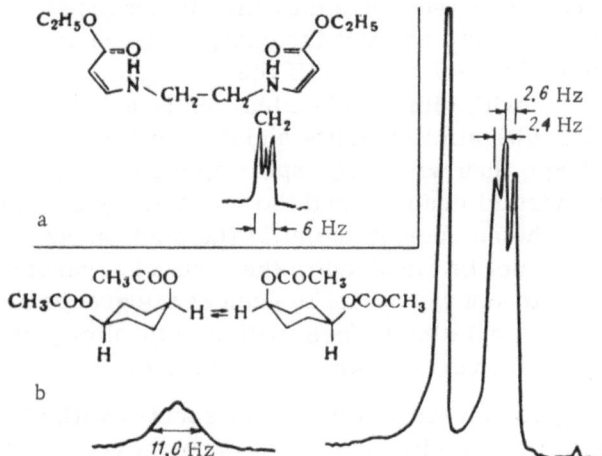

Fig. I-15. Spectra with virtual spin—spin coupling: a) N,N'-di(1-methyl-2-carbethoxyvinyl)ethylenediamine; b) cis-1,4-diacetoxycyclohexane.

solute value $|\nu_A - \nu_B|$ (and consequently, of $|\nu_B - \nu_C|$), but as they increase the intensities of the peaks are finally leveled out.

5. If $\nu_B = \nu_C$, then the quartet of A changes into a triplet because the two inner lines are separated by no more than $0.1 J_{AB}$ and cannot normally be resolved. In this case, if $J_{BC} \gtrsim J_{AB}$, the ratio of the intensities in the triplet is close to $1:2:1$ and otherwise the inner line has a lower intensity.

6. The wings converge with a decrease in $|\nu_B - \nu_C|$, and their minimal splitting is $2(\frac{1}{4} J^2_{AB} + J^2_{BC})^{1/2}$ (if they can be observed in this case). When $|\nu_B - \nu_C| > 2J_{AB}$ the distance between the wings equals approximately $2J_{BC}$. The wings depend only slightly on the value $|\nu_A - \nu_B|$.

7. The intensity of the wings increases with an increase in $|J_{AB} - J_{BC}|$ and with a decrease in $|\nu_B - \nu_C|$. When $|\nu_B - \nu_C| > 2J_{AB}$ their intensity falls to 0.05 of the intensity of one line of the quartet; in combination with rule 2 this means that the wings may also be observed if the signal of A consists of a doublet.

Virtual spin—spin coupling in a three-spin system is interesting not only in itself, but also as a model for investigating an analogous phenomenon in more complex molecules. Essentially, the main rules presented above may also be extended to complex systems if we take into account the changes which arise when a larger number of nuclei appear. If the molecule contains several nuclei A_1, A_2, ... with similar chemical shifts and several nuclei B_1, B_2, ... also with similar shifts and the shift between the nuclei A and B is large, then with large spin—spin coupling constants there may be virtual coupling and the multiplicity of each group is 2^n, where n is the number of nuclei in the other group. In practice, such spectra cannot be resolved in the form of separate lines. If the chemical shifts of the nuclei in each of the groups are the same, the spectra assume a simple form with a multiplicity of $n+1$. Naturally intermediate variants are also possible.

Above we gave several examples of spectra with virtual spin—spin coupling. In N,N'-di(1-methyl-2-carbethoxyvinyl)ethylene-diamine (Fig. I-15a) there is no direct spin—spin coupling between the NH protons and the protons of the remote CH_2 groups. As a result of the chemical equivalence of the four protons in the CH_2 groups the first order theory predicts a double signal for them with a splitting corresponding to spin—spin coupling of the NH protons and the protons of the nearest CH_2 groups. However, in actual fact a triplet is observed with a central component of low intensity due to the fact that the protons of each methylene group see not only the spin state of the nearest NH proton, but also another proton and the spin information is transferred through the CH_2 group.

cis-1,4-Diacetoxycyclohexane is a multispin system with virtual coupling (Fig. I-15b). As a result of the rapid interconversions of the two chair conformations there is averaging of the spin—spin coupling constants of the CH protons with the protons of neighboring CH_2 groups in the cis- and trans-positions. Therefore the first-order theory predicts a quintet for the CH proton and a doublet for the methylene protons split with the mean coupling constant. In actual fact, as a result of virtual coupling between remote CH and CH_2 groups the signal of the former consists of a broad band (width 11.0 Hz) while the 8 protons of the methylene groups give a triplet.

To conclude this section we should again recall that virtual coupling is not an unexpected phenomenon, but follows directly from the quantum mechanical analysis of systems of nuclear spins. Essentially, the introduction of this concept represents an attempt to present the analysis of specific complex spectra (with the constant for the coupling between remote nuclei equal to zero) in the form of simple rules without resorting to quantum mechanical calculations (see Ch. IV). This approach is justified by the fact that in experimental spectra of this type it is usually impossible to resolve the signals in the form of separate lines so that accurate calculation is practically impossible.

Signal of a Group of Equivalent Nuclei

If a molecule contains a group of chemically equivalent nuclei and spin−spin coupling of them with other nuclei of the molecule is absent or negligibly small, the signal of such a group consists of a single peak. At the same time, it is well known that there is considerable spin−spin coupling between the nuclei in the group. Thus, the coupling constant between protons in methane is 12.4 Hz, while in ethylene there are three spin−spin coupling constants of the cis, trans, and geminal protons with values of 10.5, 17.5, and 2.3 Hz, respectively, and in the H_2 molecule the spin−spin coupling between the two protons reaches 280 Hz. These constants were determined experimentally from the spectra of deuterium-substituted molecules in which the equivalence of the nuclei was disrupted while the chemical characteristics of the compound were retained, and also by comparison with a large number of substances of similar structure.

The lack of splitting of the signals of equivalent nuclei is due to the fact that in the system of spins the rf field (under conditions far from saturation) may induce only transitions in which the total spin of the system changes by ±1. This rule, which has a general physical meaning, is called the s e l e c t i o n   r u l e and plays a very important part in the analysis of complex spectra (Ch. IV). We will explain it using, as an example, a system of two equivalent nuclei.

As is well-known, each nucleus with a spin of $\frac{1}{2}$ may be in two energy states, corresponding to the spin quantum numbers $-\frac{1}{2}$ and $+\frac{1}{2}$. A system of two equivalent nuclei may have three energy states since two variants with opposite spins give states with the same energy.

The selection rule allows only transitions 1-2 and 2-3 and does

Energy, E	Spin of nucleus 1	Spin of nucleus 2	Total spin of system
1	$-1/_2$	$-1/_2$	-1
2	$+1/_2$ }	$-1/_2$ }	0
	$-1/_2$	$+1/_2$	
3	$+1/_2$	$+1/_2$	$+1$

not allow transitions 1-3 since the total spin of the system would change by two units in it. It is obvious that in the case of two equivalent nuclei the two transitions lead to the same change in the energy of the system (ΔE) as a result of which, in accordance with formula (I-3), there is the absorption of radiation with one definite frequency, i.e., only one line appears in the spectrum.

In systems of several equivalent nuclei there is the possibility of a greater number of variants, but in these cases each allowed transition leads to the same change in the energy of the system and consequently, to the appearance of a line in the same position in the spectrum.

9.   HIGH-RESOLUTION NMR SPECTROMETERS

Modern industrial high-resolution spectrometers may be classified in accordance with the following basic features: 1) the method of producing the magnetic field, i.e, with permanent and with electromagnets; 2) the magnetic field strength, which is usually expressed as the resonance frequency of protons, i.e., 40, 60, and 100 MHz; 3) the method of sweeping the spectrum, i.e., sweeping the magnetic field and a frequency sweep; 4) the purpose, i.e., for the resonance of the nuclei H^1, F^{19}, P^{31}, C^{13}, etc.

The most common instruments are 40, 60, and 100 MHz spectrometers with an electromagnet and sweeping of the spectrum through the field, which are designed for the resonance of H^1 and F^{19} nuclei. In 60 MHz instruments the resonance of fluorine is usually realized at a lower frequency (56 MHz) than proton resonance; in 40 MHz spectrometers this frequency is used for resonance of both protons and fluorine, but in the latter case it is necessary to increase the field strength somewhat.

Spectrometers with an electromagnet are fitted with a power supply for the electromagnet and a system of stabilization of the magnetic field. The stability of the magnetic field and its uniformity are decisive factors in obtaining high quality spectra.

Even very high stability of the power supply for the electromagnet does not guarantee the necessary stability of the magnetic field. In NMR spectrometers it is common to use a system of superstabilization which is based on the fact that random changes in the magnetic field induce an electric current in a special coil, which lies between the poles of the magnet. This current is amplified and fed back to the winding of the electromagnets, thus compensating for the change in the field. Superstabilization makes it possible to guarantee the stability of the magnetic field down to 10^{-7} of the nominal value and for the short period which is necessary to plot one spectrum down to 10^{-8} (0.01 ppm). The usual error of instruments fitted with superstabilization is 0.05 ppm in the measurement of chemical shifts and down to 0.5 Hz in the measurements of spin—spin splitting constants.

The latest models of spectrometers (including spectrometers produced in the USSR) are fitted with a spin stabilization system which is used particularly widely for proton magnetic resonance (proton stabilization). This system makes use of the nuclear resonance signal from a special sample, which is placed in the magnetic field (in some systems it is the signal of an internal reference, added to the sample investigated). The proton stabilization system is constructed so that a deviation of the magnetic field from the resonance conditions produces either a change in the modulation frequency of the spin generator or an electric current which corrects the magnetic field as in superstabilization. Proton stabilization guarantees the stability of the magnetic field for a long period and also high reproducibility of the spectra. With the use of proton stabilization the accuracy of measurements of chemical shifts is 0.001 ppm and that of spin—spin coupling constants, 0.1 Hz or better. The accuracy of the results may be increased still further by repeated plotting of the spectrum and averaging the data obtained in each plot.

The use of a permanent magnet simplifies the problem of stabilizing the magnetic field since in this case the stability is determined essentially only by a change in the ambient temperature and stray magnetic fields, which arise, for example, from electric lighting circuits, etc. In the spectrometer produced by the Japanese company "Hitachi" (60 MHz for proton resonance) a permanent magnet is used with complete thermal insulation and magnetic shielding and also a special system, which responds to a change in the mag-

netic field in the room. However, fields of high strength can be
produced with permanent magnets only with difficulty.

The stability of the magnetic field essentially determines the
minimal rate of sweep in the recording of the spectrum. In the
plotting of normal spectra a sweep rate of 1-3 Hz/sec is used so
that with a frequency of 60 MHz the whole of the spectrum (10 ppm)
is swept in 250-500 sec. To determine the details in multiplets it
is common to use a slower sweep of part of the spectrum and with
high-quality spectrometers it is possible to reduce the sweep rate
of 0.2-0.3 Hz/sec.

The uniformity of the magnetic field is determined by the
gradient of the field in the volume of the sample and depends largely
on the quality of the pole pieces of the magnet. In modern spectrometers the difference in the magnetic field strength in different parts
of the sample does not exceed $1 \cdot 10^{-8}$-$1 \cdot 10^{-9}$ of the nominal value,
which corresponds to 0.5-0.1 Hz at 40-100 MHz. To achieve a high
uniformity of the magnetic field, spectrometers are fitted with a
system of current shimming coils, which makes it possible to create
artificially a field gradient in different directions and thus compensate for the natural nonuniformity of the field of the magnet.
Moreover, to compensate for the nonuniformity of the magnetic
field in spectrometers it is usual to spin the sample tube. Rapid
spinning averages out the magnetic field through the volume of the
sample and this leads to a considerable improvement in the quality
of the spectrum. With a sample spinning in a nonuniform magnetic
field, additional peaks to the sides of the main signal may appear.
The intensity of these side peaks depends on the nonuniformity of
the field. The rate of spinning of the sample affects the distance
from the peaks to the main signal. To distinguish these peaks from
the signals of the substance investigated it is usual to plot the spectrum repeatedly with different spinning rates.

The uniformity of the magnetic field of the spectrometer determines the permissible dimensions of the sample investigated. In
most modern spectrometers for protons and F^{19} resonance the
sample is placed in a cylindrical thin-walled glass tube with an
external diameter of 5 mm and a length of 80-100 mm. The tube
is held in a plastic turbine, which is rotated with a stream of air.
The quality of the spectra is very dependent on the symmetry of
the tube.

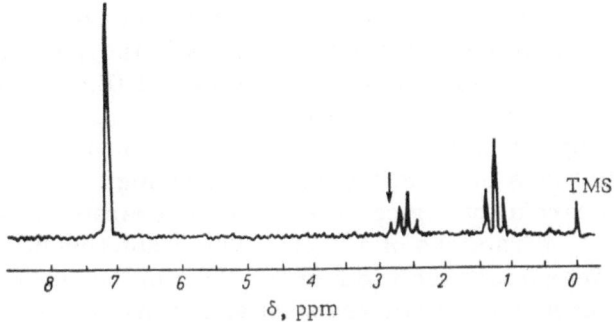

Fig. I-16. Spectrum of a 1% solution of ethylbenzene
in carbon tetrachloride.

In addition to glass, the tubes may also be made of other nonmagnetic nonmetallic materials. To plot the spectra of fluorine
compounds which attack glass, Teflon tubes are used and for small
amounts of a substance, special glass tubes which are filled with
nylon with a small spherical cavity in the region of the detector
coil [14]. Such tubes make it possible to reduce the volume of the
sample investigated from 0.5–0.1 to 0.1 cc.

The quality of the electronic part of the spectrometer, which
includes the rf oscillator, the nuclear resonance receiver, the system for amplification and detection of the signals, and the recording instrument, is equally important in obtaining a high-quality
spectrum. It should be noted that guaranteeing a sufficiently high
stability of this part of the instrument presents no serious problems to modern electronics. The stability of the spectrometer as
a whole is determined by the stability of the magnetic field. In
certain systems of NMR spectrometers the magnetic field and
resonance frequency are not stabilized separately, but the relation
of these parameters is stabilized. In such instruments it is usual
to use frequency sweeping of the spectrum. A frequency sweep is
somewhat more convenient, particularly when using the nuclear
double resonance method (see below and Ch. IV), but technically it
is more difficult to achieve.

The electronic system of the instrument determines the sensitivity of the NMR spectrometer, which is characterized by the signal-to-noise ratio in the spectra. To estimate the sensitivity of a
spectrometer, the spectrum of a 1% solution of ethylbenzene in

carbon tetrachloride is often used and in modern instruments the ratio of the amplitude of the smallest peak in the quartet of the CH_2 group to the amplitude of the noise is $4:1$ (Fig. I-16). Another characteristic of the instrument is the resolution, which is determined by the minimal distance between lines which appear as separate bands in the spectrum. Modern high-quality instruments with careful tuning can resolve into separate peaks lines at a distance of fractions of a hertz. The resolution of the instrument also determines the width of the lines in the spectrum (if they are not broadened for other reasons) and in proton resonance spectra the usual width of the line for a signal peak is less than 0.5 Hz.

Modern NMR spectrometers are often fitted with attachments, which considerably extend their possibilities. These include a system for thermostatting the sample, an integrator, a device for removing spin—spin coupling (spin decoupler), and a signal accumulator.

In principle, the sample may be thermostatted over a wide range of temperatures in NMR spectrometers. A specially designed receiver is usually used for this purpose. The rotating sample tube is placed in a tube with double walls of the Dewar vessel type, on which are the winding for producing the rf field. A stream of cooled or heated air or nitrogen with a controlled temperature is passed through the tube. By this method it is possible to thermostat in the range from -170 to $+250°C$ with an accuracy down to $\pm 1°$. It should be noted that thermostatting at temperatures which are very different from the magnet temperature leads to appreciable deterioration in the resolution of the spectrometer, even with careful design of the receiver. Therefore, when there is no need for thermostatting it is preferable to plot the spectrum under conditions of natural temperature equilibrium with the surrounding medium. Special preparation of the room in which the NMR spectrometer is placed is of great importance for the quality of the spectra: it should have a large volume and no drafts and air conditioning is very desirable. The water used for cooling the magnet should also be thermostatted. The optimal temperatures of the air and water are chosen experimentally for each spectrometer and they are usually higher than normal room temperature and the temperature of tap water. In the laboratory of the Leningrad Technological Institute, where one of the authors works, the maximum resolution on a Japanese spectrometer JNM-3 is achieved with an

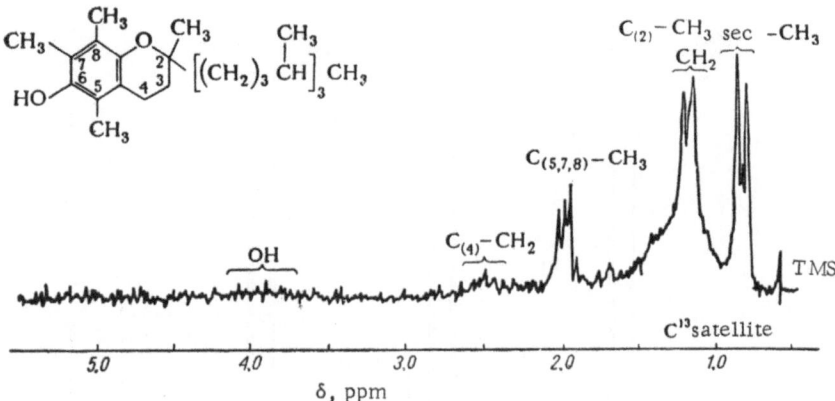

Fig. I-17. Spectrum of vitamin E (α-tocopherol) in carbon disulfide solution.

air temperature of 26-28°C and a cooling water temperature of 15-18°C.

The integrator is an electronic device for determining the areas of signals. The integration of a signal is a great help in the interpretation of spectra and may also be used for quantitative analysis.

However, it should be remembered that integrators have quite a high error (up to 10%) so that their use is of secondary value.

The device for removing spin—spin coupling (spin decoupler) consists of a calibrated audio oscillator. When connected to the nuclear resonance receiver the spin decoupler produces an additional rf field, whose frequency differs from the main field by a given value. By changing the frequency of the spin decoupler it is possible to superpose this additional field onto the signal of a particular nucleus and then the spin—spin coupling of other spins of the molecule with this nucleus is removed. The spin decoupler is more convenient with frequency sweeping of the spectrum. The disruption of spin—spin coupling (double resonance method) is examined in more detail in Ch. IV.

The signal accumulator consists of a system for storing and summing nuclear resonance signals and is used for the amplification of very weak signals which are within the limits of noise of the instrument. With repeated sweeping of the spectrum the ran-

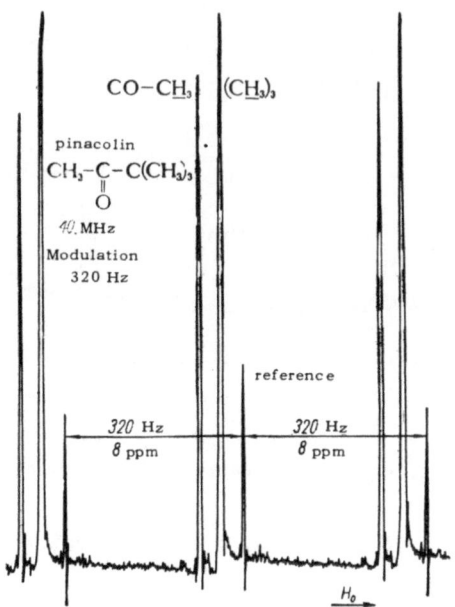

Fig. I-18. Spectrum of pinacolin
at a frequency of 40 MHz.

dom noises do not coincide in amplitude and are averaged out during summing, while signals (even very weak signals) are accumulated and thus emerge above the noise. By storage of the spectrum n times by this method it is possible to achieve a $\sqrt{n}$-fold improvement in the signal-to-noise ratio.

A signal accumulator may be used successfully only with highly stable spectrometers which do not require frequent tuning. Figure I-17 gives the spectrum of a natural product, vitamin E (α-tocopherol) in carbon disulfide solution at a concentration of $1.2 \cdot 10^{-3}$ M (0.2 mg of substance in 0.4 ml of CS_2). The spectrum was plotted on a Varian instrument, operating at a frequency of 100 MHz. Over a period of 15 h 200 scans were accumulated. This made it possible to obtain a clear spectrum of the compound. This system is one of the latest achievements in NMR spectroscopic techniques.

Calibration of NMR Spectra

In recording an NMR spectrum there is usually the problem of comparing the rate of movement of the recorder chart and the rate of sweep of the magnetic field, i.e., calibration of the spectrum. The usual calibration method is modulation of the rf field by superposition of an audiofrequency. For this purpose to the nuclear resonance receiver coils there is fed an alternating voltage with a frequency which varies from a few tens to hundreds of hertz and as a result of this, together with the main signal there appear side bands of lower intensity at a distance from the main signal corresponding to the applied modulation frequency. Figure I-18 shows the spectrum of pinacolin at a frequency of 40 MHz, plotted with modulation of the rf field with a frequency of 320 MHz. Together with the three central peaks (signals of the reference and tert-butyl and methyl groups) there appear side signals, which lie at a distance of 320 Hz from the main signals. Calibration is carried out by linear interpolation between the main and the side signals of the standard. This calibration method is known as the side band method.

When it is desirable to determine the position of the lines in the spectrum more accurately it is possible to use the side band superposition method. The modulation frequency may be selected so that the side band of the standard lies on the main signal of the line whose position is being determined. If the line is broad, the coincidence of the bands may be checked by repeated recording of the spectrum with a small variation in the modulation frequency until the intensity of the coincident signal reaches a maximum. The

TABLE I-2. Position of Lines in Spectrum of Standard Mixture for Calibration of NMR Spectrometers at 37°C [16]

Formula of component	Content, vol.%	Position of peak, Hz		Temp. co-efficient, Hz/deg*	δ, ppm
		at 40 MHz ($\pm$ 0.02 Hz)	at 60 MHz ($\pm$ 0.03 Hz)		
$Si(CH_3)_4$	3	.0	0	0	0
C_6H_{13}	2	57.32	85.98	+0.008	1.433
CH_3COCH_3	3	84.15	126.72	—0.046	2.112
CH_3CCl_3	9	109.23	163.84	—0.020	2.731
$O(CH_2CH_2)_2O$ . . .	2	144.73	217.24	—0.026	3.621
CH_2Cl_2	8	212.04	318.06	—0.038	5.301
$CHCl_3$	18	293.21	439.82	—0.052	7.330
CCl_4	to 100	—	—	—	—

modulation frequency defines the position of the band relative to the reference. The accuracy of this method is determined by the accuracy of calibration and the stability of the audio oscillator.

In spectrometers with proton stabilization of the magnetic field the calibration problem is simplified considerably since there is no temperature drift of the magnetic field. In such spectrometers it is usual to use precalibrated charts for recording the spectrum. Before the beginning of the plot, the reference signal is adjusted to the zero line on the chart and then the spectrum is swept in strict correspondence to the movement of the chart (or the pen of the recorder). To obtain more accurate results it is also desirable to use modulation in this case, at least as a periodic check of the calibration.

With frequency sweeping of the spectrum the question of calibration essentially does not arise as a separate problem, but is solved by the design of the instrument itself. The frequencies of individual spectral lines may also be measured conveniently by means of a recording frequency meter [15].

For rapid calibration of a spectrometer it is often convenient to plot the spectrum of a mixture of substances with known distances between the peaks. A convenient calibration mixture is a solution of 1% of benzene and 1% of tetramethylsilane in carbon tetrachloride; the distance between the two peaks is 436 Hz at 60 MHz or 291 Hz at 40 MHz. Another mixture contains a series of substances dissolved in carbon tetrachloride; each of the components gives a single narrow peak, which is shifted relative to the signal of tetramethylsilane. Table I-2 gives the composition of the mixture and the position of the lines in its spectrum.

For the nuclear resonance of the nuclei F^{19}, P^{31}, etc., which have large chemical shifts, the extent of the spectrum is usually considerably greater (200-400 ppm in comparison with 10-15 ppm in the case of protons). The side band method is usually used to calibrate these spectra.

10. SCOPE AND LIMITS OF APPLICABILITY OF THE NMR METHOD

The use of NMR spectroscopy for investigating organic compounds is similar to the use of infrared spectroscopy. These two methods supplement each other to a considerable extent and when used together they provide a powerful method of investigating structure, electron distribution, and reactivity of molecules. Below we list some of the fundamental differences between the NMR method and IR spectroscopy, which determine the features of their application.

1. While IR spectroscopy records the bonds between atoms, NMR spectroscopy records magnetic nuclei of a definite sort (largely protons) and the nature of the change in their state under the influence of the immediate environment. Since the molecule contains comparatively few magnetic nuclei which are recorded under given resonance conditions, NMR spectra are usually simpler to interpret and for the most part there is no need to use model compounds to analyze the spectrum.

2. NMR spectroscopy is slower in the sense that an appreciable time, measured in seconds, is required to establish energy equilibrium between a system of nuclear spins and the surrounding medium (lattice). Therefore, processes which are classified as rapid from the point of view of NMR spectroscopy (for example, internal rotation, proton exchange, etc.) may be quite slow from the point of view of infrared spectroscopy.

3. The intensities of lines (peak areas) in NMR spectra are proportional to the number of resonating nuclei, while in IR spectra they depend strongly on the nature of the bond and may change from compound to compound.

4. Nuclear resonance has a considerably lower sensitivity than IR spectroscopy. Although the sensitivity of the NMR method may be increased as the procedure and equipment are refined (dynamic polarization of nuclei, accumulation of the signal, etc.), this difference will always be maintained since the intensity of the NMR signal is determined by a very small difference in the populations of the spin states of nuclei in a magnetic field.

5. High-resolution NMR spectroscopy is only possible with the use of liquid samples of low viscosity.

These features make it possible to distinguish the spheres of application of the two methods and to determine the fields in which they can be used together most profitably.

The main application of NMR spectroscopy is in the determination of the structure of pure organic compounds. The method is particularly important for studying the configuration of the main chain, isomerism, and the stereochemistry of a molecule. The last of these applications is connected with the presence in organic molecules of magnetically anisotropic groups, whose steric position strongly affects the form of the spectrum. These groups include aromatic and three-membered rings, carbonyl groups, and acetylene and nitrile groups. The possibility of determining the stereochemistry comparatively simply has resulted in the wide application of NMR spectroscopy in the investigation of natural products. NMR spectroscopy is invaluable in the determination of cis-trans isomerism relative to a double bond, the isomerism of benzene derivatives, and the composition of a mixture of keto enols and other tautomers. The main limitations of the method are determined by the complexity of the interpretation of the spectrum when a large number of magnetic nuclei are present and also the possibility of selecting a suitable solvent (which does not absorb in the resonance region of the substance investigated). The first limitation is overcome to a considerable extent by refinement of the techniques of mathematical analysis of the spectra and by the use of special methods. The latter include double nuclear magnetic resonance, isotopic substitution, the use of instruments with a higher magnetic field strength, the investigation of resonance of C^{13} nuclei in natural compounds, etc. (Ch. IV). The second limitation is eliminated by using a set of solvents, including isotopically substituted (largely deuterated) compounds.

Functional analysis is an equally important problem for NMR spectroscopy. Functional analysis by means of NMR is simpler for groups containing magnetic nuclei. Many functional groups (amino, hydroxyl, carboxyl, carbalkoxyl, and aldehyde) and also fluorine- and phosphorus-containing groups are of precisely this type. In the case of groups which contain no magnetic nuclei (the carbonyl group, some oxygen- and sulfur-containing groups, halogens apart from fluorine, etc.) we investigate the changes which they produce in the character of the absorption of the nearest or more remote protons. The combined application of IR and NMR spectroscopy (and also mass spectrometry) for functional analysis is obviously most profitable.

We should examine in particular quantitative organic analysis by means of NMR spectroscopy. The proportionality between the peak areas and the number of nuclei resonating at a given frequency offers a method of using NMR for quantitative elementary and functional analysis. With the use of a well-calibrated integrator and calibrated sample tubes, quantitative analysis may be carried out by comparing the integral areas of the peaks in the test and standard samples separately with the same setting of the spectrometer. Another method consists of comparing areas of separate signals and the area of the peak of a standard substance, added in a definite amount directly to the sample investigated. It should be noted that the quantitative determination of fluorine by NMR in compounds with a trifluoromethyl group is almost the only method of analyzing such substances. NMR spectra are often used in combination with normal chemical methods for the quantitative determination of functional groups. Thus, for the determination of active hydrogen the substance is dissolved in heavy water (polar solvents such as acetone and pyridine are added to improve the solubility) and after exchange, the area of the H_2O (HDO) signal is determined. In this case there is no need for the exchange to proceed to completion since the signals of groups with labile hydrogen (hydroxyl, carboxyl, and amino groups) merge with the signal of water. With more prolonged exchange and the use of catalysts it is also possible to determine in this way the groups with less labile hydrogen such as acetylenyl hydrogen, methylene protons in malonic ester, protons of methyl groups in acetone, etc.

Integration of the peaks is frequently used for determination of the quantitative composition of mixtures with a small number of components, particularly in the case of difficultly separable mixtures of isomers.

The main limitation on the use of NMR for quantitative analysis is the reduction in the peak areas on saturation. To avoid saturation the quantitative analysis is usually carried out with a low level of rf field. With a low content of magnetic nuclei in the sample investigated this is contrary to the requirements for the sensitivity of the spectrometer. An increase in the rf level leads to partial saturation of the signals and the degree of saturation may be different for the test and standard peaks, leading to a considerable error.

Several methods have been proposed for avoiding errors as a result of nonuniform saturation of the peaks. One of them consists of using fast passage of the signal. In practice, with a signal width of 1 Hz and a relaxation time of 1 sec, a spectrum sweep rate of 25 Hz/sec practically eliminates the error from saturation. However, this method leads to an increase in the error of the integrator, which increases with an increase in the sweep rate. Another method of eliminating saturation consists of adding paramagnetic compounds. With small amounts of paramagnetic substances added the relaxation time is reduced so much that saturation does not occur even with a considerable rise in the level of the rf field. This method also has definite drawbacks. When paramagnetic impurities are added there is considerable broadening of the signal so that it is not possible to distinguish between separate peaks. The addition of a small amount may result in the saturation conditions for the test and standard peaks differing appreciably and this will introduce a substantial error.

In the general case, for quantitative analysis with a low intensity of the peaks it is necessary to carry out a preliminary calibration of the spectra with the substance investigated at different concentrations. The combination of this calibration with different methods of eliminating the error from saturation makes it possible to achieve the same accuracy as in normal chemical methods of quantitative analysis.

The possibility of carrying out quantitative measurements with time and under definite temperature conditions makes NMR spectroscopy useful for physicochemical investigations, primarily for studying the kinetics of chemical reactions. The work is facilitated in cases where it is possible to find a suitable criterion for following the reaction, for example, the change in the chemical shift of the intensive signal of a methyl group, etc. The NMR method is used widely for investigating the kinetics of exchange processes (proton exchange and conformational conversions), and also for studying hydrogen bonds and different forms of association. The possibility of carrying out physicochemical investigations is limited by the relatively low sensitivity of the instruments, which prevents such essential measurements at considerable dilution.

It is important to note that because of the use of glass sample tubes, making it possible to achieve complete isolation from the surrounding medium, the NMR method is convenient for the investigation of corrosive and poisonous substances. Thus, for example, NMR spectroscopy finds application in the study of the protonization of bases by strong acids (for example, trifluoroacetic, sulfuric, etc.) and in contrast to ultraviolet spectroscopy, which is used for the same purpose, when NMR is used it is possible for the most part to determine the position of addition of the proton accurately. NMR spectroscopy is evidently the most convenient method for qualitative and quantitative characterization of such poisonous compounds as esters of fluorophosphinic acids, organophosphorus insecticides, and other similar materials.

The list of different applications of NMR spectroscopy given, which is far from complete, indicates the possibilities of using this method for practical purposes. On the other hand, the parameters of NMR spectra may be correlated with the electron distribution in the molecule. For example, the chemical shifts of fluorine in meta and para derivatives of fluorobenzene correlate well with the Hammett constants and thus they may be used for determining the electronic structure of benzene derivatives. It has been established that for most compounds the spin−spin coupling constants of H^1-C^{13} nuclei connected directly are determined by the ratio of s- and p-character of the bonding orbitals; these parameters, which are important in quantum mechanical calculations on complex molecules, may be obtained by means of NMR spectroscopy. Essentially all the parameters of NMR spectra are determined largely by the electron distribution in the molecule. The establishment of these interrelations is the main scientific problem of NMR spectroscopy.

11. INDEXING OF NMR SPECTRA

NMR spectroscopy of organic compounds is firmly established as a practice in many research and industrial laboratories. As a result of the high stability of operation of modern NMR spectrometers, their reproducibility is so high that the number of high-resolution NMR spectra plotted in practice cannot be counted. As a result of this the problem of systematizing the spectra arises for generalization and simplification of the exchange of information and the accumulation of data for catalogs. Below we examine an indexing system, which has been proposed as a short notation for high-resolution NMR spectra. This system may provide a basis for processing spectra by means of computers.

Five sections are proposed and at least the first four of these must appear in the index of each spectrum.

In section I the magnetic nuclei present in the substance investigated are characterized. The observed nucleus is recorded and then the other magnetic nuclei. The atomic weight is not given. If the presence of a magnetic nucleus is not reflected in the spectrum (for example, the isotope C^{13} is present, but the splitting from this isotope is not observed), then it should not be recorded. It is recommended that only the following nuclei should be recorded by separate symbols: H, F, P. N. B, C, Hg, Tl, X (X denotes all other magnetic nuclei).

The instrument is characterized in section II. The resonance frequency (rounded off to whole numbers) is given and if double (or multiple) resonance is used, the second and subsequent frequencies are given also. The second frequency is indicated not by a number, but by the symbol of the isotope (of those given in section I) to which it corresponds. И denotes the use of an integrator.

The type of compound investigated is considered in section III. It may include up to four subsections, characterizing the skeleton of the compound:

Л : compounds with an open chain, which may include both carbon atoms and heteroatoms;

Ц : carbocyclic and nonaromatic heterocyclic compounds;

А: aromatic hydrocarbons and heterocycles;

С: complex compounds of unknown structure (which cannot be represented by a combination of the previous sections) and complex mixtures of compounds.

If the compound investigated may belong to several subsections then they are all indicated and separated by commas. If the sample is a mixture of a small number of different compounds, then the characteristics of each of them in the section are separated by the symbol +.

Each subsection contains an indication of the presence of functional groups and multiple bonds: Alk, OH, NH, NF, CHO, COOH, SH, ∂H, ∂F, =, ≡, Z. The symbol Z denotes other functional groups (it is possible to indicate only the groups which contain magnetic nuclei) and other types of bonds in addition to those listed if these are important; Э is an element. Acyl and other substituents containing a proton attached to carbon are denoted by Alk.

The complexity of the spectrum is characterized in section IV. The presence of multiplets of the following types in the spectrum is indicated in it (see Ch. IV): I, AB, A_2B, AA'B, ABX, ABC, A_2B_2, AA'XX', A_2BX, AA'BX, M.

The index I denotes simple spectra of the first order, while the index M indicates more complex multiplets. The indices are separated by a comma. For brevity, no figures are written below the line and the apostrophe is not written at all. Thus, the multiplet AA'B is written as AAB, while A_2X_2 is written as A2X2.

Some additional features of the spectrum are noted in section V.

The solvents are indicated by Ч (carbon tetrachloride), Хл (chloroform or deuterochloroform), Б (benzene), У (hydrocarbon such as cyclohexane), Ац (acetone or deuteroacetone), Д (dimethyl sulfoxide), С (alcohols), В (water or heavy water), and О (other solvents).

The use of an external reference is denoted by the index Вш. In this case, if a correction is introduced for the diamagnetic susceptibility of the sample the index П is included.

Spectra plotted with the use of a signal accumulator are indexed with a symbol Н.

If a complete computer analysis of the spectrum is carried out, then the index К is introduced. If a theoretical spectrum is constructed, then the index Т is used.

Thus, in the proposed system a total of 50 indexes is used and also three separators: . (period); , (comma), and + (plus). The main sections are separated by a period. The subsections and individual indexes of section IV are separated by a comma. The sign + is used in section III and also in sections I and II in cases where on one diagram there are several spectra of a given sample, for example, plotted under the resonance conditions of protons and fluorine. Below we examine several examples of the indexing of spectra.

The spectrum of acrylonitrile plotted in $CDCl_3$ solution with an internal reference at a frequency of 60 MHz is indexed as follows: Н.60.Л≡≡.ABC.Хл.   If in the spectrum of the same substance the satellites from spin−spin coupling with C^{13} nuclei are resolved, a complete computer analysis of the spectrum is carried out, and a theoretical spectrum is constructed, the index appears differently: HC.60.ЈI=≡.АВС. Хл KT.   If the H^1 and F^{19} spectra of 4-(2-fluorovinyl)pyridine are given on the diagram and the proton spectrum is analyzed completely by means of double proton-proton resonance (the double resonance spectra are also given) and a theoretical spectrum is constructed and also the signals of the satellites are obtained by means of an accumulator, then the index appears as follows: HFC+FHC.60C+56.A.ЈI=.AABB+ ABX. KTH.   Other examples of the indexing of spectra are given below. The system described is proposed as a basis. Naturally, in individual research groups the system of indexes may be extended in order to reflect the specific spectra features of interest to the given group. However, in the publication of spectra or information on them it is recommended that authors do not go beyond the proposed system. It should be remembered that a considerable increase in the number of indexes and complication of the system can only reduce its practical significance.

12. MANUALS AND TEXTBOOKS

ON NUCLEAR MAGNETIC RESONANCE

1. Physical Bases of the Method

A. Abragam, Principles of Nuclear Magnetism, Oxford Univ. Press, New York (1961)

I. V. Aleksandrov, The Theory of Nuclear Magnetic Resonance [in Russian], Izd. "Nauka" (1964).

E. Andrew, Nuclear Magnetic Resonance, Cambridge Univ. Press (1956).

A. Leshe. Nuclear Induction [Russian translation], IL (1963).

J. Pople, W. Schneider, and H. Bernstein, High-Resolution Nuclear Magnetic Resonance, McGraw-Hill, New York (1959).

F. I. Skripov, Course of Lectures on Radiospectroscopy [in Russian], Izd. LGU (1964).

K. B. Wiberg, Physical Organic Chemistry, Wiley, New York (1964).

2. Application of NMR Spectroscopy

in Organic Chemistry

N. S. Bhacca and D. H. Williams, NMR Spectroscopy: Steroids and Related Compounds, Holden-Day, New York (1964).

R. H. Bible, Interpretation of NMR Spectra, Plenum Press, New
    York (1964).
G. Conroy, in: Progress in Organic Chemistry, Vol. 2 [Russian
    translation], Izd. "Mir" (1964), p. 255.
J. Emsley, J. Feeney, and L. Sutcliffe, High Resolution Nuclear
    Magnetic Resonance Spectroscopy, Vols. 1, 2, Pergamon,
    London (1965).
L. M. Jackman, Application of NMR Spectroscopy in Organic
    Chemistry, Pergamon, London (1959).
G. Mavel, Theories Moleculaires de la Resonance Magnetique
    Nucleaire; Applications a la Chimie Structurale, Paris (1966).
NMR and EPR Spectroscopy [in Russian], collection of articles,
    Izd. "Mir" (1964).
B. Pesce, ed., Nuclear Magnetic Resonance in Chemistry, Academic
    Press, New York (1965).
J. D. Roberts, Nuclear Magnetic Resonance, McGraw–Hill, New
    York (1959).
H. Suhr, Anwendungen der Kernmagnetischen Resonanz in der Or-
    ganischen Chemie, Berlin (1966).
A. Weissberger, ed., Technique of Organic Chemistry, Vol. XI
    (ed. K. W. Bently), Chap. IV, Wiley, New York (1963).

3. Analysis of High-Resolution

NMR Spectra

R. H. Bible, Guide to the NMR Empirical Method, Plenum Press,
    New York (1967).
P. L. Corio, Structure of High Resolution NMR Spectra, Academic
    Press, New York (1966).
B. Dischler, Angew. Chem., Inter. Ed. in English, 5(7):623.
J. D. Roberts, Spin–Spin Splitting in High-Resolution NMR Spectra,
    Benjamin, New York (1961).
J. S. Waugh, Advances in Magnetic Resonance, Academic Press,
    New York (1965).

4. Combined Application of Spectrometric

Methods (NMR, IR, UV, and Mass Spectroscopy)

for Determining the Structure of Organic

Compounds

J. C. D. Brand and G. Englinton, Application of Spectroscopy to
    Organic Chemistry, London (1965).

T. Cairns, Spectroscopic Problems in Organic Chemistry, Vol. 1, London (1964).

D. W. Mathieson, ed., Interpretation of Organic Spectra, London (1965).

R. M. Silverstein and G. C. Bassler, Spectrometric Identification of Organic Compounds, Wiley, New York (1963).

5. Catalogs of NMR Spectra

N. S. Bhacca, L. F. Johnson, and J. N. Shoolery, NMR Spectra Catalog, Vol. 1, Palo Alto (1962).

N. S. Bhacca, D. P. Hollis, L. F. Johnson, and E. A. Pier, NMR Spectra Catalog, Vol. 2, Palo Alto (1963).

Sadtler's NMR Spectra Collection, Philadelphia (1966).

6. Tables of Calculated Spectra

R. L. Corio, Chem. Rev., 60:363 (1960).

K. B. Wiberg and B. J. Nist, The Interpretation of NMR Spectra, Benjamin, New York (1962).

7. Indexes

Formula Index to NMR Literatue Data (ed., M. G. Howell, A. S. Kende, and J. S. Webb), Vols. 1, 2, Plenum Press, New York (1964–1966).

H. M. Hershenson, NMR and EPR Spectra Index for 1958–1963, Academic Press, New York (1965).

LITERATURE CITED

1. C. I. Gorter, Phisica, 3:995 (1936).
2. I. T. Arnold, S. S. Dharmati, and M. E. Packard, J. Chem. Phys., 19:507 (1951).
3. W. G. Proctor and F. C. Yu, Phys. Rev., 77:717 (1950).
4. W. G. Proctor and F. C.Yu, Phys. Rev., 78:471 (1950).
5. H. S.Gutowsky, D. W. McCall, and C. P. Slichter, Phys. Rev., 84:589 (1951).
6. G. W.Nederbragt and C. A. Reilly, J. Chem. Phys., 24:1110 (1956).
7. É. Lippmaa, A. Olivson, and Ya. Past, Izv. Akad. Nauk ÉSSR, ser. fiz.-mat. i tekhn. nauk, 14:473 (1965).
8. M. W. Dietrich and R. E. Keller, Analyt. Chem., 36:258 (1964).
9. G. V. D. Tiers and A. Kowalewsky, Abstr. Meeting of Am. Chem. Soc., April 1960, p. 17R.
10. A. Kowalewsky, J. Biol. Chem., 237:1807 (1962).
11. L. M. Jackman, Application of NMR-Spectroscopy in Organic Chemistry, Pergamon, London (1959).

12.   F. D. Becker, J. Chem. Ed., 42:591 (1965).
13.   J. I. Musher and E. J. Corey, Tetrahedron, 18:791 (1962).
14.   I. N. Shoolery, Varian Ass. Tech. Inform. Bull., 3(3):8.
15.   A. Syugis and É. Lippmaa, Izv. Akad. Nauk ÉSSR, ser. fiz.-mat. i tekhn. nauk, 16:81 (1967).
16.   I. L. Jungnickel, Anal. Chem., 35:1985 (1963).

Chapter II

Chemical Shift

The chemical shift is the main characteristic of an atom or a group of equivalent atoms in a molecule in a high-resolution NMR spectrum. Chemical shifts were observed for the first time by Proctor and Yu [1]. However, we should regard the beginning of high-resolution NMR spectroscopy of organic compounds as the work of Arnold, Dharmatti, and Packard [2], who observed the spectrum of ethyl alcohol with three separate signals from three groups of equivalent protons for the first time in 1951.

A theoretical explanation of the phenomenon of the chemical shift was given by Ramsey [3], who showed that the difference in the resonance frequencies of nuclei of a given isotope (or, correspondingly, the difference in the magnetic field strength at a fixed frequency) is due to differences in the electronic environments of the nuclei. Lamb [4] had previously pointed out that an external magnetic field H_0 acting on atoms placed in it produces precession of the electrons of these atoms about the nuclei, analogous to the precession of nuclear spins in a magnetic field. The precession of the electrons produces a secondary magnetic field H' in the opposite direction to H_0. The field H_0 and H' are related by the equation

$$H' = -\sigma H_0 \tag{II-1}$$

where σ is the screening constant.

For isolated atoms with a spherical distribution of the electron shell (for example, for a hydrogen atom in the ^1S state) the atomic

61

diamagnetic susceptibility is the only reason for the appearance
of a local field at the nucleus, which is reduced in comparison with
the external magnetic field. As Ramsey showed by a more strict
quantum mechanical examination [3], in the general case the equation for the atomic screening constant includes two terms, a diamagnetic and a paramagnetic term, and the second of these takes
into account the wave functions of the excited states of the system
and has the opposite sign to the first. With a spherically symmetrical electron shell the paramagnetic term is reduced to zero,
but increases with distortion of the spherical distribution. Thus,
the paramagnetic part of the atomic screening is a measure of the
deviation from a spherical distribution of the electronic environment of the nucleus. It is particularly important for magnetic
resonance of heavy nuclei.

In the case where the magnetic nucleus is part of a molecule,
the electron currents and secondary fields induced by them depend
on the position of the molecule relative to the external magnetic
field. In liquids and gases there is continuous movement of the
molecule and σ, i.e., the experimentally observable values of the
magnetic screening of the atom is the result of averaging of the
secondary magnetic fields in three directions.

Thus, real magnetic nuclei surrounded by electrons are subject to the action of not the applied external magnetic field H_0, but
the local field H_{loc}, which is related to the external field by the
equation:

$$H_{loc} = H_0 (1 - \sigma) \tag{II-2}$$

$$\sigma = \sigma^P + \sigma^D \tag{II-3}$$

A. PROTON CHEMICAL SHIFTS

1. ABSOLUTE SCALE OF CHEMICAL SHIFTS
OF HYDROGEN

The theory of magnetic screening presented above was used
for calculations of the absolute chemical shift of the hydrogen
molecule. These calculations, which were based on the use of the
wave function of the H_2 molecule, were carried out by several authors. Aleksandrov [5] gave a comparative assessment of results

TABLE II-1. Absolute Proton Chemical Shifts
of Some Compounds

Compound	σ, ppm	Compound	σ, ppm
H^+	0	C_6H_{12}	26,6
H_2	26,6	C_6H_6	23,5
$Si(CH_3)_4$	35,4	$CHCl_3$	20,5
H_2O	30,2	HCl (gas)	31,2
CH_4	30,8	HF (theor.)	31,1

obtained by different routes and also carried out a calculation for
the hydrogen molecule and some others (methane, ethylene, and
acetylene) using the method of molecular orbitals. The following
results were obtained for the hydrogen molecule (in ppm):

$$\sigma^D = 28; \quad \sigma^P = -5.3: \quad \sigma = \sigma^D + \sigma^P = 22.7$$

The later calculations of Kolker and Karplus [6] for H_2 and
HF molecules using self-consistent field functions gave 26.56 and
31.1 ppm, respectively, and the difference in the chemical shifts
corresponds to experiment. These results denote that the nuclear
magnetic resonance signal of molecular hydrogen is shifted by 26–
28 ppm towards higher fields in comparison with an unscreened
hydrogen nucleus. Indirect experimental confirmation of the validity of the calculations is provided by the fact that the screening
constants obtained correspond in order of magnitude to the region
of chemical shifts of differently screened hydrogen atoms and the
difference between the calculated absolute chemical shifts corresponds to experiment.

Calculations of the absolute chemical shift for more complex
molecules have only limited value at the present time, firstly because of their laboriousness, but mainly because they are very
rough in comparison with the accuracy achieved experimentally [7].
The main result of the calculations for molecular hydrogen lies
in the possibility of constructing an absolute scale of chemical
shifts by measuring the chemical shift of molecular hydrogen relative to the generally accepted references.

Table II-1 gives the absolute chemical shifts of some compounds (in a gaseous state), which are based on an absolute chemical shift for H_2 of 26.6 ppm.

As has already been pointed out, relative chemical shifts are used in practice and these are essentially the differences in the absolute shifts of the nucleus investigated and the reference.

2. ADDITIVE COMPONENTS OF MAGNETIC SCREENING

Taking into account the effect of other atoms and molecules on the electron currents at a given nucleus and the secondary magnetic field created by them presents great difficulties within the framework of the general theory of magnetic screening. Saika and Slichter simplified the problem [8] by dividing the magnetic screening constant into approximately additive components. Below we will examine the three following groups of additive contributions to the magnetic screening:

1.   The atomic component σ_A, which is related to the electron currents of the electrons of the given atoms;
2.   the molecular component σ_M, which is due to electron currents of adjacent atoms and interatomic (intramolecular) currents;
3.   the component caused by intermolecular interactions and the sample as a whole σ'.

The total magnetic screening constant consists of the sum of the three contributions:

$$\sigma = \sigma_A + \sigma_M + \sigma' \qquad \text{(II-4)}$$

In this case the absolute chemical shift of the signal of a given nucleus (or group of equivalent nuclei) equals the screening constant σ, while the shift relative to a standard

$$\delta = \sigma_{st} \pm \sigma \qquad \text{(II-5)}$$

where the sign + or − depends on the choice of the scale of chemical shifts. When the scale of δ relative to tetramethylsilane is used (the relative shift of tetramethylsilane is taken as equal to zero and signals at a lower field have the symbol +):

$$\delta = \sigma_{TMS} - \sigma \qquad \text{(II-6)}$$

Consequently, the more the nucleus is screened, the smaller its chemical shift. Since the magnetic screening of the protons in tetramethylsilane is very great (see Table II-1), for most compounds the chemical shift remains positive. An increase in the chemical shift corresponds to a shift of the signal to lower fields.

The first two components of the magnetic screening σ_A and σ_M are related to the molecular structure of the substance investigated and are naturally characteristic of this structure. The intermolecular component σ' is determined by external factors, namely, the temperature of the sample, the aggregate state, the presence and nature of the solvent, and macroscopic magnetic properties of the molecules. As a rule, for structural investigations of compounds it is necessary to eliminate this component either by standardizing the experimental conditions or by keeping it constant. In investigations of a physicochemical nature on intermolecular interactions, the hydrogen bond, etc., the determination of σ' is often the main purpose of the experiments.

3. ATOMIC COMPONENT OF MAGNETIC SCREENING σ_A

As a rule, the paramagnetic contribution to the atomic screening σ_A^P does not play a substantial part for protons. Therefore, it is to be expected that an increase in electron density at a neighboring atom, leading to an increase in the diamagnetic contribution σ_A^D, produces an increase in the screening constant σ_A as a whole.

Experiments to study the relation of the chemical shift of protons to the electron density at an adjacent carbon atom indicates precisely this rule. Good correlation is observed particularly in the case of aromatic compounds. Spiesecke and Schneider [9] compared the chemical shifts in the tropylium cation I, benzene II, and the pentadienyl anion III.

I   II   III

These compounds make up an isoelectronic series with 6 π-electrons, which are distributed uniformly between all the carbon atoms. The π-electron density (ρ) at the carbon atoms equals $\frac{6}{7}$, 1, and $\frac{6}{5}$ of an electron, respectively. The NMR spectra of these compounds consist of single peaks in accordance with the magnetic equivalence of the protons. Comparison of the chemical shifts and the electron density gives the relation

$$\delta_A = \delta_{C_6H_6} - k(\rho - 1) \qquad (II-7)$$

where k is the empirical proportionality coefficient and in this case it equals 10.6 ppm per electron.

Equation (II-7) was found to apply to a large number of aromatic compounds, including aromatic heterocycles, and a value of k from 9.3 to 10.6 ppm per electron was used. Therefore, the chemical shifts in NMR spectra are a more reliable criterion for the determination of the π-electron density than quantum mechanical calculations [10]. However, naturally accurate results may be obtained only if careful account is taken of other contributions to the magnetic screening, namely, σ_M and σ'.

Buckingham − Musher Effect

Buckingham [11] made an attempt to explain the observed values of the chemical shifts from the point of view of the internal electric field of molecules created by the electric dipoles. Even earlier it was shown that in an isolated hydrogen atom the screening constant changes in proportion to the square of the electric field acting on the nuclei [12]. The calculations of Buckingham for the case where the hydrogen atom examined is at a fixed distance from the center of the dipole (as occurs in molecules) lead to the conclusion that the main contribution to the screening is proportional to the first power of the electric field acting in the direction of this atom, E_Z, in accordance with equation (II-8):

$$\Delta\sigma = k_E E_z - 0.738 \cdot 10^{-18} E^2 \qquad (II-8)$$

The contribution from the second term in Eq. (II-8), which is proportional to E^2, becomes substantial only with strong electric fields. The values of k_E obtained by calculation or by experiment lie in the range from $(-2$ to $-3.4) \cdot 10^{-12}$ el $\cdot$ cm $\cdot$ V/cm [13].

Equation (II-8) makes it possible to relate the rule for the change in the chemical shifts of protons to the electric charge at

the adjacent carbon atom. Let us examine the positive charge of
the ion $C_6H_7^+$, which is formed by addition of a proton to a benzene
molecule. If we ignore the disruption of the symmetry of the benzene ring, then the increase in the positive charge at each carbon
atom is $\frac{1}{6}$ of an electron or $\frac{1}{6} \cdot 4.8 \cdot 10^{-10}$ esu. In this case the
field along the axis of the C−H bond with a C−H bond length of
1.07 A changes by

$$E_z = \frac{1 \cdot 4.8 \cdot 10^{-10}}{6 \cdot (1.07)^2 \cdot 10^{-16}} = 0.70 \cdot 10^6$$

and

$$E^2 = E_z^2 = 0.49 \cdot 10^{12}$$

Taking the value $k_E = -2 \cdot 10^{-12}$ proposed by Buckingham and multiplying all the terms in equation (II-8) by 10^6 for converting into
parts per million, we obtain the change in the screening constant

$$\Delta\sigma_A = 1.40 - 0.35 = -1.75 \text{ ppm}$$

It was found experimentally that on protonation of methyl derivatives of benzene (mesitylene, 1,3,4-trimethylbenzene, durene,
and 1,3,4,5-tetramethylbenzene) at low temperatures there is distortion of the aromatic ring due to the addition of the proton. The
chemical shifts of the ring protons in the ortho, meta, and para
positions relative to the CH_2 in IV are 2.8, 2.3, and 3.2 ppm, respectively, in a weak field relative to protons attached to carbon atoms
in the sp^2 state.

H   H   −2.8 ppm   (+0.21 e)

(+)   −2.3 ppm   (+0.17 e)

IV   −3.2 ppm   (+0.24 e)

The total chemical shift of the five protons is 13.4 ppm; the increase in comparison with the value of 10 ppm per electron given
above is due to the absence of the ring current. The relation between chemical shifts of individual protons of the ring makes it
possible to determine the distribution of the charge, which agrees
closely with that calculated by Huckel's method.

We should note that even with the considerable change in the field which corresponds to the change from a neutral aromatic molecule to a positive ion, the term in equation (II-8) which is proportional to E^2 makes only an insignificant contribution to the change in magnetic screening; in most cases it may be assumed that the change in the chemical shift is related linearly to E and, correspondingly, to the charge on the neighboring carbon atom.

Naturally, correlations based on the use of equations (II-7) and (II-8) are only qualitative in the general case and good agreement may be obtained only for a series of compounds which are similar in structure. Nonetheless, they are of definite value since they make it possible to relate the chemical shift directly to the electron density.

Buckingham derived formulas for the calculation of the chemical shifts of protons of the benzene ring in monosubstituted benzene derivatives, using the dipole moment of the compound. If it is assumed that the dipole moment of compound V is determined by a point dipole lying in the middle of the $C-X$ bond, then the electric fields created by it at the protons are determined by the relations

$$E_{CH_o} = \frac{4\mu \,(10l_{CC}^2 - 8l_{CH}^2 - 2l_{CX}^2 + 10l_{CC}l_{CH} + 5l_{CC}l_{CX} + 13l_{CH}l_{CX})}{(4l_{CC}^2 + 4l_{CH}^2 + l_{CX}^2 + 4l_{CC}l_{CH} + 2l_{CC}l_{CX} - 2l_{CH}l_{CX})^{5/2}} \quad \text{(II-9)}$$

$$E_{CH_m} = \frac{4\mu \,(42l_{CC}^2 + 8l_{CH}^2 + 2l_{CX}^2 + 42l_{CC}l_{CH} + 21l_{CC}l_{CX} + 13l_{CH}l_{CX})}{(12l_{CC}^2 + 4l_{CH}^2 + l_{CX}^2 + 12l_{CC}l_{CH} + 6l_{CC}l_{CX} + 2l_{CH}l_{CX})^{5/2}} \quad \text{(II-10)}$$

$$E_{CH_p} = \frac{16\mu}{(4l_{CC} + 2l_{CH} + l_{CX})^3} \quad \text{(II-11)}$$

where l_{AB} is the length of the $A-B$ bond (in cm) and μ is the dipole moment of the compound (in esu).

For $l_{CC} = 1.39$ Å, $l_{CH} = 1.07$ Å, $l_{CX} \approx 2$Å

$$E_{CH_o} = 8.74 \cdot 10^{22}\mu \qquad E_o^2 = 76.4 \cdot 10^{44}\mu^2$$
$$E_{CH_m} = 2.32 \cdot 10^{22}\mu \qquad E_m^2 = 5.88 \cdot 10^{44}\mu^2$$
$$E_{CH_p} = 1.75 \cdot 10^{22}\mu \qquad E_p^2 = 3.07 \cdot 10^{44}\mu^2$$

The application of these relations to an isolated nitro-benzene molecule, for which $\mu = 4 \cdot 10^{-18}$ esu, using formula (II-8) gives the following results (in ppm):

$$(\Delta\sigma_A)_o = -0.70 - 0.13 = -0.83$$
$$(\Delta\sigma_A)_m = -0.19 - 0.01 = -0.20$$
$$(\Delta\sigma_A)_p = -0.14 - 0.005 = -0.145$$

As these results show, the term containing E^2 plays a secondary part except in the case of the ortho position, where the s o u r c e   o f   t h e   f i e l d is close to the proton.

Experimental data for nitrobenzene in cyclohexane (1 : 1) are as follows (shift toward lower fields relative to benzene): ortho − 0.97; meta − 0.30; para − 0.42 ppm. Better agreement may be obtained if we take into account the effect of the solvent on the shifts obtained experimentally for nitrobenzene.

Buckingham's theory of the internal electrostatic field has been confirmed experimentally in another work [14] and also by the good agreement observed between the dipole moments and chemical shifts of para-protons in monosubstituted benzenes [15]. Musher [13] reported the possibility of using equation (II-8) for calculating the chemical shifts of protons in ethane derivatives, relating the chemical shifts to the inductive effect of the substituents. However, the difficulties of an accurate determination of the electrostatic field at actual protons in a molecule limit its practical value [16]. It has also been reported that there is disagreement between experimental results and values calculated from Buckingham's formulas for 2,6-dimethyl derivatives (for example, of aniline and fluorobenzene) even with careful allowance for the other parameters in magnetic screening [17]. Therefore, the results of Buckingham have not found wide application for the calculation of chemical shifts in relation to the substituents. It will be shown

TABLE II-2. Electronegativities of Substituents [20]

Substituent	x	Substituent	x
CN	2.49	NH$_2$	2.99
COOH	2.60	Cl	3.25
COCH$_3$	2.60	OH	3.43
C$_6$H$_5$	2.75	OCOCH$_3$	3.80
SH	2.45	ONO$_2$	3.91
I . . . . . . , . . .	2.61	F	3.93
Br	2.94		

later that Buckingham's theory is of considerable value in the study of the effect of intermolecular interactions on magnetic screening.

Chemical Shift and Donor-Acceptor

Properties of Substituents

Another way of taking into account the effect of substituents consists of correlating their electronegativity with the chemical shifts of adjacent protons. The first work on this correlation was carried out by Daily and Shoolery in 1955 using an NMR spectrograph working at a frequency of 30 MHz [18]. Having determined the chemical shift of the protons in four haloethanes, the authors derived a relation between the difference in the chemical shifts of methyl and methylene groups and the electronegativity of the substituents and constructed on this basis a scale of electronegativities for more complex substituents. These data were refined later to take into account the effect of the solvents and a linear relation was obtained between the electronegativity of the substituent and the chemical shift [19]. This relation is given here in a rearranged form

$$\delta_{CH_2} - \delta_{CH_3} = 1.464x - 2.61 \qquad \text{(II-12)}$$

where x is the electronegativity of the group X in the compound CH_3CH_2X. The values of x for various groups are given in Table II-2.

A linear relation between the chemical shift and the electronegativity of the substituents has also been reported for methyl, propyl, and isopropyl compounds [20, 21] and also for derivatives of hexachlorobicyclo(2,2,1)heptane [22] and dichlorocyclopropane [23]. However, the need to construct an independent scale of electronegativities and the comparatively narrow range of compounds investigated limits the application of equation (II-12).

If the substituent is attached to an aromatic ring and also in certain other cases, similar relations in which the chemical shifts are related to the reactivity parameters σ of Hammett and Taft find practical application. A good linear correlation has been observed both for protons of the benzene ring and also for hydrogen atoms in side chains attached to the ring both through a carbon chain and through a heteroatom. The possibility of correlating the chemical shift of protons of the benzene ring with the reactivity parameters was explained by Taft [24] by the fact that the effect of polarization in the transition state is quite close to the effect in the ground state.

A particularly good linear relation is observed between the chemical shift of protons in the para position relative to the substituent and the parameters of this substituent. Taft [25] showed that for monosubstituted benzenes the chemical shift may be obtained with an accuracy down to 0.07 ppm by means of the equation

$$\delta_p = 0.40\,\sigma_I + 1.04\sigma_R^0 + 0.02 \qquad\qquad (II-13)$$

where σ_I is the Taft inductive constant and σ_R^0 is the refined resonance constant of the substituent (corrected for isovalent conjugation) [26].

The relation between the coefficients of $\sigma_I(\rho_I)$ and $\sigma_R^0(\rho_R)$ indicates how sensitive the chemical shift is to the inductive effect and the conjugation effect.

For protons of the benzene ring in the para position relative to the substituent a good linear relation is often observed with the Hammett constant σ [27]. However, in the case of substituents in the meta or ortho position the correlation is much worse [20, 28]. Somewhat better results for meta substituents may be obtained by using an equation of the type (II-13) with different coefficients of the susceptibility ρ.

Fraser [17] studied the chemical shifts of protons in derivatives of meta-xylene VI in relation to the parameters of the substituent X.

VI

If there is no appreciable steric effect for the substituents X, then a good relation of the following form is observed:

$$\delta_p = 7.03 + 0.284\sigma_I + 0.767\sigma_R$$

with a maximum error of 0.028 ppm in the case where X = I:

$$\delta_m = 7.03 + 0.284\sigma_I + 0.167\sigma_R \qquad \text{(II-14)}$$

with a maximum error of 0.04 ppm in the case where X = Br. The value δ_m was obtained by adding 0.172 ppm to the shift of the meta protons observed experimentally in order to take into account the effect of the methyl groups on the chemical shift of these protons when X = H or D.

The formulas (II-14) presented show that for protons in the meta position the inductive effects of the substituents play a much greater part, while the shift of the para protons is determined largely by the resonance constant of the substituent. This is evidently responsible for the good correlation between the chemical shifts of the para protons and the Hammett constants of the substituents.

The system of equations (II-14) makes it possible to eliminate σ_I by relating the difference in the chemical shifts of the meta and para protons (relative to any standard) to the resonance constant of the substituent. A much better correlation is observed then than when the shifts of each nucleus are used separately. A number of other studies of the chemical shift of the protons of the benzene ring also indicate that the difference in the chemical shifts of neighboring protons correlates better with the parameters of the substituents than the chemical shifts as such [20, 29]. An analogous phenomenon is observed on comparing the chemical shifts of the protons of ethyl groups and the electronegativity of substituents. This is undoubtedly explained by the fact that in this way it is possible to eliminate to a considerable extent the purely magnetic effect of the substituents, which is not taken into account by the reactivity parameters and the electronegativity. This will be examined in more detail in the next section.

As a rule, the chemical shifts of protons in the side chain of benzene derivatives correlate well with the reactivity parameters of substituents in the para position. Jackman [30, p. 57] reports that the position of the signal of the methyl group in substituted toluenes changes insignificantly and only strong electron-attracting substituents have an appreciable effect on the shift. This effect is somewhat greater for ortho substituents than for substituents in the para position.

As is to be expected, in the case of para-substituted ethylbenzenes [31], substituents have a stronger effect on the chemical shifts of protons in the α-position relative to the ring than the β-protons. It is characteristic that the halogens are exceptions to the general relationship. This peculiarity of halogens has been reported in other cases.

Cook and Danyluk [32] found that the relation between the chemical shift of the acetylenyl proton and the reactivity parameters of substituents in the para position of phenylacetylene derivatives may be represented by relations of the type [II-13], by using other coefficients of the susceptibility ρ:

$$\Delta\delta_p = 0.330\sigma_I + 0.408\sigma_R$$

The maximum deviation is observed in the case where $X = NO_2$. In the case of phenylacetylene the ratio $\rho_I : \rho_R$ is close to unity, indicating a considerably greater role of the inductive effect in the screening of acetylenic protons in comparison with protons of the benzene ring.

Somewhat poorer correlation is observed in cases where the protons screened are attached to the benzene ring through a heteroatom. In the first work in this series [33], where the chemical shifts of anisole and phenetole derivatives were examined, it was reported that with these compounds there is a correlation with the Hammett constants σ to within 15%. For the related para-substituted phenols no appreciable relation was observed between the chemical shifts of the protons of the OH group and the nature of the para substituent of the benzene ring even though the internal chemical shift of the ortho and meta protons of the benzene ring correlates approximately linearly with other physical properties of the substituents [34].

It is interesting that in the case of aniline derivatives the degree of correlation depends on the solvent in which the chemical shift is measured. Good correlation between the chemical shifts of the amino protons and the Hammett constants σ_p and σ_m was observed for 20 derivatives of aniline when acetonitrile was used for the solvent, while in carbon tetrachloride and deuterochloroform the correlation was worse [35]. It is possible that careful selection of the solvent may also improve the correlation in other cases.

A good correlation between the chemical shift of the protons of the sulfhydryl group and the Hammett constants σ_p and σ_m was observed for meta- and para-substituted thiophenols. In this case, as with phenylacetylene derivatives, the greatest deviation is given by a nitro group in the para position relative to the sulfhydryl group. However, if we use for it the refined value σ^- [36] on the grounds that there is direct conjugation of a nitro group in the para position and the sulfhydryl group, then the chemical shift and the parameters of the substituent may be related by the linear equations (the correlation coefficient $r = 0.993$) [37]:

$$\delta = 0.283\sigma + 3.24$$

The good correlation observed in the case of sulfhydryl derivatives as compared with phenols and amino compounds is evidently due to the absence of magnetic anisotropy at the sulfur atom and also the low tendency of the sulfhydryl hydrogen and the unshared pairs at the sulfur for association with the solvent or self-association. In the case where these factors play a substantial part, the correlation has more of a qualitative character. Thus, quite a rough correlation was observed for protons of the aldehyde group of benzaldehyde derivatives, whose chemical shift depends strongly on the effect of the anisotropy of the $C=O$ group [38].

In addition to aromatic compounds, quite a good linear relation between the Hammett constants σ of the substituents and the chemical shift of protons at the adjacent carbon atom was found

for a series of ethylene derivatives, namely, ω-substituted styrenes
and 1-substituted propenes [39], monosubstituted ethylenes, β-sub-
stituted acrylates, and vinyl ketones [40]. The use of the parameter recommended by Taft $(0.40\sigma_I + 0.70\sigma_R)$ (equation II-13) instead
of the Hammett constant σ does not improve the correlation. The
chemical shift of the methyl group in esters of acetic acid and the
methylene group in esters of succinic acid is in accordance with
the Taft constant σ^* of the ester radical, but there is no strictly
linear correlation, particularly in the case of branched or aromatic substitutents [41]. The chemical shifts of the protons of the OH
group in 15 aryldimethylcarbinols does not depend on the substituents in the benzene ring [42]. However, for aliphatic thiols, for
which the magnetic anisotropy of the bonds and association are insignificant, a good linear relation was observed $(r = 0.973)$ between the shift of the protons of the mercapto group and the constant σ^* of the substituent if account was taken of the degree of
branching of the radical in accordance with the equation

$$\delta = 0.77\,(\sigma^* + 0.29n) + 1.01$$

where σ^* is the Taft inductive constant and n the number of carbon —
carbon bonds. Considerable deviations of the chemical shift are
observed in cases where the substituent includes an aromatic ring.
In addition to descreening of the protons of the mercapto group due
to the magnetic anisotropy of the benzene ring, which naturally is
not allowed for by the value σ^*, here there may also be conjugation or hyperconjugation effects, leading to the withdrawal of unshared pairs from the sulfur atom and a reduction in the screening
of the proton associated with it [37].

Using substituted anilines, Dyall [35] compared results obtained by using different scales of the parameters of the substituents. For substituents with a strong negative conjugation effect
(NO_2, CN, CHO, CH_3CO) in the para position relative to a substituent with a positive mesomeric effect, the chemical shift of whose
protons is investigated (NH_2, OH, OCH_3), the best results may be
expected when using the parameters σ^- [43]. For substituents
with a positive mesomeric effect we cannot use the Hammett constants σ obtained from the dissociation constants of benzoic acids,
which make no allowance for the direct mesomeric interaction of
the substitutent with the reaction center, but it is correct to use

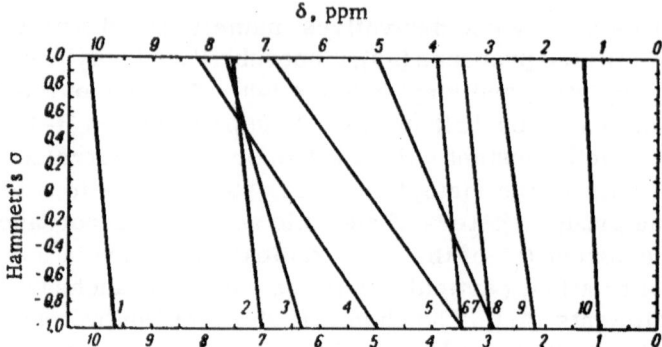

Fig. II-1. Relation of the chemical shifts of protons to the Hammett
parameters of the substituents σ: 1) protons of the aldehyde group
of meta- and para-substituted benzaldehydes; 2) protons of the
benzene ring in 2,4,6-trideutero-1-substituted benzenes; 3) para
protons in monosubstituted benzenes; 4) α-protons in β-substituted
trans-styrenes; 5) cis-protons in monosubstituted ethylenes; 6) protons of the methoxyl group in meta- and para-substituted anisoles;
7) protons of the amino group in meta- and para-substituted anilines;
8) protons of the sulfhydryl group in meta- and para-substituted thiophenols; 9) protons of the methylene group of para-substituted ethylbenzenes; 10) protons of the methyl group of para-substituted ethylbenzenes.

the values σ^0, obtained by Taft [44], starting from phenylacetic
acid. The use of the values σ_I and σ_R proposed by Taft does not
improve the correlation in the case of aniline derivatives, but it
was reported that it is useful to use these parameters in other
cases (for example, for derivatives of phenylacetylene [32]). It
should be noted that the use of the different parameters does not
lead to a substantial change in the nature of the relations.

It is interesting to compare the sensitivity of the chemical
shift of protons of different groups to the effect of substituents.
In Fig. II-1 different examples of this relation are reduced to a
single scale. The chemical shift of protons attached to a benzene
ring through one or two carbon atoms (lines 1, 9, and 10), carbon
and oxygen (6) or sulfur (8), and also protons in the meta position
relative to the substituent, shows a low sensitivity to the effect of
the substituents which does not exceed 0.5 ppm per unit of σ. The
effect of substituents on the shift of protons in the para position

relative to the substituent is somewhat greater (line 3) and according to data in [37], it is ~0.68 ppm per unit of σ. Protons of the
amino group are particularly sensitive to the effect of substituents.
For 20 meta and para derivatives of aniline in acetonitrile with
the use of values of σ^-, for 4-cyano, 4-acetyl, and 4-nitro derivatives, with the exception of the substituent 4-phenyl, for which
other factors play a substantial part, the susceptibility coefficient
$\rho = 0.975$ ppm per unit of σ was obtained. The relatively high susceptibility of the chemical shift of protons of the amino group in
aniline derivatives to the effect of substituents, the good linear
correlation (the correlation coefficient $r = 0.984$), and also the comparative accessibility of different aniline derivatives makes these
compounds particularly suitable for determining the Hammett constants σ of substituents by means of proton magnetic resonance.
For this purpose it is convenient to use equation (II-15)

$$\sigma = 1.027 \,(\delta - 3.84) \qquad\qquad (II-15)$$

where σ is the Hammett constant and δ is the chemical shift of the
amino protons of the aniline derivative in the form of a 5–15% solution in acetonitrile, relative to tetramethylsilane as an internal
reference.

It is important to note that the use of other solvents such as
carbon tetrachloride or deuterochloroform instead of acetonitrile
leads to a change in the susceptibility coefficient. This is explained
by the difference in the nature of the interaction of aniline derivatives with these solvents. The disadvantage of amines lies in the
fact that the signal of the amino protons is usually broad due to the
interaction with N^{14} nuclei and chemical exchange and this hampers
the accurate determination of the chemical shift. However, this is
compensated by the fact that the signal of the amino group is not
usually overlapped by the signals of other protons, either alkyl or
aromatic.

The chemical shift of olefinic protons in the cis position relative to the substituent (lines 4 and 5) is still more sensitive to the
effect of substituents ($\rho \sim 1.5$ ppm per unit of σ) and this is explained by the proximity of the substituent to the screened proton.
However, the same situation leads to a deviation from linearity.

For protons of the benzene ring in the ortho position relative to the substituent, which is analogous in the steric arrangement, no linear correlation is observed between the chemical shift and the reactivity parameters [20].

The good linear relation between the chemical shifts of protons and the reactivity parameters of the substituents observed in many cases is connected with the fact that they are both due to a considerable extent to the change in electron density at the atoms connecting them. However, the reactivity parameters and chemical shifts often depend on other factors, which are not directly related to the electron density. If the effect of these factors with a change from compound to compound does not change in parallel, then considerable deviations from the linear relation are observed. For the chemical shift these factors are the contributions to the screening constant σ_M and σ'. While the contribution from the intermolecular interaction and the magnetic properties of the sample σ' may be eliminated as a rule (but not always), allowing for the magnetic effect of neighboring atoms and groups in a given molecule often presents great difficulties. Below we will estimate the magnitude of this effect in different cases and show methods of allowing for it in the parameters of the substituents.

4. Molecular Component of Magnetic Screening σ_M

If we examine the effect of an adjacent atom, bond, or group of atoms on the screening constant of a nucleus which belongs to the same molecule, we see that a change in the screening constant with free movement of liquid molecules will be observed only when the magnetic field created by this neighboring group depends on its position relative to H_0. The magnetic field created by diamagnetic currents of the group A (Fig. II-2) of the given proton will increase the screening of the proton (H) when the A−H bond is parallel to H_0 and will decrease the screening when it lies perpendicular. If we assume that the secondary magnetic field which actually arises from circulation of the electrons of the group A creates a hypothetical point dipole [45], then for molecules which have axial symmetry of the group A, the contribution to the screening averaged for free movement of the individual molecules of the

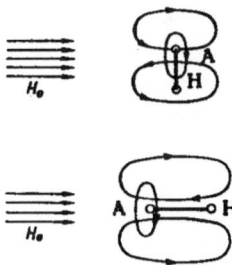

Fig. II-2. Effect of diamagnetic currents of the adjacent group A on the chemical shift of the proton H.

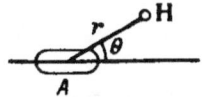

Fig. II-3. Diagram for calculation of the contribution to the chemical shift from the anisotropy from an adjacent group by formula (II-16). θ is the angle between the intercept r and the axis of symmetry of this group.

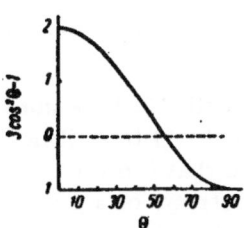

Fig. II-4. Curve of the relation $(3 \cos^2 \theta - 1)$ to θ (for calculations by equation II-16).

liquid, is determined by the McConnell equation

$$\Delta\sigma_{av}(A) = \frac{(3\cos^2\theta - 1)\,\Delta\chi}{3r^3} \quad \text{(II-16)}$$

where $\Delta\chi = \chi^{\parallel} - \chi^{\perp}$ is the difference between the longitudinal and transverse magnetic susceptibilities of the group A, r is the distance between the nucleus H and the center of the dipole of the group A, and θ is the angle between the intercept r and the axis of symmetry of this group (Fig. II-3). It is obvious that if the magnitude of the secondary field produced by the group A is independent of the arrangement of the A—H bond relative to H_0, i.e., $\chi^{\parallel} = \chi^{\perp}$, then with free motion of the molecules of a liquid (or gaseous) sample in the magnetic field $\Delta\sigma_{av}(A)$ averages out to zero and the group A thus makes no contribution to the screening of the nucleus H. In the general case when $\chi^{\parallel} \neq \chi^{\perp}$, neighboring groups may introduce both positive (screening) and negative (descreening) contributions to the magnetic screening of the nucleus examined. The absolute value of this contribution depends on the magnetic anisotropy of the group A and the magnitude of $\Delta\chi$ is a measure of this anisotropy.

Characteristic groups with a high magnetic anisotropy are halogens, a triple bond, the carbonyl group, and three-membered and aromatic rings. For such groups as double bonds and three-membered rings the use of equation (II-16) is not completely strict

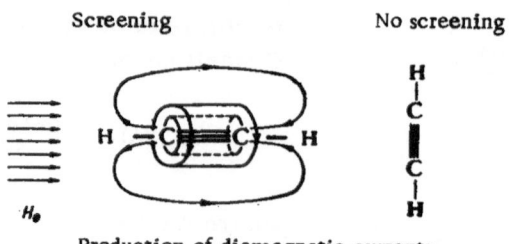

Production of diamagnetic currents

Fig. II-5. Magnetic anisotropy of the acetylene group.

since they do not have axial symmetry, but have different susceptibilities in three directions; in this case the anisotropy is determined by the equation:

$$\Delta \chi = 2\chi^{\parallel} - \chi_1^{\perp} - \chi_2^{\perp} \qquad (II-17)$$

However, the difference between $\chi_1^{\parallel}$ and $\chi_2^{\perp}$ is often unknown; in these cases equation (II-16) may be used directly for qualitative estimation with the assumption that $\chi_1^{\perp} = \chi_2^{\perp} = \chi^{\perp}$.

Another limitation on the use of equation (II-16) is due to the fact that it is based on the approximate replacement of real magnetic fields by fields from point magnetic dipoles, lying at the electronic centers of gravity of the groups. In this case, if r is comparable with the dimensions of the electron shell of the group, the use of equation (II-16) may introduce considerable errors. However, without the use of this assumption [46, 47] the calculations are much more complex.

For practical calculations using equation (II-16) it is convenient to use a graph of the relation of $(3 \cos^2 \theta - 1)$ to θ (Fig. II-4). From an analysis of the curve it follows that the function changes sign at $\theta = 55° 44'$, i.e., the same anisotropy of a group may produce screening of some protons and descreening of others, depending on their arrangement in the molecule. For linear molecules of the acetylene or hydrogen halide type one would expect an increase in the magnetic screening of protons lying along the axis of the molecule due to the production of diamagnetic currents, which are absent with a perpendicular arrangement (Fig. II-5).

The value of $\Delta \chi$ is usually expressed in cm^3 per molecule or cm^3/mole, taking into account the fact that 1 mole contains $6.025 \cdot 10^{23}$

TABLE II-3. Corrections to the Chemical Shifts of Protons
Due to Magnetic Anisotropy

Anisotropic group, A	Contributions to screening of protons, ppm*		Anisotropic group, A	Contributions to screening of protons, ppm*	
	A–H	A–C–H		A–H	A–C–H
Cl	–	+0.37	C = C = C	+100	+0.20
Br	–	+0.60	Benzene ring	from −1.5 to −2.0	−1.62
I	–	+0.87	Thiophene	−1.0	−0.40
C–C	−0.15†	–	Pyrrole	–	−0.40
C ≡ C	+4.0	+0.75	Imidazole	–	−0.40
C ≡ N	+4.25	+0.75	Cyclopropane	+1.2	−0.15

molecules. The same dimensions are obtained in calculation by formula (II-16) if $\Delta\sigma_{av}$ is given in relative units. The order of magnitude of $\Delta\chi$ is 10^{-30} cm^3 per molecule or 10^{-6} cm^3 per molecule.

The use of $\Delta\chi$ makes it possible to calculate simply the contributions to the chemical shifts of protons which are in different positions relative to the anisotropic group if the structural parameters of the molecules are known. However, it is often more convenient to use the values of the contributions to the shift of the protons obtained by direct experiment since these values are free from the errors introduced both by inaccurate determination of the molecular parameters (bond lengths and angles) and the indeterminancy in the selection of the position of the dipole in unsymmetrical groups. Data for some typical anisotropic groups are given in Table II-3.

The most important source of magnetic anisotropy is ring currents of π-electrons, which arise in aromatic compounds under the action of an applied magnetic field (Fig. II-6). When

* The sign (+) denotes a shift of the signal to higher fields and the sign (−), a shift to lower fields under the influence of the anisotropy group.
† Calculated on the basis of the value $\Delta\chi_{C-C} = 5.5 \cdot 10^{-6}$ cm^3/mole.

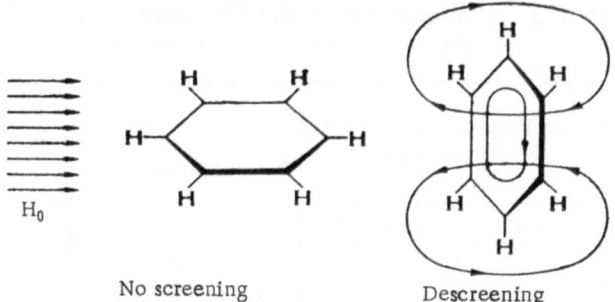

No screening   Descreening

Fig. II-6. Magnetic anisotropy of the benzene ring.

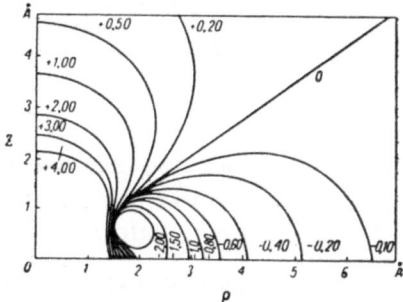

Fig. II-7. Lines of equal screening by ring
currents of the benzene ring.

the plane of the aromatic ring lies perpendicular to the field the
ring currents arising induce a secondary field, whose lines of force
are in the opposite direction to the external field above and below
the plane of the ring, while at the periphery of the ring the lines of
force of the two fields coincide. As a result, protons lying at the
periphery experience a descreening effect and their signals are
observed at lower fields than, for example, the signals of ethylene
protons. Proton-containing groups lying outside the plane of the
aromatic ring close to its center, on the contrary, experience a
screening effect from the side of this ring and their signal is
usually shifted to higher fields. When the aromatic ring lies along
the external field no appreciable electron currents arise so that
the contribution to the screening from ring currents does not disappear with random motion of molecules of liquid or gaseous aromatic compounds.

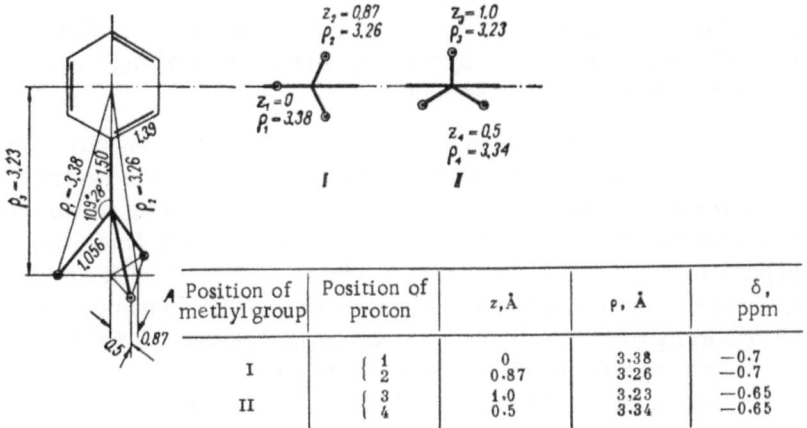

Position of methyl group	Position of proton	z, Å	ρ, Å	δ, ppm
I	1	0	3.38	−0.7
	2	0.87	3.26	−0.7
II	3	1.0	3.23	−0.65
	4	0.5	3.34	−0.65

Fig. II-8. Calculation of the contribution of ring currents of the benzene ring to the screening of methyl protons of toluene.

In calculations of the effect of the anisotropy of aromatic rings on chemical shifts it is also possible to use McConnell's equation (II-16), but appreciable errors may arise in this case, particularly for groups lying close together, connected with the fact that the distance from the center of the ring to the proton examined is comparable in magnitude to the size of the electron shell of the ring [46, 47]. Waugh and Fessenden calculated the contributions to the chemical shifts of protons close to a benzene ring without this approximation by using the empirical value of the chemical shift of the protons of benzene relative to nonaromatic compounds of similar structure, namely, cyclohezadiene-1,3, cyclooctatriene-1,3,5, and dicyclooctatetraene (an average of 1.5 ppm to lower fields). The calculation was based on the idea that the π-electron shell of benzene consists of two rings, lying on either side of the plane of the molecule symmetrically relative to the hexagonal axis and having a radius of 1.39 Å, which equals the length of the C−C bond in benzene; the length of the C−H bond is 1.08 Å. The results of the calculation are given graphically in Fig. II-7.

The figure gives the lines of equal screening in the right upper quadrant of the plane, lying perpendicular to the ring and passing through its center. The zero line generates a cone which separates the region of screening (positive contributions) from the region of descreening. The distances from the center of the ring along a radius and along the axis of symmetry are plotted along the abscissa and ordinate axes, respectively.

Example of Calculation. Let us calculate the contribution of the ring to the magnetic screening of the protons of the methyl group of toluene. Due to the fact that the methyl group in toluene rotates rapidly about the $C-C$ bond, the mean result from the superposition of the different states shown in Fig. II-8 (I and II) is observed. The coordinates ρ and z for the different positions of the methyl group are found by examining the structural parameters of the molecule (scheme A). The calculation of ρ_4 is not given on the scheme. Then from Fig. II-8 we find the contributions to the chemical shift for all four cases.

The mean contribution, allowing for the fact that in positions 2 and 4 there are two protons each and that states I and II are equally probable, is determined from the relation

$$\delta_{av} = \frac{\delta_1 + 2\delta_2 + \delta_3 + 2\delta_4}{6} = 0.67 \text{ ppm} \qquad (II-18)$$

This value agrees very well with the difference in the chemical shifts of a methyl group at an aromatic ring and at a double bond found experimentally (2.34 and ~1.7 ppm, respectively).

The calculation of the contribution for other cases, particularly for heterocyclic compounds, may be much more complex since here there are two superposed effects, namely, ring currents and the anisotropy of the heteroatom, so that the value of the molecular component σ_M is determined by equation (II-18)

$$\sigma_M = \sigma_{an} + \sigma_{ring}$$

where σ_{an} is the contribution from the anisotropy of the atoms, bonds, and groups and σ_{ring} is the contribution from the ring currents. It is difficult to allow for σ_{an} in this case since the direction of this effect relative to the structural elements of the molecule is usually unknown and it is difficult to separate from the atomic component of the chemical shift σ_A (equation II-14) caused by the electronegativity of the heteroatom. Therefore, the quantitative interpretation of chemical shifts in heterocyclic compounds

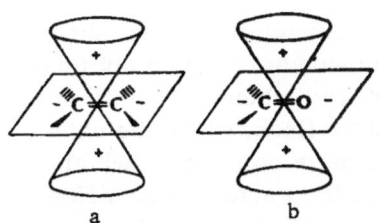

Fig. II-9. Anisotropy of ethylene (a)
and carbonyl (b) groups.

[48] is more a curiosity [49], possibly due to the fact that for the series of compounds selected the effect of the anisotropy of the heteroatom remains constant.

The method examined, which gives good results for benzene, is of limited application. Japanese investigators [50] proposed a different method, which also reduced the error due to the dipole approximation and which consists of using the McConnell equation, but with empirically chosen coefficients. For derivatives of cyclopropane the contributions due to the anisotropy of the ring may be calculated by the formula

$$\Delta\sigma_{an} = \frac{\Delta\chi}{3} \sum_{i=1}^{3} \frac{3\cos^2\theta - 1}{R_i^3}$$

where R_i is the distance from the center of each C—C bond of cyclopropane to the proton examined, and a summation of the results for all three bonds of the ring is carried out. In this formula $\Delta\chi$ is an empirical coefficient, which is taken as equal to $-20 \cdot 10^{-30}$ cm^3 per molecule instead of $-35 \cdot 10^{-30}$ cm^3 per molecule, found by use of McConnell's equation.

With the exception of benzene [52] and the simplest molecules such as H_2O and NH_3, theoretical quantum mechanical calculations of the magnetic anisotropy [51] lead to values of the same order as those observed experimentally, but they can hardly be used to find corrections to the chemical shifts due to their complexity and inaccuracy. In practice, the values obtained experimentally are usually used. No general methods have been developed for the accurate experimental determination of the values of $\Delta\chi$. The contributions from magnetic anisotropy are usually found by comparing the chemical shifts of equivalent protons in substances which contain and which do not contain the anisotropic groups. Goldstein and Reddy [53-55] proposed a method, which is based on the fact that the chemical shifts of protons are proportional to the H^1-C^{13} spin—spin coupling constants and the latter are insensitive to anisotropy. For compounds which do not contain anisotropic groups the relation of the chemical shift to $J_{H^1}-C^{13}$ is expressed by sloping straight lines, while the chemical shifts for analogous groups of protons in compounds which contain anisotropic groups do not lie within the series. The difference in the chemical shifts is ascribed to the effect of the magnetic anisotropy. The accuracy of the values obtained is checked by comparison with theoretical calculations and with results obtained by different methods (for example, the method described above for the benzene ring) and also by comparison of the effect of the given anisotropic group on different protons of the same molecule (for example, the triple bond on the methyl group and acetylenic proton in methylacetylene). Some experimental data on the effect of magnetic anisotropy on the chemical shifts of protons are given in Table II-3 (p. 81).

As the table shows, most anisotropic groups which do not have aromatic character produce an increase in the screening of protons attached to them. The double bonds $C=C$ and $C=O$ also have considerable anisotropy and this, in particular, is responsible for such a large shift to lower fields for the protons of the aldehyde group. However, it is much more difficult to estimate the effect of the groups $C=C$ and $C=O$ for several reasons. The contribution from the anisotropy of a carbon—carbon double bond apparently depends strongly on the immediate environment due to the ready polarizability of this bond by substituents. Another reason is the fact that a double bond does not have cylindrical symmetry and different directions in the plane perpendicular to the bond are not equivalent. The value $\Delta\chi = 23 \cdot 10^{-6}$ cm^3/mole found experimentally by analysis of the chemical shifts of protons in norbornene derivatives [56] has only limited value.

While the two atoms forming the bond in a $C=C$ group are more or less equivalent, the $C=O$ bond is formed by two very different atoms and has a considerable electrical dipole moment. In this case we are not certain of the position of the induced magnetic dipole or, moreover, the direction of the z axis, i.e., from C to O or vice versa [57]. It is obvious that for $C=C$ and $C=O$ double bonds it can only be stated qualitatively that the axis of the cone encompassing the region of increased screening lies in a plane perpendicular to the bond. It is possible that screening in regions above and below the plane in which the double bonds lies is less than in the plane (Fig. II-9) [58, 59].

Pritchard and Lauterbur, who investigated the anisotropy of the $S = O$ group in alkylene sulfites, consider that despite the considerable electric dipole moment of this bond, it may be assumed that the induced magnetic dipole lies approximately centrally between the O and S atoms so that the anisotropy is caused by the unshared pairs of electrons both at the sulfur and at the oxygen. The group has a relatively low anisotropy $(6 \cdot 10^{-6}$ cm^3/mole) and protons in the cis position relative to this bond (H_A) are screened less than in the trans position (H_B) (VII).

VII

The anisotropy of a $C - C$ single bond is low, but nonetheless, it makes an appreciable contribution to the chemical shift, producing decreening of the protons in the series CH_4, RCH_3, R_2CH_2, R_3CH. It is the reason why in cyclohexane and other similar rings the signals of axial protons are lower fields than equatorial protons, this being the most important rule for the stereochemistry of cyclic compounds [60].

Chemical Shifts and Aromaticity

The appearance of magnetic anisotropy under the influence of ring currents and the related contribution to chemical shifts of protons may be used to study the aromaticity of cyclic compounds. In the simplest calculation of the magnetic anisotropy of benzene Pople [61] showed that the change in the shift is determined particularly by the number of mobile π-electrons. From the point of view of nuclear magnetic resonance, aromatic compounds may be defined as compounds in which it is possible to induce ring currents [62]. A qualitative estimate of aromaticity may be carried out by simple comparison of the shifts of protons in the immediate vicinity. For example, a comparison of the chemical shifts of

benzene (7.17–7.35 ppm), the β-protons of thiophene (6.50 ppm),
furan (5.87 ppm), and pyrrole (5.85 ppm) with the shift in ethylene
(5.29 ppm) [63] indicates a decrease in aromatic character in this
series. By taking more careful account of the contributions to the
chemical shifts from the electronic environment of the protons it
is possible to make a rough quantitative calculation of aromaticity.
Elvidge and Jackman [62] compared the chemical shifts of ring
protons and protons of methyl groups in a series of methyl derivatives of pyridone-2 with the chemical shifts of analogous protons in nonaromatic heterocycles or in pyridine derivatives in
which the π-electron shell, as in benzene, is completely delocalized, and came to the conclusion that the aromatic character of
the ring of pyridone-2 is $35 \pm 5\%$ of that of benzene in terms of the
susceptibility to ring currents.

However, it is necessary to take into account the fact that
other factors, which often act in the opposite direction, may have
a greater effect on the chemical shift than magnetic anisotropy.
In particular, an increase in electron density in cyclic compounds
leads simultaneously to an increase in screening due to the atomic
components and to a decrease in the screening of the ring protons
in connection with the increased effect of the ring currents. However, as a rule, the first effect is decisive. With a change from
cyclopentadiene to cyclopentadienylsodium the signals of the ring
protons are shifted to higher fields despite the fact that the compound acquires aromatic character [64, see also 65, 66].

Long-Range Screening. Magnetic Anisotropy
and Stereochemistry

In many cases, with an appropriate configuration of a complex
molecule, anisotropic groups produce screening or descreening of
protons which are not directly attached to them, but are at a distance of two or more bonds. These cases, which are called long-
range screening, are invaluable for establishing the stereochemistry of complex organic compounds. An interesting example was
given recently of the use of the anisotropy of aromatic compounds
for establishing the stereochemistry of the diaryl spiroketones
VIII.

VIII

The signals from the protons of the methyl groups in these compounds are shifted to higher fields as compared with the normal value by up to 1.45 ppm when Ar = phenyl and 1.07 ppm when Ar = thienyl or furyl. Exact calculation of the effect of the anisotropy of the aromatic rings on the chemical shifts of these protons agrees with their relative arrangement (see VIII). In this case the anisotropic groups produce a considerable shift in the signals of protons which are at a distance of 5 bonds from them [67].

5. ADDITIVE EMPIRICAL COMPONENTS OF THE CHEMICAL SHIFT

Taking into account the contribution to the chemical shift of magnetic anisotropy, which gives valuable information on the aromaticity and stereochemistry of compounds, hampers the analysis of spectra when the latter are used for analytical purposes to determine the structure of simpler compounds. In this case it is much more convenient to use the contributions of typical substituents to the chemical shifts of adjacent protons found empirically. The contributions of various groups to the chemical shifts found experimentally include the fractions of the magnetic screening which are associated with the magnetic anisotropy of these groups. For most simple compounds the structural parameters of the substituents (bond lengths and angles) remain constant and this permits the use of additive constants. For groups which do not have appreciable magnetic anisotropy, the additive constants are determined by their electronic effect and in this case they correlate with the electronegativity, reactivity parameters, and other chemical properties of these substituents [15, 68].

A set of additive constants of the chemical shift for methane derivatives was proposed by Shoolery [69]. The chemical shifts of protons are determined by means of these constants from the formula

$$\delta = 0.23 + \sum \sigma_{eff}$$

where δ is the chemical shift relative to tetramethylsilane in an inert solvent, extrapolated to infinite dilution and σ_{eff} represents the substituent constants of Shoolery.   Table II-4 gives an expanded set of Shoolery constants.

For methyl and methylene derivatives the error does not exceed ±0.05 ppm; for trisubstituted derivatives of methane the results are much worse. An analogous set of constants for benzene derivatives has been proposed in a series of papers, which are based on the idea of the additivity of the contributions of substituents to the chemical shifts of benzene. The table gives the constants of Martin and Dailey [16] for the chemical shifts of the ring protons in polysubstituted benzenes. These constants give par-

TABLE II-4. Screening Constants for Derivatives
of Methane and Benzene

Substituent	σ_{eff}	d_o	d_m	d_p	γ
Cl	2.53	0.000	—0.065	—0.016	1.00
Br	2.33	+0.159	—0.134	—0.07	1.03
I	1.82	+0.363	—0.265	—0.07	1.10
NR_2	1.57	—	—	—	—
NH_2	—	—0.768	—0.271	—0.67	0.70
OH	2.56	—	—	—	—
OR	2.56	—0.477	—0.108	—0.41	0.67
OC_6H_5	3.23	—	—	—	—
O $\parallel$ OCR	3.13	—	—	—	—
SR	1.64	—	—	—	—
CH_3	0.47	—0.183	—0.107	—0.16	0.91
C=C	1.32	—	—	—	—
C≡C	1.44	—	—	—	—
C_6H_5	1.65	—	—	—	—
CF_3	1.14	—	—	—	—
C≡N	1.70	—0.27	+0.100	—	—
COR	1.70	—0.64	+0.091	—	—
COOR	1.55	—	—	—	—
$CONR_2$	1.59	—	—	—	—
COCl	—	+0.83	+0.156	—	—
CHO	—	+0.54	+0.195	+0.24	—
NO_2	—	+0.955	+0.155	+0.29	1.20

ticularly good results for para-substituted benzenes when the shifts
are calculated from the formula

$$\delta = d_o(R_o) + \gamma(R_o)\, d_m(R_m) + 7.17$$

in which d(R) denotes the constant of the substituent R, γ(R) is the
correction which characterizes the effect of a substituent in the
meta position, depending on which substituent is in the ortho position relative to the ring proton examined, and the letters o and m
denote the position of the substituent relative to the proton screened.
The calculated shifts agree with the experimental shifts to within
±0.015 ppm. Less accurate results are obtained for other disubstituted and trisubstituted benzenes. In this case the calculations
are carried out by the general formula

$$\delta = \sum d(R_i) + 7.17$$

and for each substitutent we use d_o, d_m, or d_p, depending on the
position of the substituent relative to the proton screened. No corrections are introduced in these cases.

Smith [68] gave several different sets of constants for ~100
substituents in para-disubstituted benzenes. In this work account
was also taken of the effect of polar solvents on the chemical shifts
of protons.

6. ISOTOPE AND CONTACT SHIFT

The isotope shift is a small shift in the resonance signal of
some magnetic nucleus when other nuclei are replaced by their
isotopes. As a rule, we are considering the shift in the signal of
protons in partly deuterated molecules in comparison with normal
compounds. The isotope shift is very small [70, 71]. Thus, for
example, with a change from styrene to α-deuterostyrene there is
a shift in the signals of both terminal protons to lower fields by
0.36 and 0.19 Hz (at 60 MHz).

In contrast to the isotope shift, the contact shift may reach
very large values in comparison with normal chemical shifts. The
contact shift arises in substances containing unpaired electrons,
for example, in complexes of transition metals. The magnitude of
the contact shift depends on the density of the unpaired electron at

the nucleus in accordance with the equation

$$\sigma_{\text{cont}} = a_i \frac{\gamma_0}{\gamma_N} \cdot \frac{g\beta S (S+I)}{3kT}$$

where a_i is the hyperfine interaction constant of the electron and the hydrogen nucleus [72], g is the g-factor [72], S is the total spin of the unpaired electrons, I is the nuclear spin, β is the Bohr magneton, k is Boltzmann's constant, and T is the absolute temperature.

For aromatic compounds the hyperfine interaction constant is determined by the relation

$$a_i = Q\rho_i$$

where Q in this case equals $-96,000$ Hz (ρ is the spin density). An analogous expression may be written for protons of the methyl group, but in this case Q is not constant. The large value of the coefficient Q indicates that even the presence of a very low density of unpaired electrons leads to considerable contact shifts. In practice, the magnitude of contact shifts is several tens of ppm.

The peculiarity of the contact shift in aromatic compounds lies in the fact that for neighboring protons of a conjugated system they are opposite in sign. This is caused by the alternation in the signs of the electron density of an unpaired electron in an aromatic ring and leads to s p r e a d i n g of the spectrum of aromatic compounds containing unpaired electrons and this may be used for a simpler determination of spin−spin coupling constants [73, 74]. It is not possible to observe a contact shift in the case of a higher density of unpaired electrons since the broadening of the spectral lines is too great.

7. COMPONENT OF MAGNETIC SCREENING σ' DUE TO INTERMOLECULAR INTERACTIONS AND MACROSCOPIC MAGNETIC PROPERTIES OF SAMPLE

In the examination of the contributions to the magnetic screening of protons σ_A and σ_M in the previous sections it was assumed that the substance investigated was ideal in the sense that it consisted of separate molecules which did not interact with each other. For real substances this condition holds with some approximation only for gases in a rarefied state. However, the possibilities of investigating substances in the gaseous state are limited both by the properties of the compounds and the sensitivity of instruments. The study of NMR spectra of gaseous substances is more of a special field and we will not examine it here in detail.

With the change of gaseous substances to the liquid state their chemical shifts may change considerably. For example, for water the difference between the shifts in the liquid and gaseous states is 4.2 ppm. The same is observed when substances are dissolved and or when the temperature of the sample is changed.

The shift in the signals of liquid substances is caused by the interaction of the molecules with each other and with the solvent. The different contributions to the magnetic screening of liquid compounds are given in equation (II-19) [75, 76]

$$\sigma' = \sigma_g + \sigma_D + \sigma_{an} + \sigma_e + \sigma_w + \sigma_c . \qquad \text{(II-19)}$$

where σ_g is the chemical shift for isolated molecules in the gaseous state, σ_D is the contribution due to the bulk diamagnetic susceptibility of the sample, σ_{an} is the contribution to the magnetic anisotropy of the solvent, σ_e is the contribution due to the electric field of adjacent molecules, which produce distortion of the electron shell of the given molecule, σ_w is the contribution due to van der Waals forces, and σ_c is the contribution due to intermolecular interaction or complex formation.

The contribution associated with van der Waals forces may be separated out by comparing the chemical shifts of substances in the gaseous state with their shifts in solutions in inert nonpolar solvents [77]. However, due to the difficulty of work with gaseous substances, in practical NMR spectroscopy the basic data for compiling tables and the parameters of chemical shifts are usually the chemical shifts of substances extrapolated to infinite dilution in inert solvents. Therefore, the screening parameters σ_A and σ_M examined previously already contain a contribution due to the van der Waals interaction as a rule. Moreover, it is more convenient to examine σ_e together with the contribution from the intermolecu-

lar interaction σ_c since the latter also includes similar elements.
On the other hand, the hydrogen bond occupies an important position among molecular interactions, particularly in proton magnetic
resonance spectra. Therefore, we will examine a somewhat different set of values determining the components of the magnetic
screening σ' (see equation II-4)

$$\sigma' = \sigma_D + \sigma_{an} + \sigma_h + \sigma_c \qquad (\text{II-20})$$

where σ_h is the contribution from the hydrogen bond, while the other
three constants are the same as in equation (II-19).

Diamagnetic Susceptibility of Sample.

External and Internal References

From the general rules of magnetostatics we know that the
field strength H inside a sample differs from the strength of the external magnetic field H_0 by a value which is determined by the bulk
magnetic susceptibility of the sample and its form in accordance
with equation (II-21) [78, p. 159]

$$H = H_0 \left(1 - \frac{2\pi}{3} \chi_v \right) \qquad (\text{II-21})$$

and the magnitude of the coefficient at χ_v results from the cylindrical form (theoretically, an infinite cylinder). The value χ_v is
dimensionless and for normal diamagnetic substance it is tenths
of a ppm. It may be determined experimentally [79]. However, in
practical NMR spectroscopy it is usually calculated by an additive
scheme, using tabular data. We then determine the molar magnetic susceptibility χ_M, which is related to the bulk susceptibility
by the equation

$$\chi_v = \frac{\chi_M d}{M} \qquad (\text{II-22})$$

where M is the molecular weight and d the density of the substance.

In using an external reference, which is placed in a capillary
arranged coaxially in the tube with the sample, it is also necessary
to take into account the magnetic susceptibility of the reference.

TABLE II-5. Pascal Constants for Elements

Element	$\chi_a \cdot 10^4$	Element	$\chi_a \cdot 10^4$
H	-2.93	F	-6.3
C	-6.00	Cl	-20.1
N		Br	-30.6
Open chain	-5.55	I	-44.6
Ring	-4.61	S	-15
Monoamide	-1.54	Se	-23
Diamide, imide	-2.11	B	-7
O		Si	-13
Alcohols, ethers	-4.16	P	-10
Aldehydes, ketones	+1.72	As	-21
Carboxyl group	-3.36		

TABLE II-6. Structural Corrections for Pascal's Scheme

Group	$\chi \cdot 10^4$	Group	$\chi \cdot 10^4$
C=C	+ 5.5	Benzene	-1.4
C≡C	+ 0.8	Cyclohexane	-3.0
C=C–C=C	+10.6	Position of carbon relative to oxygen atom	
N≡N	+ 1.85	tert-C $\alpha, \gamma, \delta, \varepsilon$	-1.3
C=N	+ 8.15	quat-C $\alpha, \gamma, \delta, \varepsilon$	-1.55
C≡N	+ 0.8	tert-C β, quat-C β	-0.5

TABLE II-7. Properties of Standards and Solvents Used Most Frequency in NMR Spectroscopy

Solvents and standards	Formula	B.p., °C	Density d_4 (t, °C)	Refractive index n_D (t, °C)	Dielectric constant ε (t, °C)	Magnetic susceptibility $\chi_M \cdot 10^6$	Chemical shift, δ
Acetone	$(CH_3)_2CO$	56.1	0.7851 (25)	—	20.47 (25)	42.5	2.09
Benzene	C_6H_6	80.2	0.8731 (25)	1.5014 (25)		68.8	7.17
Hexamethyldisiloxane	$[(CH_3)_3Si]_2O$	100.4	0.7636 (20)	1.3774 (20)	2.15 (20)	126.0	0.05
Dimethyl sulfoxide .	$(CH_3)_2SO$	85—87 (25 mm)	—	—		—	—
Dimethylformamide .	$(CH_3)_2NCHO$	153.0	0.9445 (25)	1.4269 (25)		—	—
Tetramethylsilane . .	$(CH_3)_4Si$	26.5	0.6495 (10)	1.3587 (20)		74.8	0
Chloroform	$CHCl_3$	61.1	1.4795 (25)	—	4.785 (20)	74.5	7.33
Cyclohexane	C_6H_{12}	80.7	0.7739 (25)	1.4227 (25)	2.012 (25)	83.05	1.43
Cyclopentane	C_5H_{10}	49.3	0.7404 (25)	1.4036 (25)	1.97	74.37	—
Carbon tetrachloride	CCl_4	76.7	1.5844 (25)	1.4576 (25)	2.236 (20)	83.7	—

TABLE II-8. Increments of Magnetic Susceptibility of Bonds

$$\chi_M = -10^{-6} \sum \chi_{CB}$$

Bonds	Bond increments		Bonds	Bond increments	
	of Pascal	of Dorfman		of Pascal	of Dorfman
C–C	3.7	4	C=C	1.9	3 (5.6 in ethylene)
C–H	3.85	4 (in methyl CH₃); 3.7 (in methylene CH₂*); 3.2 (in methine CH*) CH*	C≡C	–	11.8
			C=N	4.85	2
			C≡N	–	10
			C=O	2.7	3.5
			C=S	10.3	–
O–H	4.65	4.7	C–N	–	4.35
N–H	5	4.35	COOH	–	16.6
C–O	4.5	4.0 (in ethers)	NO₂	–	6.2
C–Cl	20.35	19.5 (in monochloro substituted hydrocarbons)	S=O	–	6.4
			C=C	–	5.23 (in benzene and naphthalene rings)
C–Br	29.65	–			
C–I	44.05				

In this case the expression for the chemical shift of a cylindrical sample assumes the form [80]

$$\delta_0 = \delta_{obs} + k \left(\chi_{s_0} - \chi_{s_a} \right) \qquad (II-23)$$

where δ_0 and δ_{obs} are the true and observed chemical shifts, while the expression in brackets is the difference in the bulk susceptibilities of the sample and the standard. The coefficient k, which theoretically equals $2\pi/3$, has a value between 2.3 and 3.0 in practice and is taken to be 2.6 on an average. If the sample and the standard substance are placed in spherical concentric cavities, then k = 0, i.e, the standard and substance investigated are in the same fields regardless of their susceptibilities.

There are several methods for the calculation of the magnetic susceptibility χ_M. In 1910 Pascal proposed an additive scheme for the calculation of χ_M, analogous to the well-known scheme for the calculation of molecular refractions

$$\chi_{\varkappa} = \sum \chi_a + \sum \lambda \qquad (II-24)$$

where χ_a represents the susceptibilities of individual atoms and λ represents corrections which take into account the structural characteristics of the molecule [81-84]. The atomic increments and structural corrections of Pascal are given in Tables II-5 and II-6. In Table II-8 (see p. 97) we give the increments of magnetic susceptibility for bonds, which were collected together on the basis of a large number of experimental data. This scheme makes no pretense at a strict physical basis. Pascal's increments are empirical values, which reflect the actual susceptibility of atoms or bonds only indirectly. Later investigations showed that the actual atomic susceptibilities are not constants, but often change over quite wide ranges. For example, the Pascal constant for phenolic oxygen is $-4.87 \cdot 10^{-6}$ in the case of hydroquinone and $-11.5 \cdot 10^{-6}$ in the case of o-nitrophenol [85].

Despite these drawbacks, Pascal's constants are used quite widely in the calculation of corrections for the diamagnetic susceptibility in nuclear resonance. The latter is justified for several reaons. Firstly the contribution to the chemical shift due to the difference in the magnetic susceptibilities of the substance investigated and the sample (II-23) is small as a rule so that even substantial errors in the calculation of χ_M have little effect on the accuracy of the measurement. Moreover, in accurate determinations of chemical shifts it is usual to carry out the measurements in solutions, extrapolating to infinite dilution. The bulk susceptibility of a mixture of substances approximately equals the sum of the partial contributions of each component

$$\chi_{v\ mix} = \chi_{v_1} v_1 + \chi_{v_2} v_2 \qquad (\text{II-25})$$

where v_1 and v_2 are the volume fractions of the components in the mixture. On extrapolation to infinite dilution $\chi_{v_{mix}}$ is determined essentially by the susceptibility of the solvent. In NMR spectroscopy we use a comparatively small range of solvents which are quite simple in chemical structure (Table II-7) and whose magnetic susceptibility is usually well known and this naturally reduces the error. Nonetheless, the use of Pascal constants, particularly for complex molecules containing heteroatoms and multiple bonds, should be undertaken with care.

Another approach to the calculation of χ_M is based on contemporary quantum mechanical ideas on the nature of the magnetic susceptibility. The susceptibility of molecules consists of two components, namely, a diamagnetic component produced by precession of electrons about the nuclei in the magnetic field and a paramagnetic component which is connected with deviations from the symmetrical distribution of the electron shells in the bonds:

$$\chi = \chi_d + \chi_p \qquad (\text{II-26})$$

The value of χ_d may either be calculated theoretically or be obtained by means of experimental data. Dorfman has proposed a semiempirical method of finding χ_d, based on the dependence of this quantity on the polarizability of the substance, determined experimentally [84]. The paramagnetic component for various structures is found from the known general susceptibility and the diamagnetic component.

For nonpolar ordinary covalent bonds, χ_p does not play any significant role, and χ_d makes the basic contribution to the magnetic susceptibility of the substance; χ_p increases greatly in the case of compounds possessing polar groups and bonds with an electron distribution that does not have cylindrical symmetry, for example, in the $C = C$ bond. Thus, a study of magnetic susceptibility, in particular, by the NMR method, is also of independent value, to reveal the fine structure of organic compounds.

Dorfman observed that the susceptibilities of compounds found in this way may also be represented in the form of an additive scheme. The advantage of Dorfman's increments (Table II-8) over Pascal's constants lies largely in the fact that here we are presenting real values of the susceptibilities of bonds even though they are approximate. There is no doubt that Dorfman's method is more promising since it is based on strict ideas on the nature of the magnetic susceptibility, while Pascal's empirical scheme is essentially exhausted. However, since Dorfman's method has not yet been developed adequately it still is not highly accurate. Moreover, Dorfman's data have only limited applicability to cyclic, aromatic, and unsaturated compounds.

For determination of the bulk magnetic susceptibility by the NMR method it is possible to compare the chemical shift of the substance investigated relative to an external reference with the shift relative to the same reference, but dissolved in the substance investigated. The molecules of the dissolved standard are in the same environment as the molecules of the substance investigated and therefore they all experience the action of the same magnetic field regardless of the value of the bulk susceptibility. Although this field differs from the external magnetic field, since it acts equally on the standard and on the substance investigated, the chemical shifts measured relative to the internal reference are true shifts and do not require the introduction of a correction for the bulk magnetic susceptibility. Thus, with the use of internal reference the term σ_D in formulas (II-19) and (II-20) disappears.

The advantage of internal references in the measurement of chemical shifts is generally recognized at the present time and is used widely in NMR spectroscopy. The need to add references directly to the samples investigated imposes on the substances used as internal references a series of requirements: chemical inertness, the lack of polarity and the tendency toward association, the absence of magnetic anisotropy, and a narrow intense signal which does not overlap the signals of other protons. In this respect the best standards is tetramethylsilane, which gives a narrow intense signal in the high field region of the spectrum [85]. Cyclohexane

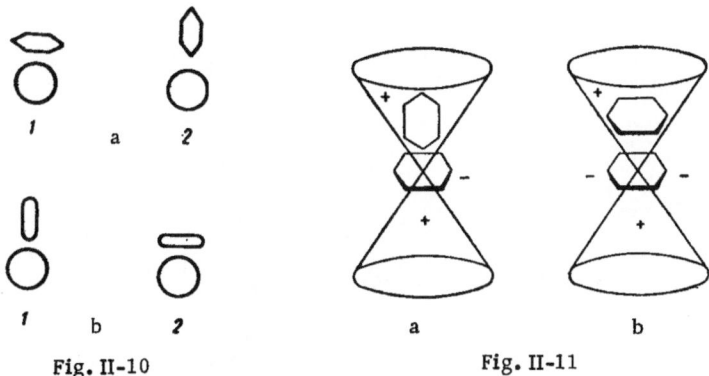

Fig. II-10   Fig. II-11

Fig. II-10. Effect of anisotropy of solvent molecules and their form on the chemical shifts σ_{an}: a) arrangement of molecules of the benzene type (dissolved substance closer to center); b) shift of signal to lower fields.

Fig. II-11. Effect of molecules of an aromatic solvent on chemical shifts: a) arrangement of molecules at an angle to each other; b) parallel arrangement of molecules.

is also a good standard. A series of other compounds which are used as internal standards was given in Ch. I (Fig. I-5).

When the purpose of the investigation is the study of intermolecular interactions and magnetic susceptibility, external references are used. External references are also used frequently in normal investigations when exact values of the chemical shift are not required. The use of external references gives a series of practical advantages; in particular, they do not contaminate the substance and are more convenient for handling. In the determination of chemical shifts with the use of an external reference it is necessary to introduce a correction for the bulk magnetic susceptibility of the sample (II-23). Water is not recommended as an external standard since its chemical shift depends strongly on temperature.

Anisotropy of the Magnetic Susceptibility
of the Solvent σ_{an}

Solvents for NMR Spectroscopy

When organic compounds are dissolved their chemical shifts (corrected for the magnetic susceptibility or measured relative to

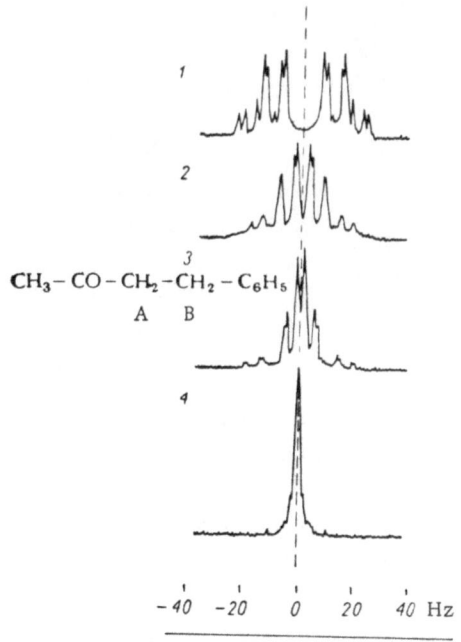

Fig. II-12. Spectra of benzylacetone in different
solvents: 1) 10 mole % in benzene; 2) without
solvent; 3) 10 mole % in carbon disulfide; 4) 10
mole % in acetone.

an internal reference) may shift either to higher or lower fields.
The change in the chemical shifts depends on the properties of the
solvent and the dissolved substance [75]. If we represent the dissolved substance in the form of spheres, then the average orientation of the molecules of a nonpolar solvent relative to it depends on
their form (Fig. II-10). Planar molecules like benzene are oriented preferentially in such a way that the dissolved substance lies
close to their center. As is well-known, this arrangement produces
an increase in the magnetic screening due to the ring currents of
the π-electrons of the ring. If the molecules of the solvents are in
the form of rods, whose anisotropy leads to increased screening
at the ends of the molecules, as for example, in acetonitrile, then
in this case the signals are shifted to lower fields due to the prefer-
ential relative arrangement of molecules of the solvent and dissolved substance (b). When methane is dissolved, its chemical
shift in benzene is shifted to higher fields by 0.15 ppm, while in

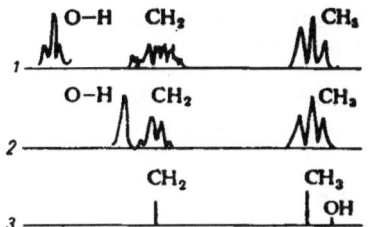

Fig. II-13. Spectra of ethanol and solutions of ethanol in carbon tetrachloride: 1) carefully purified ethanol; 2)
1.0 M solution in CCl_4; 3) < 0.01 M
solution in CCl_4 (sketch).

dicyanoacetylene $N \equiv C - C \equiv C - C \equiv N$ it is shifted to lower fields
by 0.75 ppm relative to gaseous methane [75].

If a nonpolar substance which has no magnetic anisotropy
(for example, cyclohexane) is used as the solvent, then when an
aromatic substance is dissolved in it the signal of the latter is
shifted to lower fields as a rule. In pure aromatic compounds the
relative arrangement of the molecules at an angle to each other
(Fig. II-11a) gives a substantial contribution to the chemical shift
of the protons, increasing their screening, while a parallel arrangement of the molecules (Fig. II-11b) has little effect on the
chemical shift so that as a result of the averaging out from the
motion the aromatic molecules screen each other, but this screening
is reduced with dilution. Therefore, to obtain accurately comparable
chemical shifts of anisotropic compounds and, as will be shown
later, polar molecules, it is necessary to make the measurements
in solutions of nonpolar nonanisotropic solvents (such as cyclohexane or carbon tetrachloride) with extrapolation to infinite dilution.
Nonanisotropic substances of low polarity (for example, saturated
hydrocarbons) have a low sensitivity to the effect of such solvents
and these substances may be investigated in the pure form.

If the molecules of the substance investigated have a form
which differs markedly from a sphere, then it is possible that the
effect of an anisotropic solvent is different for different protons of
the molecule. This leads to a change in the form of the spectrum
of the substance in relation to the solvent used. The effect of solvents is often used if we wish to e x p a n d or, on the contrary, to
c o n t r a c t the spectrum of a substance in order to reveal its details. Figure II-12 gives the spectrum of the protons A and B of
benzylacetone in different solvents [86]. While the protons of group
B are affected little by the solvent, the chemical shift of the protons
of group A changes appreciably so that the difference in the shifts
may be very small (spectrum 4) or may become quite appreciable

(spectrum 1) and this simplifies the determination of the spin—spin coupling constants between these nuclei (Ch. III).

The most important constants and properties of the solvent which are used most in NMR spectroscopy are given in Table II-7.

Polar Effect.   Coordination and

Hydrogen Bond

In solutions of polar compounds, in addition to the factors examined above the chemical shift of protons of a dissolved substance may be affected by an electrostatic field created as a result of the polarization of the nearest solvent molecules by molecules of the dissolved substance. The component of the external electrostatic field which is directed along the X—H bond tends to shift the pair of electrons forming the bond in the direction from H to X so that the electron density and the magnetic screening of the proton associated with it are reduced. As a result the polar effect leads to a shift in the signal of a polar compound to lower fields.

The magnitude of the shifts produced by the electrostatic field may be determined approximately from Buckingham's formula (II-8) and the electrostatic field E in this case depends on the dipole moment of the dissolved substance and the polarizability of the solvent [87] and may be determined experimentally:

$$E = \frac{2\,(\varepsilon-1)\,(n^2-1)}{3\,(2\varepsilon+n^2)} \cdot \frac{\mu}{\alpha} \qquad\qquad \text{(II-27)}$$

where ε is the dielectric constant of the medium, n is the refractive index of the dissolved substance in the form of a liquid, μ is the dipole moment of the dissolved substance in the gas phase, and α is the polarizability of the solvent.

The theory presented is called the theory of the reaction field.

If the solvent contains polar groups while the dissolved substance contains protons of an acidic character, a hydrogen bond may be formed. In this case the chemical shift of protons participating in the formation of the hydrogen bond is shifted strongly to lower fields. The hydrogen bond is responsible for the shift of 4.2 ppm in the signal of liquid water in comparison with water in

the vapor state and for the low-field shift of the hydroxyl proton in
pure alcohol as compared with a dilute solution of alcohol in carbon
tetrachloride (Fig. II-13) [88].

To interpret the chemical shift on formation of a hydrogen
bond we also use the theory of the reaction field [89]. However,
by means of theoretical calculation [90] and on the basis of infrared spectroscopic data, Aleksandrov and Sokolov showed [78]
that it is possible to explain the order of magnitude of the chemical
shift on formation of a hydrogen bond by taking into account the
effect of various factors. Let us examine them in more detail in
the case of the hydroxyl group.

Polarity of the $O-H$ Bond. The change in the polarity of the
$O-H$ bond on formation of a hydrogen bond by the hydroxyl proton
makes the main contribution to the change in the chemical shift of
this proton. The polarity of the bond is characterized by the coefficient λ, which determines the fraction of the hybrid atomic function of oxygen in the molecular orbital of the $O-H$ bond. As calculation shows, the change in the chemical shift is related to the
change in the coefficient λ by the simple equation:

$$\Delta\sigma = -1.5 \cdot 10^{-5}\,\Delta\lambda$$

Since the shift on formation of a hydrogen bond can be as
large as 5 ppm to lower fields, the change in the parameter λ must
be equal to $\Delta\lambda \simeq 0.3$, which corresponds to a decrease in electron
density at the hydrogen atom by 25%. The character of the hybridization of the orbital (the fraction of s- and p-electrons of the
atomic orbitals of oxygen) has hardly any effect on the chemical
shift of the proton.

Stretching of the $O-H$ Bond. The magnetic screening depends on the distance R between the nuclei O and H. For small
changes in the internuclear distance this relation is represented
by the formula:

$$\Delta\sigma = k \cdot 10^{-5}R\,(\text{Å}^{-1})$$

The coefficient k is close to unity and depends only slightly
on the hybridization of the atomic orbitals of oxygen. Since the
formation of a hydrogen bond leads to stretching of the $O-H$ bond,
this factor, like the polarity of the bond, leads to a decrease in the
screening of the proton. The magnitude of the contribution (1 ppm

with an increase in the internuclear distance of 0.1 Å) is comparatively small.

Effect of the Electrostatic Field of the Dipole of a Neighboring Molecule. This contribution corresponds to the Buckingham—
Musher effect of the reaction field examined previously. The linear effect of the electrostatic field leads to a change in the polarity of the bond and this has already been considered above. The
quadratic component plays a much less important role. The contribution of the quadratic component from a molecule with a dipole
moment of 1.6 D at a distance of 1.7 Å from the proton examined is
~0.6 ppm and also leads to a shift in the signals to lower fields.

Effect of the Hydrogen Bond Itself. It is interesting that the
hydrogen bond itself O...H introduces a positive contribution to the
magnetic screening constant of the proton due to the donor properties of the oxygen, which bears unshared pairs. If we assume
that the hydrogen bond is formed by the atomic orbital of hydrogen
and one 2p orbital of oxygen and the coefficient β determines the
fraction of participation of the hydrogen orbital, then with low
values of β the contribution of the hydrogen bond to the screening
of the proton

$$\Delta\sigma = (-0.8 + \beta + 3.6\beta^2) \cdot 10^{-5}$$

so that when $\beta = 0.2$ this is 2.5 ppm and corresponds to an increase
in 10% in electron density at the proton.

Effect of an Unshared Pair of p-Electrons of a Neighboring
Molecule. The p-electrons of a neighboring molecule, which is
participating in the formation of a hydrogen bond, lead to a change
in the chemical shift of the proton due to the magnetic anisotropy.
The estimation of this contribution leads to a value between 0.2
and −0.2 ppm, i.e., less than the observed value.

Thus, the main contribution to the change in the magnetic
screening of the proton on formation of a hydrogen bond is due to
the change in the polarity of the O−H bond under the effect of the
electrostatic field of a neighboring molecule. This really determines the applicability of the theory of the reaction field in the experimental investigation of the hydrogen bond. Exact calculation
of all of the parameters affecting a hydrogen bond is difficult [91].

The effect of temperature on NMR spectra is determined largely by changes in association, the strength of the hydrogen bond, etc. The chemical shift of nonpolar solutions of compounds in neutral solvents depends little on temperature. As a rule, a rise in temperature leads to a shift in the signal to higher fields for protons which are involved in the formation of a hydrogen bond, for example, for hydroxyl protons in alcohols.

As has been pointed out for a solvent, a change in temperature may also have an indirect effect on the chemical shift due to a change in the mobility of different groups in the molecule and the populations of conformations.

For the accurate determination of the chemical shift and the study of its temperature dependence modern spectrometers are fitted with special devices for thermostatting the sample. However, it should be remembered that a change in temperature also has an appreciable effect on the uniformity of the magnetic field produced by the instrument. Therefore, to achieve the highest resolution of instruments it is usual to make measurements at the temperature which results from the natural heat balance between the magnet and the surrounding medium. This is usually ~30°C, i.e., higher than the normal room temperature due to the large absorption of energy by the electromagnets. In this respect, permanent magnets which make complete thermal isolation possible have an undoubted advantage.

Proton chemical shifts are the main parameter in NMR spectra of organic molecules. This is connected with the fact that hydrogen atoms form part of an overwhelming number of organic compounds and the resonance signal of protons is most intense. In addition to the immediate electronic environment of the nuclei, other electrons in a given molecule and also perturbation due to intermolecular interactions have a great effect on the chemical shift of protons. On the one hand, this makes it possible to use proton chemical shifts for a more thorough study of organic compounds, but, on the other hand, it complicates the interpretation of the spectra and imposes special requirements on the preparation of the sample and the selection of conditions for plotting the spectrum.

TABLE II-9. Magnetic Properties and Natural Abundance
of Some Isotopes

Symbol of isotope	Spin	Relative intensity of signal in a field of 10.000 G	Resonance frequency in a field of 10,000 G (in MHz)	Natural abundance, %
H¹	$1/2$	1.00	42.6	99.98
F¹⁹	$1/2$	0.83	40.0	100
P³¹	$1/2$	0.066	17.2	100
C¹³	$1/2$	0.016	10.7	1.1
N¹⁴	1	0.001	3.1	99.635
N¹⁵	$1/2$	0.001	4.3	0.365
B¹⁰	3	0.02	4.6	18.83
B¹¹	$3/2$	0.165	13.7	81.17
Si²⁹	$1/2$	0.08	8.5	4.70
H²	1	0.01	6.5	0.0156

B. CHEMICAL SHIFTS OF NUCLEI
OF OTHER MAGNETIC ISOTOPES

The chemical shifts of nuclei of other magnetic isotopes obey
the same basic rules as proton chemical shifts. However, the
presence of a more developed electron shell in the case of these
elements introduces a whole series of peculiarities, of which the
most important are as follows:

1. The range of chemical shifts is 1-2 orders greater than
the range of proton chemical shifts.

2. For nuclei with an unsymmetrical distribution of the electron shell, for example, those containing unshared electron pairs
and covalent bonds, there is an increase in the role of the paramagnetic contribution to the atomic screening so that an increase
in electron density at the nucleus may lead to a shift in the signal
to lower fields.

3. As a rule, the molecular and intermolecular components
of the chemical shift σ_M and σ' play a much smaller role than in
proton chemical shifts.

4. The chemical shifts of the magnetic isotopes of one ele-
ment (for example, N^{14} and N^{15}, B^{10} and B^{11}) in the same chemical
compounds, expressed in parts per million relative to the same
isotopic standards, are almost exactly equal to each other. In particular, resonance at deuterium nuclei obeys the same rules as
proton magnetic resonance.

The possibilities of the experimental observation of resonance of different isotopes are determined by their magnetic properties and their natural content in the mixture with other isotopes. The
artificial preparation of isotopically substituted substances (apart
from deuterated compounds) is of very limited value in NMR spectroscopy. Table II-9 gives the properties of magnetic isotopes which play
a more important part in NMR spectroscopy of organic compounds.

The nucleus F^{19} has properties which are most favorable
(after protons) for nuclear resonance. The slight decrease in the
intensity of the signal, which is made worse to some extent by the
high atomic weight of fluorine (so that there is a smaller number
of magnetic nuclei per unit volume), is compensated by the fact
that the range of chemical shifts of F^{19} is approximately 20 times
as great as for protons. The nucleus P^{31} occupies the third place
with respect to the possibilities of investigation by NMR spectroscopy. In this case, in order to obtain a more intense signal the
volume of the sample is increased and pure liquids or very concentrated solutions are investigated. The broadening of the lines,
which is connected with the increase in volume, here is also compensated by the greater chemical shifts, whose total range is 500-
600 ppm.

Unfortunately, the direct observation of the resonance signals of other isotopes, among which the resonance of C^{13} may play
a very important part in NMR spectroscopy of organic compounds,
presents great experimental difficulties. However, the recently
discovered possibility of measuring the chemical shifts of nuclei
which have spin—spin coupling with protons by the double resonance method by observing the proton spectra of compounds [92,
93] offers the promise of considerable extension of this field of
investigation in the future.

8. CHEMICAL SHIFTS OF F^{19}

A theoretical analysis of the effect of the p-electrons of the
fluorine atom on the chemical shifts of the nucleus, which was carried out by Saika and Slichter [8], showed that the most substantial
contribution is connected with the ionic character of the A—F bond.
For a completely ionized fluorine atom F^{-}, which has a spherically
symmetrical shell of p-electrons, the paramagnetic contribution to
the screening is minimal with the result that the signal of ionic
fluorine is observed at highest fields. In the covalent molecule F$_2$

the shell of p-electrons deviates most from spherical symmetry with the result that the signal of gaseous fluorine is at lowest fields and is 625 ppm below F^-. The latter is connected with the transition from the spherically symmetrical $^1\Sigma_{g+}$ state to the $^1\Pi_g$ state, which is characteristic of covalently bound fluorine, in accordance with the equation

$$\Delta\sigma = -\frac{2}{3}\,(c^2\hbar/m^2c^2)\,(\langle 1/r^3 \rangle_{av})_{2p}\cdot 1/\Delta E \qquad (II-28)$$

where the first term in brackets represents known physical constants, ΔE is the energy of excitation of the electron in the transition between the two states given above, which equals 4.3 eV, while the expression $(\langle 1/r^3 \rangle_{av})_{2p}$ is connected with the mean distance between the nucleus and the 2p-electrons and equals $8.89/a_H^3$, where

a_H is the Bohr radius (0.53 Å), and $\hbar = h/2\pi$.

In developing these concepts, Karplus and Das [94] related the magnitude of the chemical shift to the parameters of the $A-F$ bond (hybridization or s-character s, ionic character I, and the degree of double bonding ρ) by means of the equation (II-29)

$$\Delta\sigma \simeq \sigma_0 \{-[1-(\rho_{st}/2)-s_{st}]\,\Delta I + (1+I_{st}+s_{st})\,(\Delta\rho/2)\} \qquad (II-29)$$

in which the subscript (st) indicates that the parameter refers to a standard compound of similar structure, for example, fluorobenzene for a series of derivatives of fluorobenzene, while Δ indicates the difference in the corresponding parameters for the standard compound and the compound investigated. The value of σ_0 (screening in the state $^1\Sigma$) is determined from an equation analogous to (II-28)

$$\sigma_0 = -(2c^2\hbar^2/3\,\Delta E m^2c^2)\,(\langle 1/r^3 \rangle_{av})_{2p} \qquad (II-30)$$

Further simplification and substitution of values leads to the following simple expression for derivatives of fluorobenzene:

$$\Delta\sigma = 0.765\Delta I - 0.777\Delta\rho \qquad (II-31)$$

which indicates that double bond character has a great effect on the chemical shift.

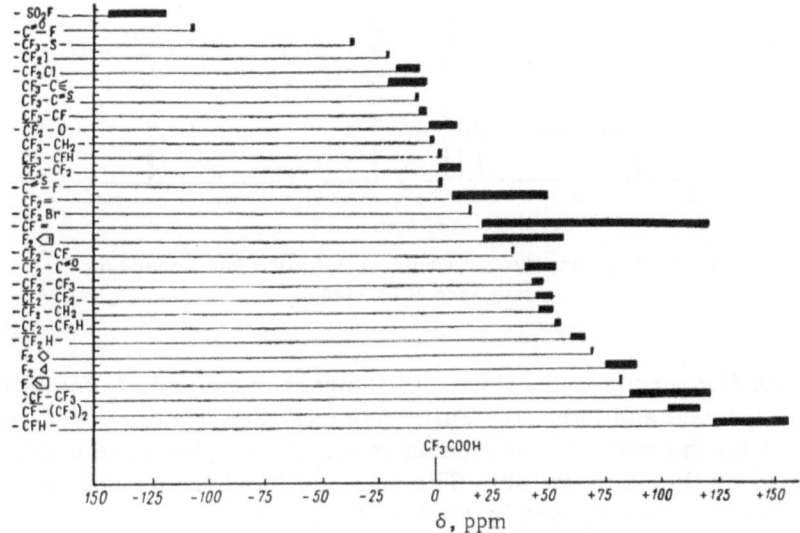

Fig. II-14. Correlation diagram of the chemical shift of F^{19}.

Calculations using formulas (II-28) and (II-31) give values which agree with experiment in an order of magnitude and this is quite satisfactory if we take into account the considerable assumptions used in their derivation. The value of these calculations lies largely in the theoretical explanation of the observed changes in the chemical shifts with a change from compound to compound.

For practical purposes the relations found experimentally are of greatest value and the most important of these is the relation between the chemical shift of F^{19} in fluorobenzene derivatives and the reactivity parameters of the substituents in the benzene ring, which was derived by Taft and his co-workers [95, 96]:

$$\int_{H}^{m-X} = -7.1\sigma_I + 0.60$$

$$\int_{H}^{p-X} = -29.5\sigma_R^0 + \int_{H}^{m-X} = -29.5\sigma_R^0 - 7.1\sigma_I + 0.60 \qquad \text{(II-32)}$$

where $\int_{H}^{m-X}$ and $\int_{H}^{p-X}$ are the chemical shifts of meta- or para-sub-

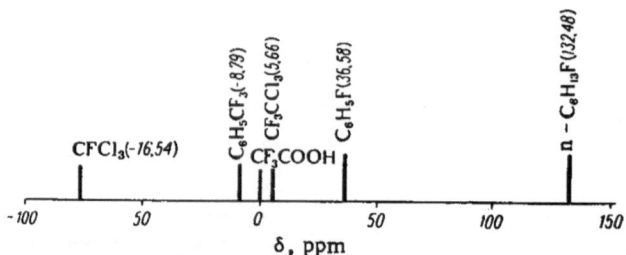

Fig. II-15. Comparison of chemical shifts of reference used for
resonance at F^{19} nuclei.

stituted fluorobenzenes relative to fluorobenzene and σ_R^0 and σ_I are
the resonance and inductive Taft constants (they should not be con-
fused with the magnetic screening parameters, which are usually
denoted by the same letter). The effect of a substituent in the o-
position is not expressed by such a strict rule.

The use of fluorobenzene as an internal standard practically
eliminates the effect of the solvent apart from cases where the substituents in the para position are groups with a +R-effect, which
tend to form a hydrogen bond, while such polar substances as 75%
aqueous methanol or trifluoroacetic acid are used as solvents. In
the other cases the chemical shifts calculated from formulas
(II-32) and those found experimentally agree within the limits of
experimental error.

For other organic compounds of fluorine the rules relating
the chemical shifts of F^{19} to structural parameters of the molecules are more qualitative in character so that for the identification of organofluorine compounds it is still preferable to use tabular values of chemical shifts for compounds of similar structure.
Figure II-14 gives the chemical shifts of F^{19} relative to trifluoroacetic acid in typical structures [97].

In addition to trifluoroacetic acid, trifluorochloromethane,
fluorobenzene, and benzotrifluoride are used as references for the
resonance of F^{19}. For conversion of the chemical shifts, Fig. II-15
gives the relative chemical shifts of the references.

Solvents have a substantial effect on the chemical shifts of
F^{19}. For example, the chemical shifts of fluorine in substituted
fluorobenzenes are shifted to lower fields by 1.4-1.6 ppm at in-

finite dilution in n-heptane [98]. However, as has already been
pointed out, this effect is relatively small in comparison with proton resonance if we compare the values of the chemical shifts of
the two nuclei; the effect of the solvent is compensated for by carrying out the investigations in dilute solutions with internal references.

9. CHEMICAL SHIFTS OF C^{13}

The low natural abundance of the C^{13} nucleus dictates the experimental approach required to determine the chemical shifts of
this isotope.

The usual method for observing the resonance of C^{13} is fast
passage of the dispersion signal with a high strength of the rf field
[99], but this method is of limited application for complex molecules. Paul and Grant [100] considerably improved the method by
observing the resonance of C^{13} at a frequency of 15.1 MHz with rotation of the sample and simultaneous irradiation at the resonance
frequency of protons. The latter eliminated the spin−spin coupling
of C^{13} nuclei with protons and also produced considerable amplification of the signal due to the nuclear Overhauser effect [101,
102]. An indirect method is also used for determining the chemical shifts of C^{13} from the frequency required for spin−spin decoupling with this nucleus as determined by observation of the C^{13}
satellites in the F^{19} and H^{1} spectrum [103, 104].

The C^{13} resonance signal in hydrocarbons is usually split
due to spin−spin coupling of these nuclei with the attached protons.
In work with unenriched samples, spin−spin coupling between C^{13}
nuclei does not appear since the probability of finding molecules
containing two or more C^{13} atoms at the same time is insignificant.
From this point it is easier to observe resonance in substances
containing several equivalent carbon atoms since the probability
of finding molecules containing the C^{13} isotope in these positions is
increased, (for example, in benzene ∼6.6% of the molecules contain the isotope C^{13}).

The rules for the chemical shifts of C^{13} nuclei are very similar to the rules derived for protons, but in this case the effects are
much stronger. Thus, the screening of C^{13} nuclei has a linear relation to the π-electron density and a change in the charge by one
electron produces a resonance shift of 160 ppm [9]. As in proton

resonance, additivity of the chemical shifts of C^{13} is observed in benzene derivatives [20, 105, 106] and in the case of these nuclei the range of investigations is extended due to the possibility of studying the effect of a substituent which is attached directly to the given carbon atom. The additivity of the chemical shifts and their relation to the electronegativity of substituents in the aliphatic series is much more approximate in character, but this may be related to the difficulties of taking account of other factors.

Figure II-16 (see insert) gives a diagram showing the structural correlation of the chemical shifts of C^{13} [107]. The regions of the chemical shifts of C^{13} are approximately the same as for protons. It is characteristic that the resonance of acetylene carbons lies between the alkane and olefin regions and this may be explained partly by the magnetic anisotropy of the triple bonds. The absolute magnitude of the paramagnetic component σ_A^p of the magnetic screening constant for sp-carbon in acetylene is considerably less than the corresponding component for olefinic sp^2-carbon and this leads to a diamagnetic shift.

In contrast to acetylene, the carbon atoms of aromatic rings give a signal in the same region as olefin carbons. The absence of an anisotropy effect in this case is undoubtedly due to the fact that the carbon atoms of the ring lie at the boundary of the region of screening produced by ring currents. The low-field position of the region of chemical shifts of a carbonyl carbon is evidently due primarily to the descreening effect of the electronegative oxygen and not the anisotropy of the $C=O$ bond. It is interesting that the signal of the central carbon atom of an allene group also lies at much lower fields than the signal of an acetylene carbon despite the fact that these atoms have the same hybridization. The study of the chemical shifts of carbon in allene and carbonyl groups illustrates the great value of spectroscopy at C^{13} nuclei in organic chemistry since the information obtained from proton resonance is very limited in these cases.

Another interesting example of this type is provided by the shifts of the tertiary carbon in carbonium ions. With a change from $(CH_3)_3C^{13}Cl$ to the tert-butyle cation $[(CH_3)C^{13}]^+$, the signal of the tertiary carbon is shifted to lower fields by 273 ppm due to the appearance of the positive charge (i.e., the decrease in screening) and also the change from sp^3- to sp^2-hybridization (see diagram

Additional material from *NMR Spectroscopy in Organic Chemistry*
ISBN 978-1-4684-1787-6, is available at http://extras.springer.com

of shifts, Fig. II-17). An analogous picture is observed on
formation of the triphenylmethyl cation from triphenylcarbinol
$(C_6H_5)_3C^{13}OH \rightarrow [(C_6H_5)_3C^{13}]^+$, but in this case the shift is much less
(129.6 ppm to lower fields) due to the fact that the positive charge is not
localized completely on the tertiary atom, but spread out due to
conjugation with the three benzene rings. It may be surmised that
only 60-80 ppm of this value is connected with the presence of the
positive charge [108]. If we compare the latter data with the results of Spiesecke and Schneider [9] it is evident that approximately
one half of the charge is concentrated on the tertiary carbon atom
and the same amount is delocalized.

A study of the chemical shifts of C^{13} in saturated hydrocarbons [109] and cis- and trans-disubstituted ethylenes [105] led
to the conclusion that in these cases the shift in the signals cannot
be explained by the anisotropy of the closest $C-C$ and $C-X$ single
bonds, though the character of the relation is the same as in proton resonance. For ethylene compounds the effect of the substituent is probably due to mesomerism involving the structures

for dihalo and dicarbalkoxy substituted ethylenes with the screening or descreening effect of the substitutents greater when they are
in the trans position since the cis arrangement may lead to disruption of the coplanarity of the molecule [110].

The effect of solvents on the chemical shifts of C^{13} has been
investigated little [111-113]. In aprotic solvents the chemical shift
of the carbonyl carbon in acetone changes insignificantly. Solvents
which tend to form a hydrogen bond produce a shift in the signal to
lower fields of from 1.6 (tert-butyl alcohol) to 37.4 ppm (H_2SO_4) and
this is due to the formation of a hydrogen bond [114] and not enolization.

The chemical shifts of C^{13} have not been studied adequately
for broader generalizations. However, even from the data presented it is obvious that this rapidly developing field of NMR spectroscopy is of great value in organic chemistry. There is no doubt

that in the near future C^{13} spectroscopy with the natural abundance
of this isotope will be next in importance to proton resonance.

10.   CHEMICAL SHIFTS OF P^{31}

The nucleus P^{31} has the properties which are most suitable
after hydrogen and fluorine for investigation by the NMR method.
With the use of modern high-resolution instruments, P^{31} spectra
are plotted at a frequency up to 40 MHz with practically the same
technique as for proton resonance or resonance of F^{19} nuclei. The
error in the measurement of the chemical shift does not exceed
±0.04 ppm [115]. The spectra are calibrated by means of the side
band technique [116] or from the splitting produced by spin—spin
coupling with protons [115] or fluorine [117].

In addition to the interest in the comparatively narrow field
of organophosphorus compounds, P^{31} NMR spectroscopy offers the
valuable possibility of studying the relation of NMR parameters
for an atom of variable valence which is capable of forming compounds of very different types. The theory of the chemical shift of
P^{31} has, however, been developed only to the first approximation
at the present time* and (as, incidentally, for other nuclei) the
empirical tables of chemical shifts are the main source of information for the identification and investigation of compounds [117,
119-121].

The chemical shifts of P^{31} in the main structures are presented in the diagram (Fig. II-17; see insert). The total scale of
chemical shifts of phosphorus is ~500 ppm if we ignore the resonance of elementary phosphorus, which gives a broad peak in the
region of +450 ppm [122]. The distribution of different compounds
on the diagram indicates that in the case of chemical shifts of phosphorus the paramagnetic term in the atomic screening is of decisive importance with the result that the introduction of electron-
donor substituents leads to a decrease in magnetic screening as a
rule. In tetra- and particularly penta-coordinate compounds of
phosphorus the paramagnetic contribution is reduced because of
the greater symmetry of the electron shell; as a result the resonance of phosphorus in such compounds is observed at higher fields.

---

* See also [118].

A theoretical analysis of the chemical shifts was carried out for tricoordinate compounds of phosphorus [116, 123] and a formula was proposed relating the chemical shifts of P^{31} in molecules of the type PX$_3$ to the characteristics of the bonds formed by it [124].

$$\delta = -230 + 29.0 \cdot 10^{-3}\varepsilon - 46.0 D$$

where δ is the shift in ppm relative to 85% H$_3$PO$_4$ and the factor D depends on this electronegativity of the atoms attached to the phosphorus and on the angle X−P−X:

$$D = \left(\frac{3}{4} - \beta^2\right) \beta^2 (1-\varepsilon)$$

where

$$\beta = -3\cos\theta/(1-\cos\theta)$$

and

$$|\varepsilon| = 0.16 |x_P - x_X| + 0.35 |x_P - x_X|^2$$

where x is the electronegativity according to Pauling [125].

Calculation by the formulas presented leads to small values of D with small X−P−X angles as, for example, in the molecule PH$_3$ with an almost pure 3p-orbital of the phosphorus, explaining the appearance of resonance of phosphines at high fields and also explaining qualitatively the appearance of signals of tetracoordinate phosphorus at higher fields than for trivalent compounds.

The usual standard for resonance of P^{31} is 85% orthophosphoric acid. A new standard was proposed recently: the oxide P$_4$O$_6$, a substance with m.p. 23.8°C, which is readily supercooled to +16°C. The advantage of this compound is the fact that it gives a much narrower peak than orthophosphoric acid (a width of 0.3 Hz at 25 MHz as compared with 3.5 Hz from H$_3$PO$_4$) with the result that the signal is observed much more readily even with much smaller amounts of the substance. The chemical shift relative to H$_3$PO$_4$ is −112 ± 0.1 ppm [126].

The chemical shifts of P^{31} change relatively little on solution in inert solvents [116] and only such solvents as methanol and par-

ticularly trifluoroacetic acid shift the signal of P^{31} of triphenylphosphine oxide to low fields due to the formation of a hydrogen bond or
the addition of a proton. When triphenylphosphine oxide is dissolved in 96% sulfuric acid (molar ratio 1 : 20) the signal of phosphorus is shifted by 25 ppm to lower fields in comparison with a
solution in dioxane or carbon tetrachloride [127].

11. CHEMICAL SHIFTS OF OTHER NUCLEI

Of the other magnetic nuclei which are of interest in organic
chemistry, substantial results may be expected from spectroscopy
of B^{11}, N^{14}, and Si^{29}. The nuclei B^{11} and N^{14} have an electric quadrupole moment, which produces broadening of the lines in the spec-
trum and hampers their detection. However, examples of the plotting of the spectrum of B^{11} in a strong magnetic field at a frequency
of 60 MHz [128, 129] and determination of the chemical shifts of
N^{14} by the double resonance method [130, 131] show that these difficulties may be overcome. The quadrupole broadening in the resonance spectra of N^{14} may be reduced considerably by using solvents
of low viscosity such as acetone or ether. Spectra of N^{14} were recorded at 3.94 MHz on a broad-line instrument [132]. The reference was a 4.5 M solution of NH_4NO_3 in 3 N aqueous hydrochloric
acid; the shifts were measured relative to the NO_3^-, line, which
lies at higher fields than the quintet of NH_4^+ by 353 ± 0.5 ppm. The
shifts of organic nitrogen–containing compounds lie in four main
groups: amines at 340, amides at 270, cyanides at 100, and nitro
compounds at 0 ppm. In each group a shift to higher fields corresponds to compounds with more electronegative substituents with
the exception of quaternary ammonium salts, in which the paramagnetic contribution is unimportant due to the symmetry of the
p-electron shell and the reverse picture is observed. For pyridine
and its derivatives substituted at the nitrogen there is the following order of changes in the chemical shifts: pyridine 60, pyridine
N–oxide 100, 1–hydroxypyridinium 150, and pyridinium 180 ppm.

An examination of most data on B^{11}, which concern boron hy-
drides, is beyond the scope of this book. From the point of view
of organic chemistry, an interesting result was obtained for para-
substituted phenylboric acids $p-Z-C_6H_4-B(OH)_2$, in which a good
linear correlation was found between the shift of B^{11} and the Hammett constants σ of the substituents with substituents of the donor
type increasing the screening of boron and electron-acceptor substituents shifting the signal to lower fields [133]. Unfortunately,

most results on the chemical shifts of N^{14} and N^{15} also concern inorganic compounds [134].

The resonance of the silicon isotope Si^{29} with the natural content (4.7%) is detected comparatively readily because of the fact that this isotope has a spin of $\frac{1}{2}$ and gives narrow signals. The chemical shifts of silicon and magnetic isotopes of the other elements of Group IV (Sn^{115}, Sn^{117}, Sn^{119}, Pb^{207}), apart from their own value, are of interest for comparison with the shifts of C^{13}. It is interesting that the order of changes of chemical shifts in related compounds of carbon and silicon are opposite and this may be connected with the presence of vacant d-orbitals in the case of silicon, which produce reversal of the rules [134]. It is characteristic that in the case of tin, although the same picture as for silicon is repeated mainly, the rules are more of an intermediate character. An analogous relation to the properties of carbon, silicon, and tin appears in other cases and this is evidently caused by the decrease in the role of the d-orbitals with a change to heavier elements of the group.

Up to the present time NMR spectroscopy of the nuclei of magnetic isotopes apart from H^1 and F^{19} is of a semiempirical nature and the examples given here are only illustrative. However, there is no doubt of the value of these methods both for identification and analysis and particularly for revealing the fine interelations in molecules of organic and heteroorganic compounds. Interesting developments in the methods of nonprotonic nuclear resonance can naturally be expected in the near future.

LITERATURE CITED

1. W. G. Proctor and F. C. Yu, Phys. Rev., 77:717 (1950).
2. J. T. Arnold, S. S. Dharmatti, and M. E. Packard, J. Chem. Phys. 19:507 (1951).
3. N. F. Ramsey, Phys. Rev., 78:699 (1950).
4. W. E. Lamb, Phys. Rev., 60:817 (1941).
5. I. V. Aleksandrov, Dokl. Akad. Nauk SSSR, 119:671 (1958); 121:823 (1958).
6. H. J. Kolker and M. Karplus, J. Chem. Phys., 41:1259 (1964).
7. N. D. Sokolov, Usp. Khim., 32:967 (1963).
8. A. Saika and C. P. Slichter, J. Chem. Phys. 22:26 (1954).
9. H. Spiesecke and W. G. Schneider, Tetrahedron Letters, 1961:468.
10. M. Grant, Ann. Rev. Phys. Chem., 15:489 (1964).
11. A. D. Buckingham, Can. J. Chem., 38:300 (1960).
12. T. W. Marshall and J. A. Pople, Mol. Phys., 1:199 (1958).
13. J. I. Musher, J. Chem. Phys., 37:34 (1962).

14. W. T. Raynes, A. D. Buckingham, and H. J. Bernstein, J. Chem. Phys., 36:3481 (1962).

15. P. Diehel, Helv. Chim. Acta, 44:829 (1961).

16. J. S. Marin and B. P. Dailey, J. Chem. Phys., 39:1722 (1963).

17. R. R. Fraser, Can. J. Chem., 38:2226 (1960).

18. B. P. Dailey, and J. H. Shoolery, J. Am. Chem. Soc., 77:3980 (1955).

19. J. R. Cavanaugh and B. P. Dailey, J. Chem. Phys., 34:1099 (1961).

20. H. Spiesecke and W. G. Schneider, J. Chem. Phys., 35:722, 731 (1961).

21. B. P. Dailey, A. Gawer, and W. C. Neikam, Discuss. Faraday Soc., 34:18 (1962).

22. K. L. Williamson, J. Am. Chem. Soc., 85:516 (1963).

23. K. L. Williamson, J. Am. Chem. Soc., 86:762 (1964).

24. R. W. Taft, Jr., J. Am. Chem. Soc., 79:1045 (1957).

25. R. W. Taft, Jr., S. Ehrenson, I. C. Lewis, and R. E. Glick, J. Am. Chem. Soc., 81:5352 (1959).

26. R. W. Taft, in: Steric Effects in Organic Chemistry [Russian translation], IL.

27. V. A. Palm, Usp. Khim., 30:1069 (1961).

28. F. Langenbucher, R. Mecke, and E. D. Schmidt, Ann. Chem., 669:11 (1963).

29. H. C. Beachell and D. W. Bestel, Inorg. Chem., 3:1028 (1964).

30. L. M. Jackman, Application of Nuclear Magnetic Resonance Spectroscopy in Organic Chemistry, Pergamon, London (1959).

31. K. L. Williamson, N. S. Jacobus, and K. T. Soucy, J. Am. Chem. Soc., 86: 4021 (1964).

32. C. D. Cook and S. S. Danyluk, Tetrahedron, 19:177 (1963).

33. C. Heathcock, Can. J. Chem., 40:1865 (1962).

34. W. G. Paterson and N. P. Tipman, Can. J. Chem., 40:2122 (1962).

35. L. K. Dyall, Austral. J. Chem., 17:419 (1964).

36. C. C. Price and G. W. Stacey, J. Am. Chem. Soc., 68:499 (1946).

37. S. H. Marcus and S. J. Miller, J. Phys. Chem., 68:331 (1964).

38. R. E. Klinck and J. B. Stothers, Can. J. Chem., 40:1071 (1962).

39. H. Kasawagi and J. Niwa, Bull. Chem. Soc., Japan, 36:405 (1963).

40. J. Niwa and H. Kasawagi, Bull. Chem. Soc. Japan, 36:1414 (1963).

41. R. O. Kan, J. Am. Chem. Soc., 86:5180 (1964).

42. M. Oki and H. Iwamura, Bull. Chem. Soc. Japan, 36:1 (1963).

43. P. R. Wells, Chem. Rev., 63:171 (1963).

44. R. W. Taft, Jr., J. Phys. Chem., 64:1805 (1960).

45. H. M. McConnell, J. Chem. Phys., 27:226 (1957).

46. J. S. Waugh and R. W. Fessenden. J. Am. Chem. Soc., 79:846 (1957).

47. C. E. Johnson and F. A. Bovey, J. Chem. Phys., 29:1012 (1958).

48. G. G. Hall, A. Hardisson, and L. M. Jackman, Discuss. Faraday Soc., 34:15 (1962).

49. K. Tori and M. Ogata, Chem. Pharm. Bull. (Tokyo), 12:272 (1964).

50. K. Tori and K. Kitahonoki, J. Am. Chem. Soc., 87:386 (1965).

51. A. A. Bothner-By and J. A. Pople, Ann. Rev. Phys. Chem., 16:43 (1965).

52. B. P. Dailey, J. Chem. Phys., 41:2304 (1964).

53. J. H. Goldstein and G. S. Reddy, J. Chem. Phys., 36:2644 (1962).

54. G. S. Reddy and J. H. Goldstein, J. Chem. Phys., 38:2736 (1963).

55. G. S. Reddy, and J. H. Goldstein, J. Chem. Phys., 39:3509 (1963).
56. H. Hogeveen, C. Massagnani, F. Montanari, and F. Taddei, J. Chem. Soc., 1964:628.
57. P. T. Narasimhan and M. T. Rogers, J. Phys. Chem., 63:1388 (1958).
58. G. Conroy, in: Progress in Organic Chemistry [Russian translation], Vol. 2, IL, 1964, p. 255.
59. J. A. Pople, Proc. Roy. Soc., A239:550 (1957).
60. R. U. Lemieux, R. K. Kullnig, H. J. Bernstein, and W. G. Schneider, J. Am. Chem. Soc., 80:6098 (1958).
61. J. Pople, W. Schneider, and H. Bernstein, High-Resolution Nuclear Magnetic Resonance, McGraw-Hill, New York (1959).
62. J. A. Elvidge and L. M. Jackman, J. Chem. Soc., 1961:859.
63. G. Martin, Chem. et Ind., 89:168 (1963).
64. G. Fraenkel, P. E. Carter, A. McLauchlan, and J. H. Richards, J. Am. Chem. Soc., 82:5846 (1960).
65. R. J. Abraham, R. C. Sheppard, W. A. Thomas, and S. Turner, Chem. Comm., 1965:43.
66. J. A. Elvidge, Chem. Comm., 1965:169.
67. H. A. P. De Jongh and H. Wynberg, Tetrahedron, 21:515 (1965).
68. G. W. Smith, J. Mol. Spectr., 12:146 (1964).
69. J. H. Shoolery, Technical Information Bulletin, Varian Ass., 2, No. 3 (1959), Palo Alto, California.
70. G. V. D. Tiers, J. Chem. Phys., 29:263 (1958).
71. G. S. Reddy and J. A. Goldstein, J. Mol. Spectr., 8:475 (1962).
72. D. Ingram, Free Radicals as Studied by Electron Spin Resonance, Academic Press, New York (1958).
73. D. R. Eaton, A. D. Josey, W. D. Phillips, and R. E. Benson, J. Chem. Phys., 39:3513 (1963).
74. D. R. Eaton, A. D. Josey, W. D. Phillips, and R. E. Benson, Discuss. Faraday Soc., No. 34:74 (1962).
75. A. D. Buckingham, T. Schaefer, and W. G. Schneider, J. Chem. Phys., 32:1227 (1960).
76. P. Diehel, J. Phys. Chim., 61:199 (1964).
77. N. Lumbroso, T. K. Wu, and B. P. Dailey, J. Phys. Chem., 67:2469 (1963).
78. I. V. Aleksandrov, Theory of Nuclear Magnetic Resonance, Izd. Nauka (1964).
79. P. Selwood, Magnetochemistry, Interscience, New York (1956).
80. C. Lussan, J. Chim. Phys., 61:462 (1964).
81. P. Pascal, Ann. chim., 19:5 (1910).
82. P. Pascal, Ann. chim., 25:289 (1912); 28:219 (1913).
83. Traité de chimie organique (Grignard), 1936, p. 3.
84. Ya. G. Dorfman, Diamagnetism and the Chemical Bond, Fizmatgiz, 1961.
85. S. K. K. Jatkar and A. J. Mukhadkar, Z. Phys. Chem., 36:221 (1963).
86. S. C. Danyluc, Can. J. Chem., 41:387 (1963).
87. P. Diehel and R. Freeman, Mol. Phys., 4:39 (1961).
88. G. Pimentel and A. McClellan, The Hydrogen Bond, Freeman, San Francisco (1959).

89.   P. J. Bercely and M. M. Hanna, J. Phys. Chem., 67:846 (1963); J. Am. Chem. Soc., 86:2990 (1964).

90.   I. V. Aleksandrov and N. D. Sokolov, Dokl. Akad. Nauk SSSR, 124:115 (1959).

91.   V. F. Bystrov, "The hydrogen bond and proton resonance," in: The Hydrogen Bond, Izd. Nauka (1964), p. 253.

92.   R. Freeman, J. Chem. Phys., 42:1199 (1965).

93.   R. Freeman and W. A. Anderson, J. Chem. Phys., 43:3087 (1965).

94.   M. Karplus and T. P. Das, J. Chem. Phys., 34:1683 (1962).

95.   R. W. Taft, E. Price, I. R. Fox, I. C. Lewis, K. K. Anderson, and C. T. Davis, J. Am. Chem. Soc., 85:709, 3146 (1963).

96.   G. E. Maciel, J. Am. Chem. Soc., 86:1269 (1964).

97.   E. G. Brame, Anal. Chem., 34:591 (1962).

98.   V. F. Bystrov, O. A. Yuzhakova, and R. G. Kostyanovskii, Dokl. Akad. Nauk SSSR, 147:843 (1961).

99.   P. C. Lauterbur, J. Chem. Phys., 26:217 (1957).

100. E. Paul and D. M. Grant, J. Am. Chem. Soc., 86:2997 (1964).

101. F. Bloch, Phys. Rev., 102:104 (1956).

102. R. Kaiser, J. Chem. Phys., 39:2435 (1963).

103. E. B. Baker, J. Chem. Phys., 37:911 (1962).

104. G. A. Olah, E. B. Baker, J. C. Evans, W. S. Tolgyesi, J. S. McIntyre, and I. J. Bastein, J. Am. Chem. Soc., 86:1360 (1964).

105. P. C. Lauterbur, J. Am. Chem. Soc., 83:1838, 1846 (1961).

106. G. B. Savitsky, J. Phys. Chem., 67:2723 (1963); J. Chem. Phys., 38:1406, 1415, 1432 (1961).

107. É. Lippmaa, A. Olivson, and Ya. Past, Izv. Akad. Nauk. Ést.SSR, Ser. Fiz.- matem. i tekh. nauk, 14:473 (1965).

108. G. A. Olah, E. B. Baker, and M. B. Comisarow, J. Am. Chem. Soc., 86:1265 (1964).

109. D. M. Grant and E. G. Paul, J. Am. Chem. Soc., 86:2984 (1964).

110. K. S. Dhami and J. B. Stothers, Can. J. Chem., 43:510 (1965).

111. J. B. Stothers and P. S. Lauterbur, Can. J. Chem., 42:1563 (1964).

112. K. S. Dhami and J. B. Stothers, Can. J. Chem., 43:479, 498 (1965).

113. D. H. Marr and J. B. Stothers, Can. J. Chem., 43:596 (1965).

114. G. E. Maciel and G. C. Ruben, J. Am. Chem. Soc., 85:3903 (1963).

115. F. Ramirez, A. V. Patwardhan, N. Ramanthan, N. B. Desai, C. V. Greco, and S. R. Heller, J. Am. Chem. Soc., 87:543 (1965).

116. N. Müller, P. C. Lauterbur, and J. Goldenson, J. Am. Chem. Soc., 78:3557 (1956).

117. J. F. Nixon and R. Schmutzler, Spectrochim. Acta, 20:1835 (1964).

118. J. H. Letcher and J. R. Van Wazer, J. Chem. Phys., 44:815 (1966).

119. J. R. Van Wazer, C. F. Callis, J. N. Shoolery, and R. C. Jones, J. Am. Chem. Soc., 78:5715 (1956).

120. H. Finegold, Ann. N. Y. Acad. Sci., 70:875 (1958).

121. M. L. Nielsen, J. V. Pustinger, Jr., and J. Strobel, J. Chem. Eng. Data, 9:167 (1964).

122. H. S. Gutowsky and D. W. McCall, J. Chem. Phys., 22:162 (1954).

123. J. R. Parks, J. Am. Chem. Soc., 79:757 (1957).
124. J. R. Van Wazer, Phosphorus and Its Compounds, Interscience, New York (1959).
125. L. Pauling, The Nature of the Chemical Bond, Cornell Univ. Press, Ithaca (1960).
126. A. C. Chapman, J. Homer, D. Mowthore, and R. T. Jones, Chem. Comm., 1965:121.
127. G. E. Maciel and R. V. James, Inorg. Chem., 3:1650 (1964).
128. R. L. Pilling, F. N. Tebbe, M. F. Hawthorne, and E. A. Pier, Proc. Chem. Soc., 1964:402.
129. R. L. Pilling, M. F. Hawthorne, and E. A. Pier, J. Am. Chem. Soc., 86:3568 (1964).
130. E. W. Randall and J. D. Baldeschwieler, Proc. Chem. Soc., 1961:303.
131. E. W. Randall and J. D. Baldeschwieler, J. Mol. Spectr., 8:365 (1962).
132. D. Herbuson-Evans and R. E. Richards, Mol. Phys., 7:515 (1963/1964).
133. H. C. Beachell and D. W. Bestel, Inorg. Chem., 3:1028 (1964).
134. P. C. Lauterbur, in: Determination of Organic Structures by Physical Methods, Vol. 2, Academic Press, New York-London (1962), p. 465.

Chapter III

Spin-Spin Coupling

The fine structure in high-resolution NMR spectra produced by
spin−spin coupling* between magnetic nuclei provides a very important source of information on the structure of complex organic
molecules and the distribution of the electron cloud in them.

The splitting of NMR signals as a result of spin−spin coupling
was observed in 1951 during the very first experiments on high
resolution in work with liquid samples [2-4]. Spin−spin coupling
may appear between nuclei of different sorts (for example, H^1 and
P^{31}) or nuclei of the same magnetic isotope. The fine structure in
spectra may be produced by spin−spin coupling of nuclei connected
either directly together or separated by several covalent bonds.
Spin−spin coupling between protons separated by more than 3 covalent bonds is called l o n g - r a n g e c o u p l i n g.

The absolute values of the spin−spin coupling constant vary
from zero to several thousand hertz with the lower observable limit
determined essentially by the resolution of the instruments. Moreover, in accordance with experiment the theory predicts that the
coupling constants may be positive or negative. The latter arise
when in a complex molecule the parallel orientation of two nuclear
spins is more favorable energetically than the antiparallel orientation. This may be represented clearly using as an example two

---

* We have adopted the term used in Abragam's monograph [1]. The terms "pairing,"
"association," "interaction," "splitting," and "indirect interaction" are also used
for spin−spin coupling.

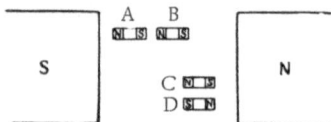

Fig. III-1. Test magnets in a magnetic field: the appearance of spin—spin coupling constants with opposite signs.

pairs of similar magnets, which are oriented differently in an external magnetic field (Fig. III-1). While for magnets A and B their orientation in the external field and their relative orientation are compatible (this state may be regarded as corresponding to a negative constant for the coupling between magnets A and B), with the more favored relative orientation of magnets C and D the position of one of them relative to the external field is unfavorable (this state may be compared with a positive coupling constant). However, experimentally we can determine only the relative signs of coupling constants; the determination of the absolute signs of the constant is based on the very well substantiated hypothesis that the coupling constants between directly connected hydrogen and C^{13} nuclei are positive (see the recent work [77-79]). A detailed account of methods of determining the signs of coupling constants experimentally is given in Ch. IV.

The most common theory of spin—spin coupling in molecules is based on the hypothesis that in liquids and gases the direct dipole—dipole interaction of nuclear spins which is observed in solids (its value is of the order of thousands of hertz) is averaged out to zero as a result of the rapid reorientation of the molecules due to Brownian movement. The fine structure in the high-resolution spectra of liquids and gases is produced by the indirect magnetic interaction of covalently bound nuclei through the electron shells of the common electrons. The magnetic fields of the nuclei create local magnetic fields around them and these induce some magnetic moment in the electron shell of the molecule; in its turn, this magnetic moment acts on the magnetic nuclei, producing splitting of their energy levels. Transitions of the molecules from one energy state to another under the influence of rf radiation produce lines, i.e., the fine structure in the nuclear resonance spectra [5]. Within the framework of the quantitative theory of spin—spin interaction [1, 6, 7] this induced magnetic moment is divided into separate contributions, due, in particular, to the interaction of nuclear spins with orbital motion of the electrons, dipole—dipole interaction of nuclear and electron spins, and the so-called

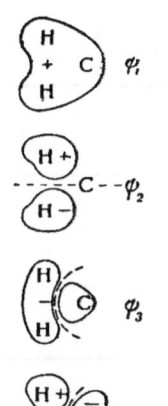

Fig. III-2. Mo-
lecular orbitals
of the fragment

$$C\!\begin{array}{c}\nearrow H\\\searrow H\end{array}.$$

contact or Fermi interaction of nuclear spins with electrons. As calculations
show [8, 9], the contact interaction makes the
main contribution to the spin–spin coupling
constant, at least in the case of proton–proton
interaction. A calculation of the coupling constant in the hydrogen molecule (~200 Hz), which
was based on the above consideration, agrees
well with the value found experimentally (the
coupling constant in the HD molecules is 43 ± 1
Hz [10] and conversion for the H_2 molecule
gives J_{HH} 280 Hz).

However, when the theory is used for calculating the coupling constants in more complex
molecules difficulties arise because the Fermi
interaction is zero if the electron wave function
of the molecule reverts to zero close to the nucleus. Only the wave functions of s–electrons
are nonzero close to the nucleus, while the orbitals of p– and d–electron and others and also
the hybrid orbitals responsible for the forma-
tion of chemical bonds in the molecule equal zero at the point of the nucleus. To explain the appearance of spin–spin coupling in more complex molecules, particularly in systems with conjugated π-electrons, the concept of the so-called configurational interaction,
which takes into account excited states of the molecule, was introduced [11]. The subsequent development of the theory involved the
use of the methods of molecular orbitals and valence bonds for calculation of the coupling constants. The value of these methods [12]
lies mainly in the possibility of a qualitative explanation of experimental results and also quantitative calculations for a series of
compounds of similar structure with the use of empirical parameters. The exact a priori quantitative calculation of coupling constants for more complex compounds presents considerable difficulties at the present time.

1. SPIN – SPIN COUPLING BETWEEN GEMINAL PROTONS [13, 76]

Spin–spin coupling between two protons attached to the same
carbon atom may vary over a wide range of more than 60 Hz but

TABLE III-1. Contribution to the Spin—Spin Constant
of Geminal Protons

Transition	Nature of transition	Sign of product of AO coefficients	Sign of contribution to J_{HH}
$\psi_2 - \psi_3$	Antisymmetrical—symmetrical	−	+
$\psi_2 - \psi_4$	Antisymmetrical—antisymmetrical	+	−
$\psi_1 - \psi_3$	Symmetrical—symmetrical	+	−
$\psi_1 - \psi_4$	Symmetrical—antisymmetrical	−	+

is usually (from -21.5 Hz in cyclopentadienone I to $+42.4$ Hz in formaldehyde) within the limits of 20 Hz in absolute magnitude

I

The study of geminal coupling constants provides valuable information both on the structure and conformations of a substance and on fine electronic interactions so that, in particular, it is possible to detect the difference between an inductive effect and hyperconjugation. Attempts to calculate geminal constants using the method of valence bonds or molecular orbitals, taking into account the mean value of the excitation energy [14-16], lead to inaccurate results since they predict positive values for all geminal coupling constants. Pople and Bothner-By [13] examined spin—spin coupling between geminal protons from the point of view of the LCAO MO-method, taking into account the electronic excitation energies of all triplet states. The theory they developed does not make it possible to make a quantitative calculation of the absolute values of the coupling constants, but it does explain the tendency for a change in their values and signs, depending on the substituents. A brief account of this theory is given below.

For the four-electron fragment of the molecule $C\begin{smallmatrix} \diagup H \\ \diagdown H \end{smallmatrix}$ it

TABLE III-2. Some Spin–Spin Coupling Constants of Geminal Protons, Hz

A. Trigonal CH₂ group

1) $H_2C=CH_2$	+2.3	6) $H_2C=CHMgBr$.	+7.4
2) $H_2C=O$ . . . . from	+40.2	7) $(H_2C=CH)_3P$ . .	+2.02
	to +42.42	8) $H_2C=CHCH_3$ . .	+2.08
	(depending on	9) $H_2C=CHSCH_3$ . .	−0.3
	the solvent)	10) $H_2C=CHCl$ . .	−1.4
		11) $H_2C=CHOCH_3$. from −2.0 to −2.2
3) $H_2C=N-C_4H_9$-tert	+16.52	12) $H_2C=CHF$ . . .	−3.2
4) $H_2C=C=C(CH_3)_2$ . .	−9.0	13) $H_2C=CF_2$	−4.8
5) $H_2C=C=O$	−15.8		

B. Tetrahedral CH₂ group

1) CH_4 . . . . ± 12.4
2) CH_3OH . . . . −10.8
3) $(CH_3CH_2O)_2SO$ . . . . . . −9.3
4) CH_3CN . . . . ± 16.9
5) $C-CH_2-C$ . . . . ± 20.4

6) N —CH₃ . . . . . . ± 14.5

7) structure: HC=C(O)... CH₂ . . . . −21.5

8) H_3C CH_3 / H_2C——CH_2 . . . . −4.5

9) Cl Cl / H_2C——CH_2 . . . . −6.0

10) NH / H_2C——CH_2 . . . . +2.0

11) S / H_2C——CH_2 . . . . . . ± 0.4

12) O / H_2C——CH_2 . . . . . . +5.5

is possible to describe the following set of molecular orbitals (Fig. III-2): ψ_1: bonding, occupied, symmetrical; ψ_2: bonding, occupied, antisymmetrical; ψ_3: antibonding, unoccupied, symmetrical; ψ_4: antibonding, unoccupied, antisymmetrical.

The energies of the molecular orbitals increase in the direction 1 → 4. The molecular orbitals are made up of linear combinations of atomic orbitals with definite coefficients. The spin–spin coupling constant, which is determined by the value of the 1s-orbital of the hydrogen atoms at the nucleus, depends on the coefficient of mutual polarization of the two 1s-orbitals of hydrogen, which is determined in its turn by the product of the coefficients of the two atomic orbitals. As a result, the sign of the coupling constant depends on the electronic excitation in the methylene group as shown in Table III-1.

The introduction into the methylene group of substituents which attract electrons from the orbital ψ_1, i.e., normal substituents with a negative inductive effect such as fluorine, leads to the product of the coefficients becoming more negative for the transition $\psi_2-\psi_3$ and less negative for $\psi_1-\psi_4$, leaving the transitions $\psi_2-\psi_4$ and $\psi_1-\psi_3$ unchanged. Since the energy of the transition $\psi_2-\psi_3$ is much less than for $\psi_1-\psi_4$, the change in the product of the coefficients in the latter case predominates so that the effect as a whole leads to a change in the coupling constant toward p o s i t i v e values.

An analogous, but reverse situation arises if the attraction of electrons from the methylene group proceeds by a hyperconjugation mechanism in the case of such substituents as a phenyl group. In this case the attraction of the electrons is from the anti-symmetrical bonding orbital ψ_2 and this leads to less negative coefficients for $\psi_2-\psi_3$ and more negative coefficients for $\psi_1-\psi_4$, producing a change in the coupling constant toward n e g a t i v e values. In the case of electron-donor substituents the change is in the opposite direction.

The conclusions of the theory agree completely with available experimental data on the magnitudes and signs of the coupling constants of geminal protons. Some of these data are presented in Table III-2.

The replacement of one methylene group in ethylene by more electronegative oxygen leads to a considerable increase in the coupling constant of the geminal protons. The effect is increased as a result of the reverse shift of the unshared pair of oxygen to an antisymmetrical bonding orbital II by a mechanism analogous to hyperconjugation.

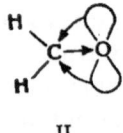

II

As a result the coupling constant in formaldehyde is greatest both in absolute magnitude and also taking into account the sign of all the known proton-proton coupling constants apart from that for the H_2 molecule [17]. The introduction of a nitrogen atom gives an analogous, but smaller effect. The theory predicts for imines, oximes, and other compounds of type 3A (see Table III-2) a positive value of J_{HHgem}.

In the case of allene compounds and ketene (compounds 4A and 5A) one would expect strong attraction of the electrons from the antisymmetrical bonding orbital by a hyperconjugation mechanism and for these compounds the theory predicts a negative sign for the constant. The prediction is confirmed by the fact that for ketene, in which the effect is obviously stronger, J_{HHgem} is greater in absolute magnitude than for dimethylallene.

The introduction of electronegative substituents into the β-position relative to a methylene group leads to a change in the coupling constant in the reverse direction since the MO theory predicts that in this case the attraction of electrons is largely from the antisymmetrical orbital. Electronegative substituents (Table III-2, compounds 9-13A) lead to a negative constant, while donor substituents in the β-position produce an increase in the positive value of J_{HHgem}.

In the case of tetrahedral hybridization of the methylene group, the character of the change in the coupling constants is analogous, but J_{HHgem} changes much less in absolute value. For most compounds one would expect a negative coupling constant for geminal protons. Available experimental data confirm this conclusion. The shift toward positive values of the coupling constants in three-membered rings (Table III-2, compounds 8-12B) is connected with the fact that the hybridization of carbon is intermediate in character between sp^3 and sp^2.

In molecules of saturated compounds the effect of a substituent depends on its steric position relative to the pair of geminal protons. Thus, for molecules of a monosubstituted ethane with the staggered conformation III

III

one would expect that the electronegative substituent X would make a negative contribution to the spin—spin constant of the protons H—H' and a positive contribution to the protons H—H" or H'—H'. The accumulation of experimental data and the refinement of methods of determining the values and signs of the spin—spin coupling constants of geminal protons makes it possible to use them more widely in conformational analysis.

2. SPIN − SPIN COUPLING BETWEEN VICINAL PROTONS [18-20, 76]

The spin−spin coupling constants of protons attached to neighboring carbon atoms vary over a narrower range (4-20 Hz). Theoretical analysis by the valence bond method predicts a positive value of the coupling constant for such protons. The positive value of these constants is taken as a basis in the determination of the absolute sign of spin−spin coupling constants for other nuclei.

Theoretical calculations and the results of many experiments indicate that vicinal constants depend strongly on the configuration of the molecule. Calculation of the contact interaction by the valence bond method, taking into account the σ-electrons, leads to the following approximate relation between the constant $J_{HH_{vic}}$ and the dihedral angle between the planes $H_A C_1 C_2$ and $C_1 C_2 H_B$* (Fig. III-3):

$$J_{HH_{vic}} = A + B \cos \theta + C \cos 2\theta \qquad (III-1)$$

The coefficients in equation (III-1) in their turn depend on various factors, namely, the length of the $C−C$ and $C−H$ bonds, the angle $H−C−C$, and the polarity of the bonds, i.e., they may be related to structural and electronic characteristics of the given molecule. Thus, for ethane calculation leads to the coefficients $A = 4.22$, $B = −0.5$, and $C = 4.5$ Hz [19]. Examination of ethylene with sp^2 hybridized carbon atoms leads to calculated coupling constants $J_{HH_{cis}}$ 6.1 Hz and $J_{HH_{trans}}$ 11.9 Hz. The results for ethane and ethylene agree well with experiments and this confirms the validity of the mechanism of proton−proton spin−spin interaction examined at the beginning of the chapter.

The angular dependence of the spin−spin coupling constant of vicinal protons plays an important part in the investigation of the conformations of cyclic compounds and ethane derivatives. In practice, instead of equation (III-1) the earlier results of Karplus are used frequently and these are given in equation (III-2).

$$J_{HH_{vic}} = \left[\begin{array}{l} K_1 \cos^2 \theta + T \ (\text{For angles} \ \theta \ \text{from 0 to 90°}) \\ K_2 \cos^2 \theta + T \ (\ " \quad " \quad \theta \ \text{from 90 to 180°}) \end{array} \right] \qquad (III-2)$$

---

*Calculation by the molecular orbital method leads to an analogous result.

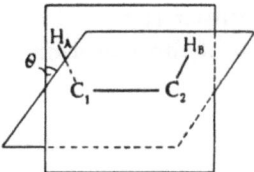

Fig. III-3. Analysis of spin —
spin coupling of vicinal protons.

In the case of ethane K_1, K_2, and T equal 8.5, 9.5, and -0.28 Hz, respectively. Because of the small value of the parameter T it is neglected in practical calculations.

The effect of substituents on the coupling constants of vicinal protons is much less than in the case of geminal protons. For a large number of ethane derivatives the mean value of the coupling constant $J_{HH vic}$ varies in relation to the substituent over the range of 4-8 Hz. Abragam and his co-workers found that for ethane derivatives the following relation of the mean coupling constant to the electronegativity of the substituent holds with an accuracy down to ± 0.3 Hz:

$$(J_{HH\text{-}vic})_{av} = 14.5 - 0.8 \sum_4 E \qquad \text{(III-3)}$$

where $\sum_4 E$ is the sum of the electronegativities of the first carbon atoms of the four substituents at the fragment $\rangle CH-CH\langle$ on Huggins' scale [22]. Apart from purely practical interest, this relation is important in that it confirms the electronic mechanism of spin—spin coupling since the electronegativity of the elements is undoubtedly due to their electronic properties.

The effect of substituents on the spin—spin coupling constants of protons in other systems has not been expressed by actual relations. However, it is definite that in all cases the configuration of the molecule plays an important part so that a comparison of the constants may be made only when it is certain that the relative steric position of the protons is exactly the same in the pairs compared. Calculations showed that a change in the π-character of the carbon—carbon bond does not have a substantial effect on the spin—spin coupling of vicinal protons [23]. Only a small increase in the coupling constants was observed for the ortho protons in benzene derivatives from adjacent substituents [76]. For most benzene derivatives these constants vary over a narrow range of 1.5 Hz [24]. The decrease in the coupling constant of vicinal protons with a change from olefins to aromatic compounds is evidently due to purely structural factors and not a change in aromaticity. In particular, the calculation of Karplus for the fragment H−C=C−H (cis) in ethylene leads to the following relation of the coupling constant to the length P of the C = C bond (in Hz):

$$J_{HH_{cis}} = 1 - 2.9\,(P - 1.35) \qquad\qquad \text{(III-4)}$$

for P within the range of 1.35-1.55 A.

The coupling constants for typical groups in organic molecules are given in Table I-1. The ease of experimental measurement of the vicinal constants makes them a very important NMR parameter for determining the structure and stereochemistry of organic molecules.

3. SPIN — SPIN COUPLING BETWEEN REMOTE PROTONS (LONG-RANGE COUPLING) [25]

In individual favorable cases splitting of proton signals is observed in NMR spectra as a result of spin—spin coupling with other protons which are separated by three, four, and sometimes more bonds. The magnitude of the splitting may sometimes reach 10 Hz; on the other hand, in other cases the spin—spin coupling is so slight that instruments are unable to resolve the splitting and the interaction between the nuclei appears as broadening of the lines in the spectrum. It is usual to measure the so-called half-width of the lines, i.e., the width at half the height, in comparison with the width of the line of a standard substance. Naturally an estimate of spin—spin coupling from the width of lines can only be qualitative in character and this approach should be used with caution as the width of lines depends on the relaxation time of the protons, the viscosity of the sample, individual characteristics of the instruments, the plotting conditions, and other factors.

Long-range spin—spin coupling appears most often in structures containing multiple bonds, particularly allene and acetylene bonds, though considerable interaction has also been observed in saturated compounds including those containing heteroatoms. Typical structures in which considerable spin—spin coupling between remote protons is observed are given in Table III-3.

TABLE III-3. Long-Range Spin—Spin Coupling Constants in Typical Structures

Structure	Systems participating	Coupling constants $\pm J_{HH}$, Hz
H–C=C–C–H	Allyl	1-4
H–C–C=C–C–H	Homoallyl	0.5-5
H–C=C=C–H	Allene	6-7
H–C=C=C–C–H		3-4
H–C=C=C=C–C–H		up to 1
H–$(C\equiv C)_n$–H	Acetylene	~2 (n=2)
H–$(C\equiv C)_n$–C–H		2-3
H–C–$(C\equiv C)_n$–C–H		(n=1)
H–C=C–C=C–H	1,3-Diene and polyene	{ up to 2
H–C=C–C–H		up to 1
(aromatic structures)	Aromatic (including heterocyclic) { meta / para	1-3 / up to 1
(aromatic structures with C–H side chain)	Aromatic ring and side chains or different ring of polynuclear aromatic hydrocarbons and heterocycles	up to 1.5
H–C–C–C–H (with =)	Saturated, including sp^2-hybridized carbon atoms	0.5-5
H–C–C=C–H	Saturated	0-8
H–C–X–C–H etc.	Including a heteroatom	0-3

The theory of long-range spin—spin coupling has been developed in detail only for unsaturated systems. Calculation by the valence bond method, using experimental values of hyperfine splitting constants in the corresponding free radicals and taking into account the Fermi interaction by the mechanism of configurational interaction [26], gives the following predictions for long-range spin—spin coupling constants.

1. Cisoid and transoid coupling constants in allyl and homo-allyl systems must be similar in magnitude (in contrast to the coupling constants of cis- and trans-protons in olefins).

2. The magnitude of the spin—spin coupling constants must depend on the angle between the bond connecting the proton to an sp^3-hybridized carbon atom and the plane of the multiple bond.

3. The coupling constants of protons separated by an odd number of bonds must be positive and by an even number of bonds, negative.

4. The replacement of $=C-H$ by $=C-CH_3$ must lead to a change in the sign, but should not affect the magnitude of the coupling constants of the coupling of protons through a π-system.

The theoretical relations derived are confirmed completely by the results of experimental work.

For a qualitative estimate of the coupling constants of different protons in one compound or several compounds of similar structure it is often worthwhile in practice to examine the possibility of $\sigma-\pi$ overlapping of the orbitals in these compounds. As a rule, an increase in c o n j u g a t i o n between protons in the usual chemical sense leads to an increase in the absolute value of the constant of spin—spin coupling constant between them.

The most complete study has been made of spin—spin coupling between allyl protons, for which J_{HH} is negative in all examples investigated. In contrast to vicinal protons, in this case $J_{transoid}$ (1-4 Hz) is usually somewhat lower in absolute magnitude than the coupling constants of cisoid protons IV J_{cisoid} (2-4 Hz) with the difference in each pair equal to 0.5 Hz.

IV

In the investigation of cyclic systems with a known configuration it was established that allyl spin—spin coupling reaches a maximum when $\theta = 90°$ and a minimum when $\theta = 0$ or $180°$. This relation becomes understandable when we examine the overlapping of the orbitals of the allyl system.

The maximum overlapping is reached precisely when $\theta = 90°$. In accordance with this, a decrease in the spin—spin coupling constants of allyl protons is often due to steric hindrance to effective $\sigma—\pi$ overlapping. A characteristic example is provided by derivatives of cholestenone V, in which there is appreciable spin—spin coupling between protons in positions 1 and 6 only if the latter are axial.

The participation of the double bond of an allyl system in conjugation with a benzene ring, for example, in methylstyrenes, does not substantially affect the magnitude of the spin—spin constants. In cases where the allyl protons belong to a cyclic system we often observe anomalously small or, on the contrary, anomalously large absolute values of the spin—spin coupling constants of the allyl protons, regardless of the steric hindrance discussed above, so that the conclusions drawn for such systems from an examination of the allyl constants should be used with caution. It is interesting to note that delocalization of the electron shell, as, for example, in the negative ion of dimedone VI, may lead to the disappearance of appreciable spin—spin coupling between allyl protons. At the same time, in the acetyl derivative VII normal coupling is observed with the constant $J_{26} = 1.2$ Hz.

As Table III-3 shows, the spin—spin coupling constants for homoallyl protons lie over a wider range than for allyl protons and this agrees completely with the predictions of the theory since in this case the magnitude of the constant depends on the change in two angles, namely, θ_1 and θ_2. The best-known example is spin—spin coupling between protons of vicinal methyl groups in different derivatives of pseudobutylene $CH_3-CH=CH-CH_3$, with constants from 1.2 to 1.8 Hz. Spin—spin coupling with a transoid arrangement of the groups is usually somewhat higher (by ~ 0.5 Hz) than with the cisoid arrangement. Somewhat higher coupling constants are observed with cyclic compounds and this may be explained by the more favorable steric arrangement due to the absence of free rotation. It has been reported [27] that in futronolide VIII

VIII

the coupling constant between the homoallyl protons $J_{7-11} = 5$ Hz.

The strongest spin—spin coupling between remote protons is observed in allene and acetylene systems. In available examples of the experimental determination of the relative signs of the constant there is complete agreement with the predictions of the theory. Little is known of the steric dependence of the coupling constants in such systems since they have a rigid configuration and there are no differences between cisoid and transoid arrangements of the protons in allenes. The NMR spectra of allene compounds confirm this conclusion, which was reached previously by other methods. On the other hand, in the cumulene compound IX in which the methyl groups are not equivalent relative to the proton in the α-position, the coupling constant $J_{H\alpha-CH_3} \simeq 1.2$ Hz,

differs somewhat for the two methyl groups, judging by the form of the spectrum [28]. No strict theory of spin—spin coupling in acetylene, allene, and cumulene systems has been developed up to the present time.

As a rule in conjugated 1,3-diene and enyne systems the spin–spin coupling constants are somewhat lower than for coupling between the corresponding protons in acetylene and allene compounds. It is characteristic that the coupling constants between the terminal trans–trans protons in butadiene (1.30 Hz) and the terminal trans and acetylene proton in vinylacetylene (0.8–0.9 Hz) are higher than for coupling between the corresponding protons in the cis-configurations [29, 30]. Spin–spin coupling between meta or para protons in aromatic hydrocarbons or heterocycles is still less marked in comparison with protons in acyclic nonaromatic systems separated by the same number of bonds. It is obvious that in the cases presented a part is played by delocalization of the electron cloud analogous to VI. In contrast to allene systems, it is assumed that the coupling constant between meta protons in benzene is positive.

Spin–spin coupling constants of approximately the same magnitude as in benzene derivatives are observed for protons in heterocyclic and polynuclear aromatic and heteroaromatic compounds separated by the same number of bonds. It is possible that spin–spin coupling through π-electrons, which does not play a substantial part for protons lying close to each other, is substantial in the case of a long-range interaction [31].

Spin–spin coupling between benzyl protons and ortho protons of the benzene ring is usually very slight (of the order of 0.5 Hz), but the presence of coupling is indicated by broadening of the signal of the benzyl protons. Somewhat greater coupling between ortho protons of the ring and α-protons of the side chain is observed in five-membered heterocycles. In 4-methylthiazole X ($J_{CH_3-H_5} = 2\,Hz$) [32] it is similar to coupling in an acyclic allene system.

The usual mechanism of spin–spin coupling with the participation of π-electrons is also possible for protons separated by single bonds in systems containing sp^2-hybridized carbon atoms or heteroatoms with unshared p-electrons. In this case there is the possibility of overlapping of the orbitals of the type in XI with the use of one part of the π-orbital and two C−H σ-bonds. Typical systems of this type are acetone (J_{HH} 0.54 Hz, determined from the spectra of C^{13} satellites [33]) and also various aldehydes, ketones, and unsaturated compounds. The coupling constants are somewhat higher if the system contains several hybridized carbon atoms and in this case the coupling constants reach 2 Hz and above, as, for example, for coupling between the protons H_A and H_X in bromobenzoquinone XII. No spin−spin coupling is observed between the protons H_A and H_X in bromobenzoquinone XII. No spin−spin coupling is observed between the nuclei H_B and H_X. The interaction between the protons H_A and H_X is similar in character and magnitude to spin−spin coupling of the meta protons in benzene [25]. Spin−spin coupling in systems containing a heteroatom with an unshared pair of p-electrons has an analogous character.

Long-range interaction of considerable magnitude through 4 (and sometimes more) σ-bonds has been observed in the systems which do not have unsaturation. Such systems are characterized by a direct dependence of the spin−spin coupling constants on the configuration of the molecule and the relative arrangement of the protons. Thus, for example, in 2,2,1-bicyclohexanes XIII, strong coupling (J = 7−8 Hz) is observed between the α and α' protons, while no appreciable interaction has been observed between the protons $\alpha - \beta'$ or $\beta - \beta'$.

XIII

An analogous relation is also observed for many other alicyclic and also saturated heterocyclic compounds. Although strict calculations were unable to give a theoretical explanation for an interaction of this type, the idea of overlapping of small parts of sp^3-hybridized orbitals of carbon of the t a i l − t o − t a i l type in the fragment XIV seems very attractive.

XIV

With precisely this steric arrangement of the protons, spin—spin coupling between them reaches very high values. Unfortunately, there is as yet insufficient experimental material, including data on the relative signs of the constants, for a more strict discussion of the mechanism of spin—spin coupling in saturated systems. The practical study of an interaction of this type also plays an important part in the stereochemistry of complex cyclic molecules.

There are examples of long-range spin—spin coupling in systems containing heteroatoms, namely, oxygen, sulfur, nitrogen, etc. In contrast to saturated carbon systems, in these molecules unshared pairs of p-electrons of the heteroatoms may participate in the transmission of the spin—spin interaction. The signs of the coupling constants have not been studied in such systems.

4. H^1 – C^{13} SPIN – SPIN COUPLING

The presence in natural carbon of 1.08% of the isotope C^{13}, which has a magnetic moment and spin of $\frac{1}{2}$, results in the appearance of weak additional signals in the main proton spectrum. These signals may be detected in many cases and are produced by splitting of the signals of protons connected directly to this isotope or separated from it by a few bonds and are the so-called C^{13} satellites. The study of C^{13} satellites plays an important part in the NMR spectroscopy of organic molecules since the electronic structure of carbon is studied most fully in organic chemistry and here we can expect the establishment of fine and theoretically based correlations.

Since the magnitudes of spin—spin coupling constants are determined mainly by the contact interaction of electrons and nuclei (p. 126), one would expect that $J_{H^1-C^{13}}$ would be affected to a con-

TABLE III-4. Effect of Hybridization of Carbon
on the Spin—Spin Coupling Constants $J_{H^1-C^{13}}$

Compound	Hybridization state	s-Character, %	J, Hz	J_{calc}, Hz
Methane	sp^3	25	125	125
Benzene	sp^2	33,3	159	167
Methylacetylene	sp	50	248	250

siderable extent by the character of the hybridization of the carbon orbital in the direction of the C—H bond. In actual fact, this relation was observed even in early work on the analysis of spectra of C^{13} satellites [34–36] (Table III-4).

The last column of the table gives calculated values of J_{HC}, starting from the hypothesis that they are proportional to the s-character of the carbon orbital and taking as the starting value the constant J_{HC} in methane, for which p u r e sp^3-hybridization is most probable

$$J_{HC} = J_{HC}(CH_4)\ \frac{\text{s-character}}{25} \qquad\qquad \text{(III-5)}$$

or in a somewhat different form

$$J_{HC} = 500\alpha_H^2 \qquad\qquad \text{(III-6)}$$

where α_H^2 is the fraction of 2s-character of the hybrid carbon orbital in the direction of the C—H bond, which appears in the wave function of this orbital:

$$\psi_{CH} = \alpha_H s_{CH} + \sqrt{1 - \alpha_H^2} \cdot p_{\sigma\,CH} \qquad\qquad \text{(III-7)}$$

In the case of methane $\alpha_H^2 = {}^1/_4$ and this makes it possible to calculate the coefficient in (III-6).

Fromula (III-6) makes it possible to explain the dependence of the constant J_{HC} on the substituents in methane derivatives CHXYZ. As Malinowski [37] showed, the following relation holds quite well for methane derivatives:

$$J_{HC}(CHXYZ) = \zeta_X + \zeta_Y + \zeta_Z \qquad\qquad \text{(III-8)}$$

Here ζ_X, ζ_Y, and ζ_Z are the additive contributions of the substituents X, Y, and Z to the coupling constant J_{HC}, determined from the relation

$$\zeta_X = J_{HC}(CH_3X) - 83.4 \qquad\qquad \text{(III-9)}$$

where the constant 83.4 Hz in this formula is determined from the constant J_{HC} in methane (125 Hz):

$$2\zeta_H = \frac{1}{3} J_{HC}(CH_4) = 83.4 \qquad \text{(III-10)}$$

Juan and Gutowsky [38, 39] pointed out the fact that with a change in the substituents at carbon the different parameters characterizing the C—H bond compensate each other to a considerable extent; as a result, between certain limits J_{HC} is independent of the degree of ionic character of the bond and is directly proportional to the s-character, which is expressed by the value α_H^2. In this sense the values of ζ_X essentially characterize the effect which the substituent X has on the hybridization of the carbon orbital. Replacement of hydrogen by a more electronegative atom leads to rehybridization of the orbitals so that the remaining C—H bonds have a greater proportion of 2s-character [36]; this also justifies the use of formula (III-6) for the interpretation of the additivity of the contributions of the substituents to J_{HC}.

In cases where there are two or three electronegative substituents bearing unshared pairs of electrons there is a positive deviation from additivity of the contributions, which reaches 24 Hz in the case of trifluoromethane. These deviations may be due to the considerable contribution of the ionic structure XV,

$$\overset{+}{X}=\overset{\overset{\displaystyle H}{|}}{C}-H$$
$$X^-$$
$$\textbf{XV}$$

in which the C—X bond has a considerable degree of π-character due to conjugation between the substituents at the carbon so that the fraction of s-character in the C—H bonds is increased and this leads to a positive deviation of J_{HC} [40].

To allow for these deviations in the case of disubstituted methanes empirical corrections have been proposed for the so-called p a i r   i n t e r a c t i o n of the substituents [41], but this is obviously only of limited value since it requires a large number of experimental data.

Table III-5 gives the empirical increments of the substituents ζ for calculating J_{HC} by Malinowski's additive scheme (equation

TABLE III-5. Increments of Substituents for Calculation of J_{HC} for Methane Derivatives

Substituent	J_{HC} (CH,X)	ζ *	ρ †
H	125	41.7	0
F	149	65.6	24
Cl	150	66.6	25
Br	152	68.6	27
I	151	67.6	26
OR, OH	143	59.6	15
S(O)CH$_3$, SCH$_3$, SH	138	54.6	13
NH$_2$	133	49.6	8
NHCH$_3$	132	48.6	7
N(CH$_3$)$_2$	131	47.6	6
NO$_2$	147	63.6	22
CN	136	52.6	11
C≡CH	132	48.6	7
COOH	130	46.6	5
CHO	127	43.8	2
COCH$_3$	126	42.6	1
C$_6$H$_5$	126	42.6	1
CH$_3$	126	42.6	1
CCl$_3$	134	50.6	9
Si(CH$_3$)$_3$	118	34.6	—7
C(CH$_3$)$_3$	124	40.7	—1

*For calculations by (III-8).
†For calculations by (III-11).

III-8) and also the increments for Douglas' scheme, in which J_{HC} is found by means of equation (III-11):

$$J_{HC} = 125 + \rho_1 + \rho_2 + \rho_3 + g \qquad \text{(III-11)}$$

The increments ρ in this scheme are determined from the constants in monosubstituted methanes

$$\rho_X = J_{HC}(CH_3 X) - 125 \qquad \text{(III-12)}$$

and the constant 125 (in Hz) in equations (III-11) and (III-12) is J_{HC} in unsubstituted methane. The increment for hydrogen $\rho_H = 0$.

In disubstituted methanes and also in trisubstituted derivatives which contain no more than two polar substituents (for example, fluorine or an alkoxyl group) it is possible to improve the results of calculation by means of a correction for pair interaction g (Table III-6).

TABLE III-6. Corrections for Pair Interaction
in Methane Derivatives

Substituents		g	Substituents		g
F	F	13	Cl	Cl	3
F	Cl	9	NR$_2$	NR$_2$	0
F	CN	4	NO$_2$	NO$_2$	0
OR	OR	6	Cl	COOH	2

The additivity of the contributions of substituents to J_{HC} of methane derivatives is due to the small contribution of structures with π-bonds to the molecular orbitals of these compounds. As a result of this the constants J_{HC} make it possible to determine experimentally the s-character of both the C – H bond and also of the hybrid orbitals directed toward the other substitutents since the following normalization condition must apply:

$$a_X^2 + a_Y^2 + a_Z^2 + a_H^2 = 1 \tag{III-13}$$

The additive relations derived for methane derivatives also apply to compounds with a carbon atom in the sp^2-hybridization state, though in these cases the correlation is less strict, evidently due to the more considerable π-contribution to the molecular orbitals. If we assume that this contribution is constant, then for ethylene derivatives we may obtain the following additive relation for the contributions:

$$J_{HC}(CH_2{=}C^{13}HX) = 157 + \frac{4}{3}\rho_X \tag{III-14}$$

In this formula ρ is taken from Table III-5; the value 157 Hz represents J_{HC} for ethylene [39].

An analogous relation was obtained for the spin – spin coupling constant J_{HC} in aldehydes, which also contain a carbon atom in a state of hybridization close to sp^2:

$$J_{HC}(XCHO) = 5.3 + 4.01\zeta_X = 172.4 + 4.01\rho_X \tag{III-15}$$

In this equation the constants were obtained by the method of least squares from data for a series of aldehydes; the value 172.4 coincides with J_{HC} in formaldehyde (172 Hz) [42, 43].

To the extent to which J_{HC} in aldehydes is dependent on the s-character of the C—H bond, this constant may be related to various scales of electronegativities, which in their turn, are determined to a considerable extent by the fraction of s-character in the hybrid orbitals. Below we give some relations between J_{HC} (XCHO) in aldehydes and the electronegativities of the first atoms in the substituents X on different scales [44]:

$$E_G = 0.0140 J_{HC}(XCHO) + 0.25 \quad [45] \tag{III-16}$$

$$E_{C, D} = 0.0129 J_{HC}(XCHO) + 0.48 \quad [46] \tag{III-17}$$

$$E_{H, W, J} = 0.0165 J_{HC}(XCHO) - 0.83 \quad [47] \tag{III-18}$$

The relations obtained for J_{HC} in different compounds are empirical or semiempirical in character since they include initial experimental values of J_{HC} in various compounds. An objective, independent confirmation of the possibility of using relations of this type was provided by theoretical calculations of the constants J_{HC} by the molecular orbital method, starting directly from Ramsey's equation [48, 49]. Although the results obtained correspond only roughly to the experimental values of J_{HC}, they still make it possible to confirm in principle the validity of the conclusions drawn on the basis of empirical premises.

Thus, the study of spin—spin coupling constants is a convenient method of studying the fine structure of the electron shells of molecules and may be used to obtain parameters which were obtained previously by laborious calculations that are difficult to check. However, this method should be used with caution for compounds with an unusual molecular structure (for example, carbonium ions and other charged groups and aromatic heterocycles with a large number of heteroatoms) since the experimental spin—spin coupling constants J_{HC} in compounds of this type [50, 51] often show poor correlation with the probable hybridization of the orbitals, possibly due to other contributions to J_{HC} in addition to the contact interaction.

The spin—spin coupling constants of protons with C^{13} nuclei separated by two or more bonds are considerably lower than between nuclei connected directly. For compounds of the type

$(CH_3)_3CC^{13}$ they are 3.6-6.0 Hz with sp^3-hybridization of the C^{13} carbon atom, 3.7-6.4 Hz with sp^2-hybridization, and 5.38 Hz in the nitrile of trimethylacetic acid $(CH_3)_3CCN$, which contains an sp-hybridized carbon atom [52].

The signs of the $H-C^{13}$ spin−spin coupling constants apparently obey the same rules as for J_{HH}. It is known that J_{HC} is positive for atoms connected directly; as was shown on the example of propyne [53], in this case $J_{H^1CC^{13}}$ is negative, while $J_{H^1CCC^{13}}$ is again positive.

5.   SPIN − SPIN COUPLING OF PROTONS WITH THE NUCLEUS F^{19}

The spin−spin coupling constants of protons and fluorine are found both in proton magnetic resonance spectra and in spectra of F^{19} nuclei and they play an important part in establishing the structure of compounds containing these two nuclei. The absolute magnitudes of the spin−spin coupling constants of protons and fluorine in molecules of different structures are given in Table III-7.

As a rule, the $H-F$ spin−spin coupling constants are greater in absolute magnitude and vary over a wider range than the corresponding proton−proton constants. In many cases this makes it possible to correlate the constants J_{HF} better with the structure of the molecules. However, up to now there has actually been no satisfactory theoretical interpretation of these constants. It is possible that the additional contribution to J_{HF} is due to the dipole−dipole interaction of nuclear spins, which is not averaged out completely due to the anisotropy of the magnetic screening of the nuclei and also the spin−orbital and dipole−dipole interaction of the nuclei and the electrons surrounding them [6].

An experimental investigation of the effect of substituents on the constants J_{HCCF} in ethane derivatives [21] indicates that the mechanism of spin−spin coupling in these compounds is evidently very similar to the mechanism of $H-H$ interaction.

In a study of the average coupling constants in ethane derivatives it was observed that as in the case of proton−proton interaction, J_{HF} in compounds of this type correlates linearly with the sum of the Huggins' electronegativities of the atoms of the four

TABLE III-7. Spin–Spin Coupling Constants of H^1 and F^{19} Nuclei

Structure of group	Type of compound	J_{HF}, Hz
H–F	Hydrogen fluoride	615
H–C–F	Geminal coupling	
	Fluoroolefins	
	$-CF_2H$	57.2-52.4
	$-CFH_2$	47.3-45.5
	Fluoroolefins $=C\langle^{F}_{H}$	72-84.7
H–C–C–F	Vicinal coupling	
	Fluoroalkanes (coupling averaged over conformations)	3-25.5
	Fluoroolefins	
	cis	1-20
	trans	12-52.4
H–C–C–C–F	ortho-H–F in benzene derivatives	6-10
H–C–C–C–C–F	meta-H–F in benzene derivatives	6-8
	para-H–F in benzene derivatives	2-3
	ortho-$(CH_3)_3C -C_6H_3(OH)F$	1
H–P–F	$CH_3-P(O)F_2$	~120
H–C–P–F	$=C\langle^{H}_{P(O)F_2}$	6.6
	—	0
H–C–C–C–P–F	$CH_3-C\equiv C-P(O)F_2$	1

remaining substituents in the fragment $\diagdown$CF—CH$\diagup$ (cf. equa-

tion III-3)

$$(J_{HF})_{av} = 53.03 - 3.38 \sum_4 E \qquad \text{(III-19)}$$

with a mean deviation from a straight line of 1.3 Hz over a total range of changes in J_{HF} of 22 Hz. Had a mechanism different from the mechanism of H—H coupling played a substantial part in H—F spin—spin coupling, then one would have expected substantial deviations from the additive formula derived. As will be shown below, in the case of F—F interaction these deviations are more substantial.

It is obvious that stricter conclusions on the mechanism of spin—spin coupling of protons and fluorine may be drawn after the determination of the relative and absolute signs of the spin—spin coupling constants of these nuclei in molecules of different types and also their changes in relation to the structure and the character of the substituents. The signs of J_{HF} have been studied only in very recent years. It was shown that in fluorinated ethanes J_{HFgem} and J_{HFvic} are identical in sign, while in trifluoroethylene $J_{HFtrans}$ and J_{HFgem} have the same sign, but are opposite in sign to J_{HFcis} [54]. A complete analysis of the spectrum of vinyl fluoride [55] made it possible to establish that in this compound all three constants J_{HF} have the same sign as J_{HHcis} and $J_{HHtrans}$. Since it is very probable that the last two constants are positive, the data obtained make it possible to establish the absolute sign of J_{HF} in this compound. Unfortunately, available data do not yet make it possible to give general rules for the absolute signs of J_{HF}, particularly since the results given above for two compounds of the same structure, $CF_2 = CHF$ and $CH_2 = CHF$, differ from each other. Evidently the coupling constants for H—F in the geminal position, in contrast to the corresponding H—H constants, are positive in both saturated and ethylenic compounds.

A theoretical analysis of the J_{HF} constants in benzene derivatives [56] indicates that either all three J_{HF} constants in benzene are positive or one of them, namely, $J_{HF\,para}$, is negative, while the other two are positive. An experimental check on this is difficult due to the low absolute magnitude of J_{HFpara}.

TABLE III-8. Spin—Spin Coupling Constants J_{FF}

Structure of group	Type of compound	J_{FF}, Hz
F—C—F	Geminal atoms in saturated and cyclic compounds	142-343
	Terminal fluorine atoms of a vinyl group	6.6-87
F—C—C—F	Fluorine derivatives of benzene	0-18.8
	Fluoroolefins { cis	112-117
	FC = CF { trans	24-33
	Ortho derivatives of benzene	~20
F—C—C—C—F	Saturated fluorohydrocarbons	7-10
	Meta derivatives of benzene	0.7-5
F—C—C—C—C—F	Saturated and unsaturated fluorohydrocarbons	2-7
	Para derivatives of benzene	7.4-16

An experimental investigation of long-range spin—spin coupling of protons and fluorine, separated by 5-6 bonds, indicates a steric dependence of J_{HF}, which may be due to the transmission of the interaction through the chemical bonds and not through space. Thus, in 6β-fluoro steroids there is splitting of J_{HF} of up to 7 Hz through 5 bonds between the protons of the angular methyl group and fluorine if the vectors in the C—H and C—F directions intersect (i.e., essentially, if the vector in the C—F direction intersects the surface of the cone formed by the C—H bonds on rotation of the methyl group about the C—C bond). Otherwise no interaction is observed between these nuclei.*

6. SPIN — SPIN COUPLING BETWEEN

FLUORINE NUCLEI J_{FF}

The principles of spin—spin coupling between F^{19} nuclei separated by two or more bonds differ substantially from the principles characteristic of H—H and H—F interactions both in the absolute magnitudes and in the relative and absolute

---

* H—F spin—spin coupling in fluoro steroids is examined in detail by Bhacca and Williams, NMR spectroscopy: Steroids and Related Compounds, Holden-Day, New York (1964).

signs of the spin—spin coupling constants. Up to now no satisfactory general theory has been developed for F—F interaction.
According to the so-called theory of direct interaction
through space, which is promoted most actively, it is surmised that due to the great spread of the 2p-orbitals of fluorine
there is the possibility of direct overlapping of the electron shells
of the two nuclei directly through space, leading to a one- or two-
electron interaction of these magnetic nuclei [57]. The direct
interaction mechanism may be particularly important when
the interacting nuclei are separated by a large number of bonds
(4-7), but lie close together in space. Though the theory of direct
interaction through space explains satisfactorily many cases of
large values of J_{FF}, it still is not generally accepted and has been
criticized seriously [21].

In practice, empirical rules for changes in the constants and
signs of spin—spin coupling J_{FF} and their correlation with various
parameters of the substituents play an important part in the study
of F—F interaction. Experimental values of J_{FF} in various structures are given in Table III-8.

The spin—spin coupling constants J_{FF} lie over a very wide
range (according to some data, they reach 400 Hz) and they often
change between wide limits with a change in substituents even in
compounds of the same type. The ranges of variation in J_{FF} often
overlap and this reduces the possibility of using experimental results for correlation with molecular structures.

The signs of F—F spin—spin coupling constants have been
studied little. The results of measurements of the relative signs
of constants in compounds of various structures also indicate a
considerable difference in the principles of F—F interaction and
H—H or H—F interaction. Spin—spin coupling between geminal
fluorines in halogen derivatives of ethane is opposite in sign to
vicinal interaction and in this case J_{FFgem} is evidently positive
and J_{FFvic} negative. An investigation of fluorinated propanes led
to a result which agrees with these data: while coupling between
two pairs of vicinal fluorines (J_{12} and J_{23} in the structure $\underset{1}{C}-\underset{2}{C}-\underset{3}{C}$)
has the same sign and is evidently negative, J_{13} is opposite in sign
and has the same sign as J_{gem} with both of these constants evidently positive. The proposed absolute signs for the spin—spin

coupling constants are based on a comparison with compounds in which it is possible to observe both H— F and F—F interaction and also spin—spin coupling between two protons, where the absolute sign of J_{HH} may be derived from theoretical arguments. The data obtained for F—F coupling agree well with the absolute values of the corresponding constants since it is extremely probable that values of J_{FF}, which are of large magnitude, must be positive. The arguments presented also agree with data on the signs of J_{FF} in olefins, for which it was found that J_{FFgem} and $J_{FFtrans}$ have the same sign (+), while J_{FFcis} is opposite to them in sign. The following relative signs of J_{FF} in saturated and ethylenic compounds of various structures have been given [54]:

$$
\begin{array}{ccc}
\underset{\pm}{\overset{F}{\underset{F}{>}}C-C} &
\underset{\mp}{\overset{F}{>}C-C\overset{F}{<}} &
\underset{\pm}{\overset{F}{\underset{}{>}}C-C-C\overset{F}{<}} \\[2em]
\underset{\pm}{\overset{F}{>}C=C\overset{F}{<}} &
\underset{\pm}{\overset{F}{\underset{F}{>}}C=C} &
\underset{\mp}{\overset{F}{>}C=C\underset{F}{<}} \\[2em]
\underset{\mp}{\overset{F-C}{\underset{F}{>}}C=C} &
\underset{\pm}{\overset{F-C}{>}C=C\overset{F}{<}} &
\underset{\mp}{\overset{F-C}{>}C=C\underset{F}{<}}
\end{array}
$$

The difference in the signs of the constants $J_{FFortho}$ and J_{FFmeta} in fluoro derivatives of benzene, which is observed experimentally, and also the comparatively low absolute value of the meta constants in comparison with the ortho and para constants are in agreement with theoretical predictions that the ortho and para constants must be positive and only the meta constant may have a negative sign [56].

The spin—spin coupling constants of geminal fluorine atoms vary over a very wide range and reach very high values with the ranges for saturated and vinyl compounds differing markedly. For compounds of the type

$$
\overset{F}{\underset{Q}{\overset{F}{>}}}C-C\overset{F}{\underset{Cl}{<}}H
$$

(where Q represents substituents containing heteroatoms) an inverse relation was observed between the electronegativity of the

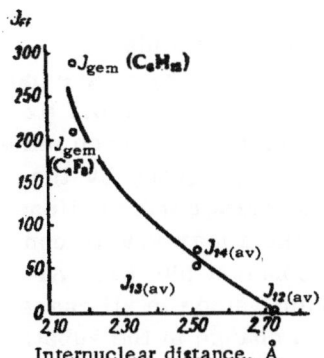

Fig. III-4. Relation of the spin—spin coupling constants of F^{19} nuclei to the internuclear distance.

heteroatom in Q and the spin—spin coupling constant J_{FFgem}

$$J_{FF_{gem}} \approx 540/x \quad \text{(III-20)}$$

where x is Pauling's electronegativity of the heteroatom. If it is assumed that an increase in electronegativity leads to a parallel fall in electron density at the carbon atom, then it is obvious that the coupling constant is directly proportional to the electron density between the interacting nuclei [58]

An examination of the geminal coupling constant of vinyl fluorines leads to approximately the same conclusion. For compounds of this type a linear relation was observed between the coupling constants and the sum of the chemical shifts of the geminal fluorine atoms of the type

$$J_{FF_{gem}} = K(\delta_1 + \delta_2) - C \quad \text{(III-21)}$$

and for different groups of compounds, somewhat different values of the constants K and C were obtained (K = 1.4, C = 217 and K = 0.938, C = 130.4; δ represents the chemical shifts [in ppm] relative to CCl_3F) [59]. The existence of a relation between the spin—spin coupling constants and the chemical shifts of the F^{19} nuclei indicates that the main contribution to spin—spin coupling is made by a purely electronic mechanism of interaction of the nuclei.

However, in the case of vicinal fluorine atoms this picture becomes unsatisfactory. In 1,2-difluoro derivatives of ethane the spin—spin coupling constant of the vicinal fluorine atoms correlates only roughly with the sum of the electronegativities of the four substituents of the fragment $\rangle CF - CF \langle$ [21]

$$(J_{FF_{vic}})_{av} = 91.4 - 6.15 \sum_4 E \quad \text{(III-22)}$$

(E is Huggins' electronegativity), though in the case of H−H and H−F interactions an analogous correlation holds with high accuracy. The low values of J_{FF} in the groups CF_3-CF_2- are particularly surprising: in perfluoropropane this constant is 0.7 Hz, while in perfluoroethane it falls to zero. Attempts have been made to explain this fall in J_{FFvic} in the perfluoroethyl group by the fact that the values of J_{FFvic} in transoid and skew conformations have opposite signs; even small changes in them may lead to considerable variations in the observed mean constant [60, 61]. Although this explanation now seems improbable, since in all cases the character of the change in J_{FFvic} with a change in the substituent is the same for J_{HHvic} [cf. formulas (III-3) and (III-22)], for which it has been shown that the coupling constants in the transoid and skew conformations are positive, at the same time it cannot be rejected completely without an unequivocal independent determination of J_{FF} in different conformations. It should be remembered that with a change in temperature there is the possibility of a change in the mean coupling constant both as a result of differences in the populations of the conformations and also due to changes in J_{FF} which are characteristic of the separate conformations and reach values of 1-2 Hz [21].

The nature of the changes in vicinal spin—spin coupling of fluorine nuclei may be explained by assuming that for compounds of this type the mechanism of direct interaction through space is important. In actual fact, the internuclear distance between vicinal fluorine nuclei (Fig. III-4) may be greater than in compounds of another type and this leads to a relative decrease in J_{FF}. However, it is quite possible that the low value of J_{FFvic} in the perfluoroethyl group is due to the high electronegativity of the fluorine atoms, leading to the attraction of the electron cloud from the C−C bond and a corresponding decrease in the transmission of spin information between the fluorine atoms so that there is no need to introduce for the explanation any type of mechanism which differs from the mechanism of H−H or H−F spin—spin coupling [21].

Nonetheless, the idea of a direct interaction through space by vicinal fluorines is very attractive if we include olefinic compounds in the examination. It is interesting that in this case there is a substantial difference from the nature of the coupling constants between protons in analogous structures. While the coupling constants between fluorine nuclei in the cis-position are large in magnitude and positive, the values of $J_{FFtrans}$ have the opposite sign and are comparatively small in absolute magnitude and this may be due to the steric remoteness of these nuclei and the resulting importance of the electronic mechanism of spin—spin interaction.

Direct spin—spin coupling of fluorine nuclei through space is evidently particularly important for atoms which are separated by four or more bonds since in these cases an interaction through electron clouds may make only a very small contribution to the value of the constant. In actual fact, coupling constants of chemically remote fluorine nuclei of considerable magnitude appear as a rule when the interacting groups are close in space, while J_{FF} may fall to zero when they are remote from each other [57]:

XVI   XVII   XVIII   XIX

$J_{FF}=10.9$ Hz $J_{FF}=12.8$ Hz $J_{FF}=10-12$ Hz $J_{FF}=0$

At the same time, the character of the change in the coupling constants in isomeric di- and polyfluorobenzenes cannot be explained from this point of view and here it is obvious that the coupling of sterically remote fluorine atoms in the para position involves the π-electron cloud of the benzene ring.

A more exact mechanism of spin—spin coupling of fluorine nuclei in molecules of various structures may evidently be established as a result of further correlation of the magnitudes and signs of the constants with other parameters of the molecules. Valuable information may be obtained by a comparison with the nature of the changes in the coupling constants of protons and fluorine with other nuclei or other magnetic nuclei with each other.

7. SPIN — SPIN COUPLING OF PROTONS
AND FLUORINE WITH MAGNETIC NUCLEI
OF GROUP IV B ELEMENTS

Group IV B elements (apart from the isotope C^{13}, which has already been mentioned) have a considerable natural content of isotopes, many of which have a spin of $\frac{1}{2}$ and are suitable for study by the NMR method (Table III-9).

TABLE III-9. Magnetic Isotopes of Group IV B Elements

Isotope	Atomic No.	Natural abundance %	Spin	Magnetic moment
C^{13}	6	1.08	$^1/_2$	0.702
Si^{29}	14	4.70	$^1/_2$	−0.555
Ge^{73}	32	7.61	$^9/_2$	−0.877
Sn^{115}	50	0.35	$^1/_2$	−0.913
Sn^{117}	50	7.67	$^1/_2$	−0.995
Sn^{119}	50	8.68	$^1/_2$	−1.041
Pb^{207}	82	21.11	$^1/_2$	0.584

In the proton and F^{19} nuclear magnetic resonance spectra of compounds of these elements there usually appear satellites caused by spin—spin coupling of protons or fluorine with magnetic isotopes. The spin—spin coupling constants with a direct chemical bond H—M or F—M (M represents magnetic isotopes of Group IV B elements) vary from hundreds to thousands of hertz and are observed quite readily. As would be expected, the spin—spin coupling constants of hydrogen or fluorine with different magnetic isotopes of the same element (for example, $H-Sn^{117}$ and H^1-Sn^{119}) in such molecules are proportional to the magnetic moments of the isotopes.

The spin—spin coupling of protons and fluorine with heavy elements of Group IV B is interesting in comparison with H^1-C^{13} spin—spin coupling, which was described previously, and $F^{19}-C^{13}$ spin—spin coupling since these elements are similar to carbon in chemical behavior and essentially have the same outer electron shell, while the structure of the inner electron layers is different. As has been observed, the coupling constants of $F^{19}-C^{13}$ and also H^1-M and $P^{31}-M$ show considerably less correlation with the electronegativity of the substituents at the central tetrasubstituted atom, they do not obey additive relations well, and they also show much less obvious connections with the state of hybridization of the element than in the case of H—C constants. This is evidently due to the contribution to spin—spin coupling caused by the polarity of the H—M and F—M bonds and also the participation of electrons of inner layers in the transmission of spin information. Common rules were observed for all Group IV B elements in similar chemical structures, indicating that the mechanism of spin—spin coupling is the same for these nuclei and H^1-C^{13} [62] and, in particular, that the contact term in the Ramsey equation is important.

Since the magnitudes of the constants in the case of isotopes of one element (with identical spins) are determined by the magnetic moments of these isotopes μ, it is natural to compare the ratio of J_{HM} or J_{FM} to the magnetic moment of the nucleus M for different elements. It was found [62] that these ratios correlate strictly with the atomic number of the element M both in the case of a direct chemical bond $H-M$ or $F-M$ or with interaction of these nuclei through a carbon atom:

$$\sqrt{|J_{HM}/\mu_M|}=0.676Z_M+8.0 \tag{III-23}$$

$$\sqrt{|J_{HCM}/\mu_M|}=0.107Z_M+1.95 \tag{III-24}$$

$$\sqrt{\left|\frac{J_{FM}/\mu_M}{J_{FC^{13}}/\mu_{C^{13}}}\right|}=0.0245Z_M+0.84 \tag{III-25}$$

The rules derived may originate from the fact that the density of the electron cloud in the region of the nuclei for elements of one subgroup varies according to approximately the same law, while the other parameters which determine the magnitude of J_{HM} or J_{FM} change insignificantly.

Thus, the results of investigating spin–spin coupling of protons and fluorine with magnetic nuclei of Group IV B indicate that the interaction mechanism in this case is generally the same as for H^1-C^{13} or H^1-H^1 interaction, but the presence at these nuclei of additional filled and unfilled electron orbitals complicates the rules for the changes in the coupling constants under the action of substituents. Nonetheless, the spin – spin coupling constants with Group IV B nuclei may be used successfully both for analytical purposes and for studying fine electron displacement in compounds of these elements.

8. SPIN – SPIN COUPLING INVOLVING P^{31} NUCLEI

As has already been mentioned, P^{31} nuclei, whose isotopic abundance in natural phosphorus is 100%, are next to protons and fluorine in convenience for investigation by NMR spectroscopy since

TABLE III-10. H‒P and F‒P Spin‒Spin Coupling Constants

Structure of group	Type of compound	J_{HP}	J_{FP}
X‒P	Compounds of tricoordinate phosphorus	170‒230	570‒1420
	Phosphoryl compounds	500‒800	980‒1190
	Thiophosphoryl compounds	490‒630	‒
	Penta- and hexacoordinate compounds	943	965‒1100
X‒C‒P	Tricoordinate compounds	13‒20	90
	Phosphoryl and thiophosphoryl compounds and phosphoranes	10‒27	156
H‒C‒C‒P	‒‒	10‒30	‒
H‒C‒O‒P	‒	5‒11	
H‒C‒S‒P			‒

they have a spin of $\frac{1}{2}$ and a considerable magnetic moment. Nonetheless, direct resonance at P^{31} nuclei is complicated by the fact that the intensity of P^{31} signals is only 6.6% of the proton signal (in the same field) and it is also difficult to observe this signal due to the relatively low content of these nuclei in normal organic molecules in comparison with protons. The latter situation often leads to a considerable multiplicity of the resonance lines of P^{31} [for example, in the P^{31} NMR spectrum of trimethyl phosphate $(CH_3O)_3PO$ the signal of phosphorus is split into ten components] due to spin‒spin coupling with a large number of protons and this also hampers the use of direct resonance of phosphorus nuclei, particularly in the investigation of dilute solutions. Therefore, the study of spin‒spin coupling of phosphorus nuclei with hydrogen and fluorine assumes an important role in the investigation of organophosphorus compounds since the latter may be observed and identified readily in the normal spectra of resonance at these nuclei.

A characteristic of spin‒spin coupling with the P^{31} nucleus is the considerable change in magnitude and often in sign of the coupling constants with a change in the valence state of the phosphorus atom. The $H^1‒P^{31}$ and particularly $F^{19}‒P^{31}$ coupling constants reach very considerable values (more than 1400 Hz) and often appear with a considerable distance between the interacting nuclei. Data on the spin‒spin coupling constants of phosphorus with protons and fluorine in various structures are summarized in Table III-10.

TABLE III-11. Spin−Spin Coupling Constants J_{HP}
for Nuclei at Different Distances [63]

Compound	No. of bonds	J_{HP}, Hz	r_{HP}, Å	$J \cdot r^3$, Hz · Å^3
HPO(OC$_2$H$_5$)$_2$	1	687	1.4	1885
CH$_3$PO(OC$_2$H$_5$)$_2$	2	17.5	2.4	241
HC≡CPO(OC$_2$H$_5$)$_2$	3	13.9	4.05	923
CH$_3$C≡CPO(OC$_2$H$_5$)$_2$ . . .	4	7.4	5.0	920
CH$_3$C≡CC≡CPO(OC$_2$H$_5$)$_2$	6	2.15	7.6	950

In compounds containing allene and acetylene groups in hydrocarbon radicals attached to a phosphorus atom there is appreciable
long-range spin−spin coupling of protons with phosphorus and in a
phosphonate with a methyldiacetylenyl radical the constant J_{HP} of
nuclei separated by six bonds still has a considerable magnitude
(2.15 Hz) [63, 64].

The spin−spin coupling constants of phosphorus with protons
in the β-position often substantially exceed the spin−spin coupling
constants with α-protons. Together with a decrease in the absolute
value of $J_{H\alpha P}$ there is often a change in the relative sign of this
constant. Some data on the signs of spin−spin coupling constants
for tricoordinate organophosphorus compounds are given below
[65, 66]:

The absolute signs of the coupling constants given were
derived on the assumption that spin−spin coupling between protons
in these organophosphorus compounds obeys the same rules as in
hydrocarbons.

TABLE III-12. Additive Components of $F-P$
Spin–Spin Coupling Constants

Substituent	η, Hz	Substituent	η, Hz
NaO	391	HO	458
CHO	474	C_6H_5O	499
F	528	iso-C_3H_7O	481
$(CH_3)_2N$	462	C_2H_5 av	585 ± 10
iso-C_4H_9O	484	$R—CH=CH$ (av)	550 ± 25
n-C_3H_7O	482	C_6H_5 (av)	555 ± 2
n-C_4H_9O	483	CH_3 (av)	556 ± 5
n-$C_6H_{13}O$	484	Cl	588
C_2H_5O	482	$CH_3—C≡C$	454

The signs of the spin–spin coupling constants in tetracoordinate compounds of phosphorus have been studied much less. It has
been reported that in alkyl phosphates the constants $J_{H\alpha P}$ and $J_{H\beta P}$
have the same sign, in contrast to phosphites [66]. In the acid fluoride of methylacetylenephosphinic acid $CH_3-C\equiv C-P(O)F_2$ the spin–
spin coupling constants J_{HP} and J_{FP} have opposite signs.

Available data on the magnitudes and relative signs of the constants for phosphorus compounds indicate that spin–spin coupling
between phosphorus and protons may occur by two different mechanisms, one of which is important for directly connected H and
P atoms and the other appears at a greater distance. In the case
of protons separated from phosphorus by two bonds, the superposition of the two mechanisms may explain the reduction in the absolute value of J_{HP} and the change in the sign of the constant. In this
respect it is interesting to compare the values of J_{HP} for nuclei
separated by different numbers of chemical bonds. In order to compensate for the effect of the distance between the nuclei, in Table
III-11 we compare the products of the absolute values of the coupling
constants and the distance between them to the third power.

The table shows that at great distances the magnitude of the
constant J_{HP} is determined almost exclusively by the distance between the interacting nuclei, while for nuclei separated by one or
two bonds, other effects are obviously important. However, it is
impossible at the present time to state definitely precisely the mechanism of spin–spin coupling J_{HP} in different cases. It is only

certain that the contribution from the contact interaction of the nuclei is not the only one in this case.

Spin–spin coupling of phosphorus and fluorine has been studied in more detail for nuclei connected directly and it has been found that there is additivity of the contributions of different substituents to the value of J_{FP} in phosphoryl compounds [67], making it possible to use these constants for analytical purposes. Table III-12 gives the increments of substituents, by means of which it is possible to calculate the spin–spin coupling constant J_{FP} in compounds

of the type $\begin{smallmatrix} X \\ \\ Y \end{smallmatrix} \!\!\diagdown\!\! P \!\!\diagup\!\! \begin{smallmatrix} O \\ \\ F \end{smallmatrix}$ by means of the formula (III-26):

$$J_{FP} = \eta_X + \eta_Y \qquad \qquad \text{(III-26)}$$

From an examination of the increments of substituents given it is difficult to derive a general rule for the changes in them. Analysis of the constants J_{FP} is also hampered by the uncertainty on the role of the vacant d-orbitals of phosphorus, which undoubtedly play a substantial part in tetracoordinate compounds with such electronegative substituents as fluorine, oxygen, nitrogen, and the $C \equiv C$ group. However, it is noticeable that alkyl and alkenyl substituents lead to an increase in the value of J_{FP}. Determination of the absolute sign of the constant J_{FP} (it is very probable that these constants are negative relative to J_{HC}) may make it possible to draw definite conclusion on the mechanism of this interaction.

Homonuclear spin–spin coupling between phosphorus nuclei has been studied little due to the difficulty of observing it. In some organophosphorus compounds it has been possible to determine the magnitudes and relative signs of J_{PP} from the proton spectrum, whose form depends on these parameters. The magnitudes of J_{PP} vary over a wide range, depending on the structure of the molecules. For directly connected phosphorus nuclei they vary over the range of 18.7-220 Hz and are apparently negative (absolute sign) [68]. It is interesting that for more remote phosphorus nuclei J_{PP} varies over a still wider range (480 Hz in the anion $[PO_3PHO_2]^{3-}$ and 17 Hz in the pyrophosphate anion $[O_3P-O-PO_3]^{4-}$).

The study of the constants J_{PP} is particularly interesting in NMR spectroscopy since this is apparently the only heavy polyvalent element for which it has been possible to observe experimentally homonuclear spin—spin coupling without the use of enrichment.

9. SPIN — SPIN COUPLING OF PROTONS

AND FLUORINE WITH OTHER

MAGNETIC NUCLEI

There are many other magnetic nuclei, spin—spin coupling with which is of great interest to chemists (for example, N^{14} and N^{15}, B^{10} and B^{11}, Li^7, Cl^{35}, S^{33}, Br^{79} and Br^{81}, etc., which are often found in organic compounds). Unfortunately the experimental observation of the spin—spin coupling constants involves considerable experimental difficulties, largely due to the fact that most of these isotopes have a spin greater than $\frac{1}{2}$ and an electrical quadrupole moment which produces broadening of the lines that exceeds the magnitude of the coupling constants. Nonetheless there is a series of examples of the successful observation of spin—spin coupling with heavy nuclei and these are examined below.

The investigation of ammonia with the nitrogen isotope N^{15} (spin $\frac{1}{2}$) and its deuterated derivatives [69] indicates quite strong spin—spin coupling of this nucleus with protons $J_{H^1-N^{15}}$ (61.2 ± 0.9 Hz). It is characteristic that a comparison of the ratios J_{HX}/μ_X

TABLE III-13. Constants for H^1-Tl^{205} and H^1-Hg^{199} Spin—Spin Coupling [74, 75]

Compound	Proton group	Spin—spin coupling constants			
		R_3Tl	R_2Tl^+	RTl^{2+}	$RHgBr$
CH_3—CH_2—X	CH_2	−198	−340	—	—
	CH_3	+396	+628	—	—
CH_2=CH—X	gem	—	+842	+2004	291
	cis	—	+805	+1806	331
	trans	—	+1618	+3750	658
C_6H_5X	o	+259	+451	+948	—
	m	+80	+139	+356	—
	p	+35	+52	+123	—

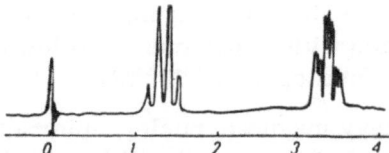

Fig. III-5. Proton resonance spectrum of
tetraethylammonium iodide in D_2O. Index
HN. 60. ЛAlk. IB. There is splitting of
the proton signals of the CH_3 group due to
spin—spin coupling with the N^{14} nucleus.
The scale of chemical shifts is given as
ppm relative to the internal standard HDO.

for ammonia and phosphorus shows that for nitrogen the absolute
value of this ratio is 1.335 times that for phosphorus. It is interesting to note that the constant J_{HH} of the geminal protons in ammonia (10.35 ρ 0.80 Hz) is close in absolute magnitude to the constant in saturated hydrocarbons.

In the proton spectrum of ammonia with a predominance of
the isotope N^{14} [70] it is also possible to distinguish spin—spin
coupling of protons with nitrogen. The signal of the protons in this
case appears in the form of a diffuse triplet (1 : 1 : 1), caused by the
nuclear spin of the isotope N^{14}, which equals 1. Spin—spin coupling
of protons attached directly to the nucleus N^{14} in organic amines
appears as a rule as broadening of the signal and it is usually impossible to observe the fine structure. In perfluorinated amines
$R-NF_2$ the constants for $N^{14}-F^{19}$ spin—spin coupling reach 117
Hz [71].

It is not usually possible to observe spin—spin coupling of
nitrogen with more remote protons in organic amines. However,
in isonitriles and quaternary ammonium salts it has been possible
to observe appreciable spin—spin coupling of nitrogen N^{14} with protons attached to a β-carbon atom (Fig. III-5) with J_{NH} =1.8 Hz.
The symmetrical arrangement of the electron cloud at the nitrogen
atom in these compounds increases the quadrupole relaxation time
so that it is possible to observe spin—spin coupling. Disruption of
the symmetry in tetraalkylammonium salts due to the introduction
of different substituents, particularly those containing a sp^2-hybrid-
ized carbon atom, reduces the value of J_{NH}. It is characteristic

that in these cases the absolute magnitude of the spin—spin coupling constants for nitrogen with α-protons is so low that it is impossible to detect them in normal spectra [72, 73].

Among the heavy magnetic nuclei, spin—spin coupling with protons has been studied in more detail for the isotopes Tl^{203}, Tl^{205}, and Hg^{199}, which have a spin of $1/2$. Table III-13 gives the characteristic values of the constants for $H^1—Tl^{205}$ and $H^1—Hg^{199}$ spin—spin coupling in various organic compounds.

Unfortunately the number of organic compounds with heavy metal atoms bound by covalent bonds is very limited. Nonetheless, the data presented already are of considerable interest for elucidating the mechanism of spin—spin interaction. The table shows that the character of the change in the values (and where they have been determined, the signs) of the spin—spin coupling constants is very similar to the character of the change for proton—proton interaction. This evidently indicates the importance of the contact Fermi interaction of the nuclear spins and the appearance of the electron density in the region of the nucleus may be connected with the participation of the 6s-orbitals of these atoms. The increased absolute values of the constants may be due to the high effective charges of the nuclei of these heavy atoms [74]. In the future it will be interesting to determine the magnitudes and signs of the constants for longer-range spin—spin coupling of protons and other magnetic nuclei with nuclei of these heavy metals for a more thorough elucidation of the mechanism of this coupling.

In this chapter we examined the experimental data and foundations for the theoretical concepts on the mechanism of spin—spin coupling of various magnetic nuclei. The results obtained up to the present time make it possible to use the values and signs of the constants both for purely analytical purposes and for correlation with fine changes in the electronic structure of organic molecules. The further development of this important trend in NMR spectroscopy will evidently involve more accurate determination of the signs of spin—spin coupling constants and the study of long-range interaction with the use of a wider range of magnetic isotopes and also the generalization of these experimental results from the point of view of quantum chemistry.

An important problem connected with the analysis of spectra arises in the study of spin—spin coupling in NMR spectra. The simple rules for determining the values of the coupling constants from the spectra, which were given in Ch. I, only apply to spectra of the first order. Analysis of complex nuclear resonance spectra with a large number of magnetic nuclei may be extremely difficult and in individual cases it may be a practically insoluble problem. In the next chapter we examine methods for mathematical analysis of complex NMR spectra and also methods for the experimental simplification of complex spectra in order to determine the main parameters, namely, the chemical shifts of magnetic nuclei and the magnitudes and signs of the spin—spin coupling constants between them.

LITERATURE CITED

1. A. Abragam, Principles of Nuclear Magnetism, Oxford Univ. Press, New York (1961).
2. W. G. Proctor and F. C. Yu, Phys. Rev., 78:471 (1950).
3. H. S. Gutowsky and D. W. McCall, J. Chem. Phys., 22:162 (1954).
4. E. L. Hahn and D. E. Maxwell, Phys. Rev., 84:1246 (1951).
5. F. I. Skripov, Course of Lectures on Radiospectroscopy, Izd. LGU (1964), p. 144.
6. J. Pople, W. Schneider, and H. Bernstein, High-Resolution Nuclear Magnetic Resonance, McGraw-Hill, New York (1959).
7. I. V. Aleksandrov, Theory of Nuclear Magnetic Resonance, Izd. Nauka (1964).
8. N. F. Ramsey, Phys. Rev., 91:303 (1953).
9. N. F. Ramsey and E. M. Purcell, Phys. Rev., 85:143 (1952).
10. H. Y. Carr and E. M. Purcell, Phys. Rev., 88:415 (1952).
11. D. Ingram, Free Radicals as Studied by Electron Spin Resonance, Academic Press, New York (1958).
12. N. D. Sokolov, Usp. khim., 32:967 (1963).
13. J. A. Pople and A. A. Bothner-By, J. Chem. Phys., 42:1339 (1965).
14. H. M. McConnell, J. Chem. Phys., 24:460 (1956).
15. H. S. Gutowsky, M. Karplus, and D. M. Grant, J. Chem. Phys., 31:1278(1959).
16. H. M. McConnell, J. Chem. Phys., 24:460 (1956).
17. B. L. Shapiro, R. M. Kopchik, and S. J. Ebersole, J. Chem. Phys., 39:3154 (1963).
18. M. Karplus, J. Chem. Phys., 30:11 (1959).
19. M. Karplus, J. Am. Chem. Soc., 85:2870 (1963).
20. R. J. Abraham and K. G. R. Pachler, Mol. Phys., 7:165 (1964).
21. R. J. Abraham and L. Cavalli, Mol. Phys., 9:67 (1965).
22. M. L. Huggins, J. Am. Chem. Soc., 75:4123 (1953).
23. N. Jonathan, S. Gordon, and B. P. Dailey, J. Chem. Phys., 36:2443 (1963).
24. G. W. Smith, J. Mol. Spectr., 12:146 (1964).
25. S. Sternhell, Rev. Pure Appl. Chem., 14:15 (1964).
26. M. Karplus, J. Am. Chem. Soc., 82:4431 (1960); J. Chem. Phys., 33:1842 (1960).

27.   H. H. Appel, R. P. Bond, and K. H. Overton, Tetrahedron, 19:635 (1963).

28.   E. M. Kosower and T. S. Sorensen, J. Org. Chem., 28:687 (1963).

29.   K. T. Hobgood, Jr., and J. H. Goldstein, J. Mol. Spectr., 12:76 (1964).

30.   E. I. Snyder, L. J. Altman, and J. D. Roberts, J. Am. Chem. Soc., 84:2004 (1962).

31.   R. A. Hoffman, Arkiv Kemi, 17:1 (1961).

32.   A. Taurins and W. G. Schneider, Can. J. Chem., 38:1237 (1960).

33.   J. R. Holmes and D. Kivelson, J. Am. Chem. Soc., 83:2959 (1961).

34.   N. Muller and D. E. Pritchard, J. Chem. Phys., 31:768, 1471 (1959).

35.   J. N. Shoolery, J. Chem. Phys., 31:1427 (1959).

36.   H. A. Bent, Chem. Rev., 61:275 (1961).

37.   E. R. Malinowski, J. Am. Chem. Soc., 83:4479 (1961).

38.   C. Juan and H. S. Gutowsky, J. Am. Chem. Soc., 84:307 (1962).

39.   C. Juan and H. S. Gutowsky, J. Chem. Phys., 37:2198 (1962).

40.   S. G. Frankiss, J. Phys. Chem., 67:752 (1963).

41.   A. W. Douglas, J. Chem. Phys., 40:2413 (1964).

42.   E. R. Malinowski, L. Z. Pollara, and J. P. Larmann, J. Am. Chem. Soc., 84:2649 (1962).

43.   R. M. Hammaker, Can. J. Chem., 43:2916 (1965).

44.   R. M. Hammaker, J. Chem. Phys., 43:1843 (1965).

45.   W. Gordy and J. O. Thomas, J. Chem. Phys., 24:439 (1956).

46.   J. R. Cavanaugh and B. P. Dailey, J. Chem. Phys., 34:1099 (1961).

47.   J. Hinze, M. A. Whitehead, and H. H. Jaffe, J. Am. Chem. Soc., 85:148 (1963).

48.   J. H. Goldstein and R. T. Hobgood, Jr., J. Chem. Phys., 40:3592 (1964).

49.   D. S. Bartow and J. M. Richardson, J. Chem. Phys., 42:4018 (1965).

50.   J. S. McIntyre and I. J. Bastein, J. Am. Chem. Soc., 86:1360 (1964).

51.   P. C. Haake and W. B. Miller, J. Am. Chem. Soc., 85:4644 (1963).

52.   G. J. Karabatsos and C. E. Orzech, Jr., J. Am. Chem. Soc., 87:560 (1965).

53.   H. Dreeskamp, E. Sackmann, and G. Stegmeier, Ber. Bunzenges. Phys. Chem., 67:860 (1963).

54.   D. F. Evans, S. L. Manatt, and D. D. Elleman, J. Am. Chem. Soc., 85:238 (1963).

55.   C. N. Banwell and N. Sheppard, Proc. Roy. Soc., A263:136 (1961).

56.   G. A. Williams and H. S. Gutowsky, J. Chem. Phys., 30:717 (1959).

57.   S. Ng and C. H. Sederholm, J. Chem. Phys., 40:2090 (1964).

58.   J. Dyer, Proc. Chem. Soc., 1963:275.

59.   J. Reuben, Y. Shvo, and A. Demiel, J. Am. Chem. Soc., 87:3995 (1965).

60.   L. Carpo and C. H. Sederholm, J. Chem. Phys., 33:1583 (1960).

61.   R. K. Harris and N. Sheppard, Trans. Faraday Soc., 59:606 (1963).

62.   L. W. Reeves and E. J. Wells, Can. J. Chem., 41:2698 (1963).

63.   B. I. Ionin, V. B. Lebedev, and A. A. Petrov, Dokl. Akad. Nauk SSSR, 152:1354 (1963).

64.   C. Carrier, W. Chodkiewicz, and P. Cadiot, Bull. Soc. chim. France, 1966:1002.

65.   S. L. Manatt, L. Juvinall, and D. D. Elleman, J. Am. Chem. Soc., 85:2664 (1963).

66.   E. Duval, J. Ranft, and C. J. Béné, Mol. Phys., 9:427 (1965).

67. V. F. Bystrov, A. A. Neimysheva, A. I. Stepanyants, and I. L. Knunyants,
     Dokl. Akad. Nauk SSSR, 156:637 (1964).
68. R. K. Harris and R. G. Hayter, Can. J. Chem., 42:2282 (1964).
69. R. A. Berheim and H. Batiz-Hernandez, J. Chem. Phys., 40:3446 (1964).
70. R. A.Ogg and J. D. Ray, J. Chem. Phys., 26:1515 (1957).
71. R. Ettinger and C. B. Calburn, Inorg. Chem., 2:1311 (1963).
72. P. G. Gassman and D. C. Heckert, J. Org. Chem., 30:2859 (1965).
73. J. M. Lehn and M. Frank-Neunamm, J. Chem. Phys., 43:1421 (1965).
74. J. P. Maher and D. F. Evans, J. Chem. Soc., 1965:637.
75. P. R. Wells,W. Kitching, and R. F. Henzell, Tetrahedron Letters, 1964:1029.
76. R. C. Cookson, T. A. Caoblo, J. J. Frankel, and J. Hudec, Tetrahedron, Suppl.,
     No. 7:353 (1966).
77. E. L. Mackor and C. MacLean, J. Chem. Phys., 44:64 (1966).
78. S. L. Smith and R. H. Cox, J. Chem. Phys., 45:2848 (1966).
79. S. L. Smith and A. M. Ihring, J. Chem. Phys., 46:1187 (1967).

Chapter IV

Analysis of Complex Nuclear Magnetic Resonance Spectra

In previous chapters we examined the connection between the structure of organic molecules and the parameters of the NMR spectra,
namely, the chemical shifts and the spin—spin coupling constants.
The determination of these parameters is quite simple when the
substance investigated gives a simple spectrum of the first order.
Spectra of the first order (Ch. I) arise when the differences in the
chemical shifts of groups of equivalent nuclei considerably exceed
the constants for spin—spin coupling between these nuclei. Since
an increase in the strength of the magnetic field of the spectrometer (with a corresponding increase in frequency) leads to a pro-
portionate increase in the chemical shifts, but leaves the spin—
spin coupling constants unchanged (if they are both expressed in
frequency units, Hz), plotting the NMR spectra on spectrometers
with a high magnetic field strength leads as a rule to simplification
of the spectra and makes them closer to spectra of the first order.
Figure IV-1 gives spectra of acrylonitrile plotted at frequencies
of 40, 60, and 100 MHz. While the spectrum of acrylonitrile at 40
MHz is very complex, in the spectrum at 60 MHz it is possible to
pick out the region of resonance of the nucleus in the α-position
relative to the nitrile group and though this spectrum cannot be
regarded as a first-order spectrum, its analysis is still simpler,
as will be shown below. In the spectrum at 100 MHz the chemical
shifts already considerably exceed the spin—spin coupling constants
and this spectrum may be analyzed by first-order rules.

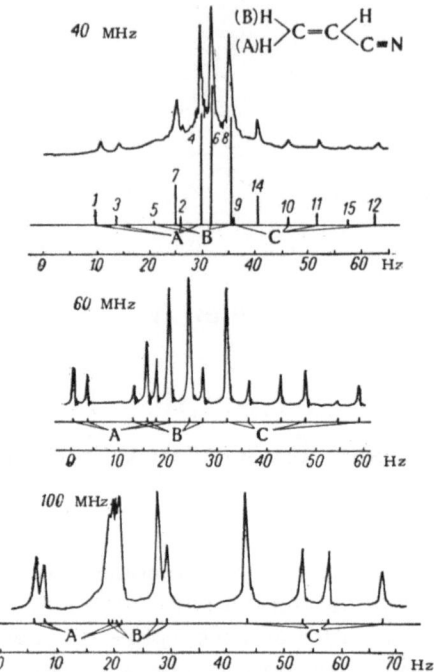

Fig. IV-1 Spectra of acrylonitrile.

The use of spectrometers with a high magnetic field strength is a great help in the investigation of the nuclear resonance of substances which give complex spectra. However, it is quite understandable that this approach is limited by technical possibilities. Moreover, increasing the field strength has little effect in cases where complex spectra arise due to the magnetic nonequivalence of chemically equivalent nuclei, i.e., when nuclei of one group which are chemically equivalent have different spin−spin coupling constants with some third nucleus of the spin system. Thus, for example, while 1,1-difluoroallene I gives a simple first-order spectrum, the spectrum of 1,1-difluoroethylene II is a complex spectrum since here the two protons are not arranged in equivalent positions relative to the two fluorine nuclei.

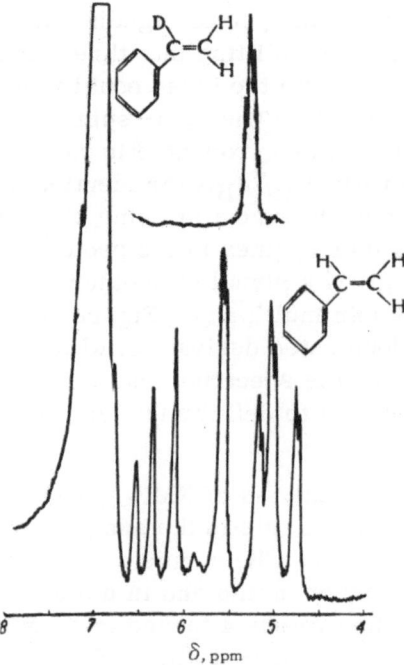

Fig. IV-2. Spectra of styrene and the vinyl
protons of deuterostyrene at 40 MHz. Index
of styrene spectrum A=.ABC + M.

It should be noted that the appearance of complex spectra is
of definite value since these spectra give valuable additional information and, in particular, they often make it possible to determine the relative signs of the spin–spin coupling constants and to
determine the coupling constants between chemically equivalent
nuclei. In the present chapter we examine methods for mathematical analysis of complex spectra and also the nuclear magnetic
double resonance method (NMDR), in which nuclear resonance is investigated by means of special apparatus. For experimental chemists the use of compounds in which some of the magnetic nuclei
have been replaced by their isotopes may be of great help in the
analysis of spectra in many cases.

Partly or completely deuterated substances are the isotopically substituted compounds used most frequently. Deuterium has a
spin $I = 1$ and a magnetic moment associated with it so that it may

also produce splitting of the proton signals due to spin—spin coupling. In this case there is splitting into three components of equal
intensity in accordance with the three equally probable nuclear
spin states (+1, 0, and—1). The spin—spin coupling constant J_{HD}
is reduced due to the lower gyromagnetic ratio of the deuterium
nucleus so that the ratio J_{HH}/J_{HD} for identical structures equals
6.51. Therefore the replacement of some of the protons by deuterium reduces the number of lines in the proton resonance spectrum
because deuterium gives a signal at another frequency and also
contracts the spin—spin multiplets. Figure IV-2 gives the spectra
of styrene and its deuterated derivative, which show that deuterium
substitution simplifies the spectrum and makes it possible to assign
the lines of the spectrum to definite nuclei of the molecule more
simply.

    Another form of analysis of NMR spectra with the use of iso-
topes is of great value and this is the analysis of C^{13} satellites in
proton resonance spectra. The magnetic isotope of carbon C^{13} constitutes 1.1% of all carbon atoms and in molecules containing nuclei of this isotope the protons attached to them give a split signal
in the spectrum with the constant J_{HC} 125-160 Hz. As a result
additional signals appear on either side of the main signal of the
protons and these are called C^{13} satellites. The form of the satellites differs from the form of the main signal since the effect of
splitting for protons attached to a C^{13} nucleus is equivalent to the
appearance of an additional chemical shift at them [1]. It is often
possible to find spin—spin coupling between these protons attached
to a neighboring C^{12} carbon atoms as, for example, in dioxane,
whose main signal consists of a narrow line. Analogous satellites
are observed in the spectra of organosilicon and tin compounds.

    The use of data for substances of similar structure is also
of great help in the analysis of complex spectra. The construction
of a theoretical spectrum from known parameters is a much
simpler problem than the determination of these parameters from
an experimental spectrum. Therefore, before beginning to analyze
an actual experimental spectrum it is convenient to reproduce this
spectrum from parameters obtained for substances of similar
structure. This procedure forms the basis of methods of analyzing
spectra by means of digital computers. The usual programs for
analysis of spectra by means of computers [2, 3] include the construction of theoretical spectra from given parameters, comparison

of these spectra with the experimental spectra, subsequent refinement of the starting parameters, and a repeat of the cycle to obtain the best agreement between the theoretical and experimental spectra.

In subsequent sections of this chapter we will examine methods of analyzing complex spectra without the use of computers. This calculation of spectra is quite simple for most typical cases. When the system of nuclear spins contains not more than two groups of magnetically equivalent nuclei, the relation between the position of the lines in the spectrum and its parameters may be expressed in the form of tables or diagrams [4] with the ratio $J/\Delta\nu$ as the independent variable. It is also necessary to assimilate methods for manual calculation of spectra to understand the analysis of spectra with computers, whose use is particularly valuable when the system investigated contains three or more nonequivalent nuclear spins.

Classification of NMR Spectra. The magnetic nuclei in a molecule of the substance investigated are usually denoted by capital letters of the Latin alphabet. Modified designations have been used subsequently [5] and in these chemically equivalent nuclei are denoted by the same letters, nuclei with similar chemical shifts by adjacent letters, and those nuclei between which the chemical shifts (expressed in hertz) substantially exceed the spin−spin coupling constants between them, by letters from other parts of the alphabet. If among a group of chemically equivalent nuclei we find magnetically nonequivalent nuclei, i.e., those with different coupling constants with some other nucleus of the molecule, they are distinguished by the number of primes. Thus, for example, the spectrum of the molecule CH_2Cl-CF_2Cl should be denoted by A_2X_2 since the magnetic nuclei in each of the two groups are both chemically and magnetically equivalent due to rapid rotation about the $C-C$ bond. The spectrum of difluoroallene I should also be denoted in the same way, while the spectrum of difluoroethylene II is denoted more correctly by AA'XX'. Four different spin−spin coupling constants appear here in the general case.

It should be noted that the selection of a particular designation may often be arbitrary. This is particularly concerned with the question of the relation between the magnitudes of the chemical shifts and the spin−spin coupling constants. Strictly speaking the chemical shifts may be regarded as l a r g e only when the nuclei belong to different magnetic isotopes. However, in practice, for nuclei of one sort with chemical shifts which are 5-10 times as great as the spin−spin coupling constants the spectra differ little from those calculated on the assumption of a large chemical shift. The same compound may have l a r g e and s m a l l chemical shifts, depending on the working field strength of the spectrometer and on whether or not the signals of C^{13} satellites are taken into account, for example. Finally, the spectrum of a substance may be of a different type, depending on the resolution of the spectrometer: with an increase in resolution there may appear in a first-order spectrum additional lines which cannot be explained by simple rules. In practice the spectra are often denoted in one way or another, depending on the method used for analyzing them.

Equally subjective are the terms s t r o n g and w e a k spin−spin coupling or s t r o n g l y c o u p l e d and w e a k l y c o u p l e d spectra, which we will use to denote spectra with s m a l l and l a r g e chemical shifts, respectively, relative to the spin−spin coupling constants.

In practice, particularly in older work, the primes indicating magnetic nonequivalence of the nuclei are often omitted in the designations. This is permissible for simplicity in writing in cases where the structure of the molecules investigated unequivocally indicates the form of the spectrum.

A. COMPLETE ANALYSIS OF SPECTRA OF COMPLEX SYSTEMS OF NUCLEAR SPINS

In the present section we examine mathematical methods of analyzing complex NMR spectra which are based on the quantum mechanical analysis of a system of nuclear spins. These methods are essentially a complete theory of complex spectra, which relates the parameters of the spectrum to the number, position, and intensity of the lines. The complete mastery of this theory is difficult for chemists who have no special physicomathematical training. The purpose of the present section is not a strict account of the theory of complex spectra, but only practical recommendations for analysis of the most important and most frequently encountered cases. Theoretical problems are discussed in a series of monographs [6, 7], the review by Corio [4], and in a more popular form, in Roberts' book [8].

By a system of nuclear spins we mean a group of magnetic nuclei between which there is spin−spin coupling. It is possible that any particular nucleus of the system interacts only with some and not all of the spins. The molecule may also contain other magnetic nuclei which do not belong to the given system if they do not interact with any one nucleus of the given system or if the spin−spin coupling between them is so weak that it may be neglected. It is possible that a molecule may contain two independent systems of nuclear spins and their signals may overlap as a result of a small difference in the chemical shifts of nuclei of the two systems. The simplest example of this case is the spectrum of a dilute solution of ethanol in deuterochloroform (see Fig. A-10). The ethyl group of the compound forms an A_3X_2 system, while the signal of the hydroxyl proton, which does not belong to this system in the given case, gives a separate peak, lying in the same region of the spectrum. In a special case the molecule may contain two equivalent spin systems, whose chemical shifts coincide. An example of such a system is provided by diethyl ether (see A-4), in which there is no appreciable interaction between the spins through the oxygen bridge, or naphthalene, in which the interaction between protons belonging to different nuclei may be neglected when the resolution is not too high. The two systems are analyzed independently and in the case of equivalent systems, they are analyzed as if only one such system was present.

Systems with strong spin−spin coupling behave as if the resonance of each spin were connected with a change in the energy state of all other spins of the system (in quantum mechanical terminology this denotes a shift in the wave functions of the individual nuclei). A successful analysis of the spectra of strongly coupled systems may be carried out only if the energy states of the system are considered as a whole. Therefore it is more convenient to express the chemical shifts of the nuclei and the spin−spin coupling constants in comparable units and frequency units, hertz, are usually used for this purpose.

Subsequently ν_1, ν_2, etc., will denote the positions of individual lines in the spectrum in hertz relative to a particular point in the spectrum (for example, relative to the signal of a reference substance or the center of the spectrum), ν_A, ν_B, ν_X, etc., will denote the chemical shifts δ_A, δ_B, δ_X, etc., in these spectra; the signs

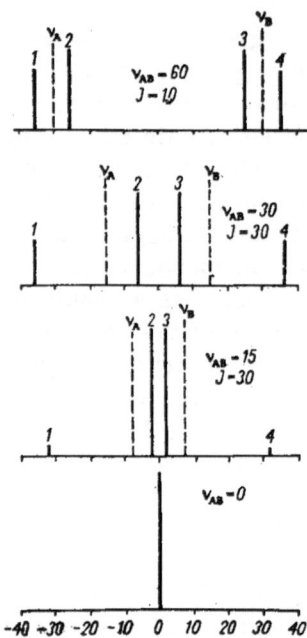

Fig. IV-3. Theoretical spectra
of the system AB with different
relations between J_{AB} and ν_{AB}.

$\nu_{1,2}$, ν_{AB}, δ_{BA} will denote the differences in the values $\nu_1 - \nu_2$, $\nu_A - \nu_B$, $\delta_B - \delta_A$; the absolute values of the differences will be distinguished by vertical lines in the generally accepted way: $|\nu_{1,2}|$.

1. SYSTEM OF TWO NONEQUIVALENT NUCLEI AB

A system of two spins is one of the simplest and most readily interpreted systems. In Ch. I we examined a particular case of this system, namely, an AX spectrum in which the chemical shift between the nuclei A and X considerably exceeds the splitting due to spin−spin coupling. The spectrum of the system AX consists of two symmetrical doublets, the distance between the components of which gives the spin−spin coupling constant, while the distances from the centers of the doublets to the signal of the reference gives the chemical shifts. Another particular case of an AB system is two equivalent nuclei (an A_2 system). As has already been pointed out (Ch. I), the coupling constant between equivalent nuclei does not appear in the spectrum and the spectrum consists of a single line.

TABLE IV-1. Position and Intensity of Lines in the Spectrum
of an AB Two-Spin System

Line No.	Frequency	Relative intensity
1, 4	$\frac{1}{2}(\nu_A + \nu_B) \pm \frac{1}{2}\left[(\nu_{AB}^2 + J_{AB}^2)^{1/2} + J_{AB}\right]$	$1 - \dfrac{J_{AB}}{(\nu_{AB}^2 + J_{AB}^2)^{1/2}}$
2, 3	$\frac{1}{2}(\nu_A + \nu_B) \pm \frac{1}{2}\left[(\nu_{AB}^2 + J_{AB}^2)^{1/2} - J_{AB}\right]$	$1 + \dfrac{J_{AB}}{(\nu_{AB}^2 + J_{AB}^2)^{1/2}}$

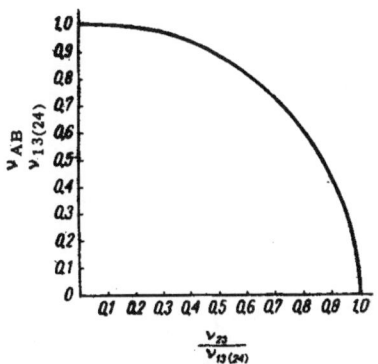

Fig. IV-4. Diagram for analysis of spectra of the AB system.

In the general case the spectrum of an AB system consists of four peaks with the two central peaks more intense than the two side peaks. The spectrum is symmetrical relative to the center and therefore, in many cases it is more convenient to measure the distance between the reference and the line of symmetry of an AB spectrum, while the chemical shifts of the nuclei A and B are already measured relative to this line (Fig. IV-3).

The spectrum of an AB system is independent of the sign of the coupling constant and therefore we assume for simplicity that J_{AB} is positive.

Quantum-mechanical analysis of the system of spins AB leads to the relations of the position and intensity of the lines in the spectrum to the coupling constants and chemical shifts of the nuclei given in Table IV-1.

The coupling constant J_{AB} may be obtained directly from the spectrum as the distance between lines 1-2 or 3-4. To determine the chemical shift ν_A or ν_B it is necessary to measure the distance between lines 2 and 3. This distance equals $\sqrt{\nu_{AB}^2 + J^2} - J$, and hence the difference in chemical shifts:

$$\nu_{AB} = \sqrt{(\nu_{2,3} + J)^2 - J^2} \qquad \text{(IV-1)}$$

It is convenient to analyze a spectrum of type AB by means of a simple diagram (Fig. IV-4), which consists of a quarter of a circle. Along the abscissa and ordinate axes are plotted the ratios $\nu_{2,3}/\nu_{1,3}$ and $\nu_{AB}/\nu_{1,3}$. Having determined from the spectrum the distances 2-3 and 1-3 (or, what is the same, 2-4) and taken their ratio, we determine from the diagram what fraction of the distance 1-3 is ν_{AB}. In the limit with large chemical shifts $\nu_{2,3} \approx \nu_{1,3}$ and then ν_{AB} may be determined directly from the distance 1-3. This procedure means that the analysis of the AB spectrum was carried out by the AX approximation.

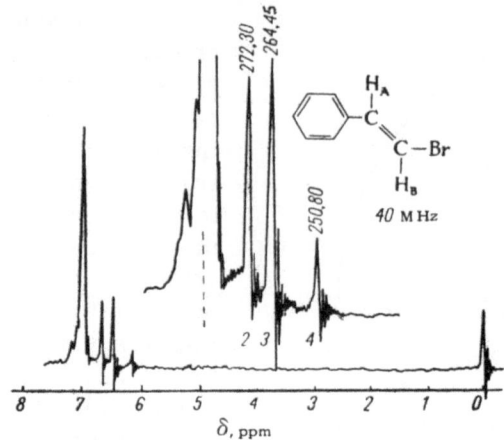

Fig. IV-5. Spectrum of trans-ω-bromostyrene, whose
vinyl protons form an AB system. Index H. 40.A
$=$.AB $+$ M.

To determine the chemical shifts of the nuclei A and B rela-
tive to the standard the value of $\frac{1}{2} \nu_{AB}$ must be calculated and added
to the distance from the center of the AB spectrum to the reference
(in Hz) and for conversion to parts per million the result is divided
by the working frequency of the spectrometer in MHz. As has already been pointed out, a spectrum of type AB itself does not make
it possible to determine which of the chemical shifts obtained refers to the nucleus denoted by the letter A and which to B.

The analysis procedure presented does not require experimental determination of the intensity of the lines. This approach
is quite sound since in the determination of intensities the accuracy
is considerably lower than in the establishment of the position of
lines. Nonetheless, for a rapid approximate estimate of the chemical shifts it is possible to use the following procedure. The ratio
of the intensities of two lines, for example, 1 and 2, is determined.
The resonance frequency of the nucleus A lies between the lines 1
and 2, closer to the inner line (more intense), with the distances to
the two lines inversely proportional to their intensities, in other
words, it lies at the c e n t e r o f g r a v i t y of these two lines.

In strongly coupled AB spectra the outer lines of the quartet
are often of such low intensity that they are lost in noise or are
overlapped by other lines of the spectrum. In this case the chemi-

cal shifts may also be determined by the usual method by taking
for J_{AB} any (to a certain extent) value which is used (according to
literature or experimental data) for compounds of similar structure.
Even with considerable errors in the selection of the value of J_{AB},
this leads to comparatively small errors in the magnitude of the
shifts. For example, for the case where $\nu_{AB} = 20$ Hz and $J_{AB} = 10$
Hz, an error in the selection of J of ±3 Hz gives quite acceptable
accuracy in such a rough determination of the chemical shift of
±2 Hz (0.05 ppm at 40 MHz).

In individual cases the coupling constant may be estimated
from the intensity of the inner lines in comparison with the intensity of the lines of other nuclei. If we assume that the intensity of
the unsplit signal of one nucleus equals 2, then the coupling constant
J_{AB} depends in the following way on the integral intensity P of lines
2 and 3 and the distance between them:

$$J_{AB} = \frac{P-1}{2-P} \nu_{2,3}$$ (IV-2)

However, this method should be used with care since it may
lead to considerable errors as a result of inaccurate determination
of the intensity.

Example of Analysis of an AB Spectrum

Figure IV-5 gives the spectrum of trans-ω-bromostyrene as
a pure liquid (without solvent), plotted at 40 MHz with tetramethylsilane as an internal reference. The upper part of the figure gives
the part of the spectrum which interests us with the position of the
lines relative to the standard (in Hz). The spectrum is complicated
by the fact that one of the signals of the AB part of the spectrum
overlaps an intense band of protons of the benzene ring. Nonetheless, complete analysis is quite possible. The spin—spin coupling
constant is determined directly from the spectrum from the distance between lines 3 and 4. The magnitude of this constant (13.65
Hz) indicates that the two protons are in the trans position. For
the determination of the chemical shifts of the nuclei A and B we
first find $\nu_{2,3}$, $\nu_{2,4}$, and their ratio. These values equal 7.85, 21.50
Hz, and 0.365, respectively. From the diagram in Fig. IV-4 we
find that in this case $\nu_{AB}/\nu_{2,4} = 0.93$, whence $\nu_{AB} = 0.93 \cdot \nu_{2,4} = 20$ Hz.
By taking half of the sum of the frequencies of lines 2 and 3 we find

the position of the center of the spectrum $\frac{1}{2}$ $(\nu_2 + \nu_3) = 268.37$ Hz and from this $\nu_A = 258.37$ Hz or $\delta_A = 6.47$ ppm; $\nu_B = 278.37$ Hz or $\delta_B = 6.97$ ppm.

From the analysis of the spectrum it is not clear which of the protons of ω-bromostyrene should be considered as H_A and which as H_B. However, it may be assumed with certainty that resonance at lower fields corresponds to the proton lying closer to the benzene ring as a result of the effect of the magnetic anisotropy of the latter. This is also confirmed by comparison with the spectra of other styrene derivatives.

It should also be noted that the position and intensities of the lines in the spectrum do not change if the spin—spin coupling constant is positive or negative so that the sign of the constant cannot be determined from the spectrum.

2. SYSTEM OF THREE NONEQUIVALENT NUCLEI

The spectrum of a system of three spins differs fundamentally from the spectrum of a two-spin system.

1. In the general case, the distribution of the line intensities in the spectrum depends on the relative signs of the spin—spin coupling constants. The spectrum does not change with a change in the signs of all three coupling constants. A change in the sign of one or two constants leads to a substantial change in the appearance of the spectrum.

2. The spectrum may contain up to 15 lines, i.e., 3 lines more than predicted by first-order theory. These three lines are called combination lines and are produced by a simultaneous change in the energy state of the three nuclear spins.

Therefore before analyzing the spectrum of an actual compound it is useful to examine data for a substance of similar structure to make a preliminary estimate of the possible magnitudes of the chemical shifts and coupling constants and the relative signs of the latter. This preliminary estimation both facilitates the analysis and also helps to avoid chance errors in quite laborious calculations.

Before analyzing a pure ABC system, we will examine several particular cases. Spectra of the systems A_2X and AMX are trivial and are analyzed readily by means of the rules for first-order spectra. Analysis of the system A_2B is more complex, but due to the small number of the parameters (δ_A, δ_B, and J_{AB}), it may be carried out by means of simple procedures.

A_2B Spectrum

An A_2B spectrum consists of nine lines which are usually numbered in the order shown in Fig. IV-6. As a rule, the ninth line, which is connected with a combination transition, has a low intensity or does not appear at all. Lines 5 and 6 are often poorly resolved and overlap.

TABLE IV-2.　Position and Intensity of Lines in A_2B Spectrum

Line No.	Origin	Frequency	Relative intensity
1	B	$\nu_B - \frac{1}{2}\left(\nu_{BA} - \frac{3}{2}J - R'\right)$	$\dfrac{1 - \sqrt{2}\,Q'^2}{1 + Q'^2}$
2	B	$\nu_B - \left[\nu_{BA} - \frac{1}{2}(R + R')\right]$	$\dfrac{[\sqrt{2}\,(Q - Q') - 1]^2}{(1 + Q^2)(1 + Q'^2)\,Q'^2}$
3	B	ν_B	1
4	B	$\nu_B - \frac{1}{2}\left(\nu_{BA} + \frac{3}{2}J - R\right)$	$\dfrac{(\sqrt{2}\,Q + 1)^2}{1 + Q^2}$
5	A	$\nu_A - \left[\nu_{BA} + \frac{1}{2}(R - R')\right]$	$\dfrac{[Q(1 + 2\sqrt{Q'}) + \sqrt{2}]^2}{(1 + Q^2)(1 + Q'^2)}$
6	A	$\nu_A - \frac{1}{2}\left(\nu_{BA} - \frac{3}{2}J + R'\right)$	$\dfrac{(Q' + \sqrt{2})^2}{1 + Q'^2}$
7	A	$\nu_A - \left[\nu_{BA} - \frac{1}{2}(R - R')\right]$	$\dfrac{[Q'(\sqrt{2}\,Q - 1) + \sqrt{2}]^2}{(1 + Q^2)(1 + Q'^2)}$
8	A	$\nu_A - \frac{1}{2}\left(\nu_{BA} + \frac{3}{2}J + R\right)$	$\dfrac{(Q - \sqrt{2})^2}{1 + Q^2}$
9	Combination	$\nu_B - \left[\nu_{BA} + \frac{1}{2}(R + R')\right]$	$\dfrac{[Q(\sqrt{2} - Q') - \sqrt{2}\,Q']^2}{(1 + Q^2)(1 + Q'^2)}$

In this table　$R = \nu_{BA}^2 - \nu_{BA}J + {}^9/_4 J^2$　　$Q = \dfrac{2J}{\nu_{BA} - {}^1/_2 J + R}$

$R' = \nu_{BA}^2 + \nu_{BA}J + {}^9/_4 J^2$　　$Q' = \dfrac{2J}{\nu_{BA} + {}^1/_2 J + R'}$

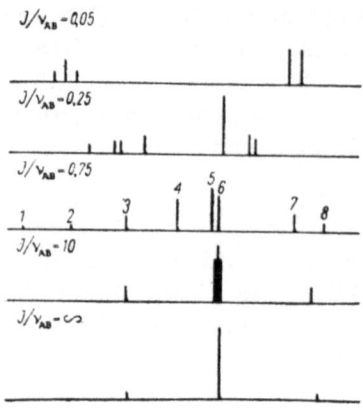

Fig. IV-6. Theoretical spectra of the
A₂B system.

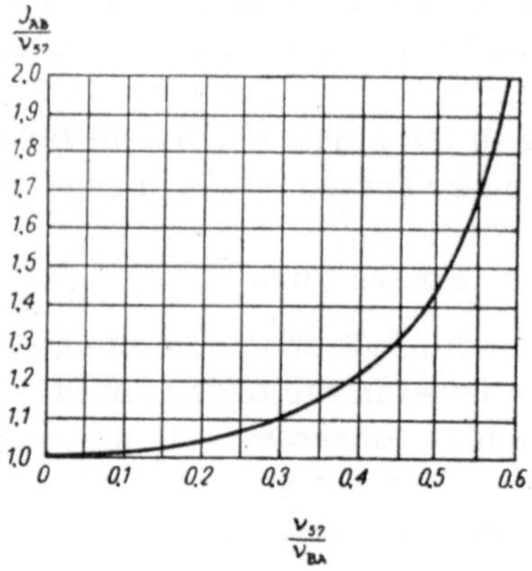

Fig. IV-7. Graph for determination of
J_{AB} in spectra of the system A₂B.

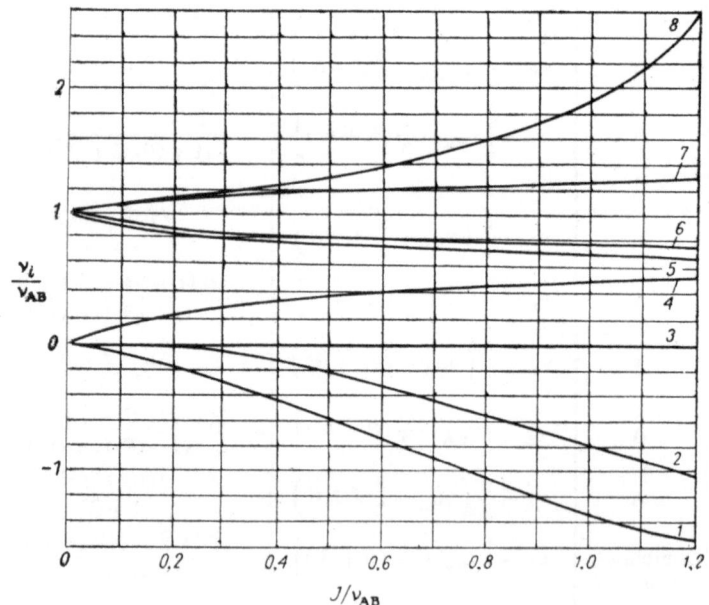

Fig. IV-8. Frequencies of lines in an A₂B spectrum
(relative to the value ν_{AB}).

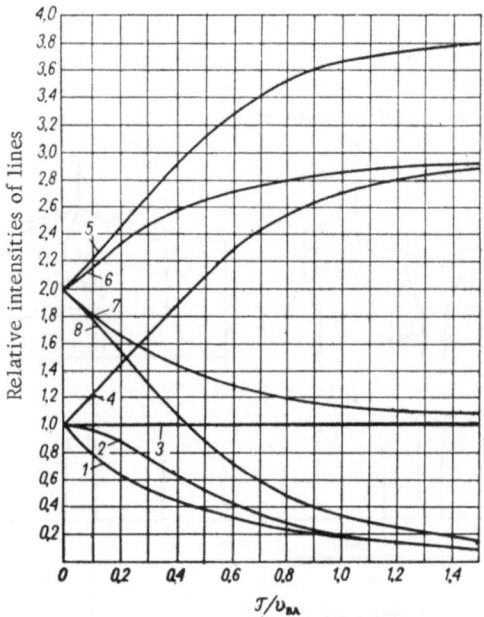

Fig. IV-9. Relative intensities of the lines in an
A$_2$B spectrum.

The chemical shift of one of the nuclei (B) is determined
from the position of line 3: $\nu_B = \nu_3$. The resonance frequency of
the nuclei A corresponds to the position halfway between lines 5
and 7, $\nu_A = \frac{1}{2} (\nu_5 + \nu_7)$. The same two lines (of which line 5 is the
most intense in an A$_2$B spectrum) are used to determine J_{AB} by
means of the curve in Fig. IV-7. In the limit, when $J_{AB} \ll \nu_{AB}$,
this distance exactly equals the coupling constant (A$_2$X spectrum).

By using the data in Table IV-2 it is possible to construct a
theoretical A$_2$B spectrum. A comparison of the theoretical and experimental spectra provides a check on the validity of the assignment of the lines in this case and this is particularly important if
part of the spectrum is blacked out or if some of the lines are
poorly resolved.

For rapid construction of a theoretical spectrum it is convenient to use the diagrams in Figs. IV-8 and IV-9, which give the
graphical relations of the position and intensities of the lines in an
A$_2$B spectrum to the ratio J/ν_{BA}.

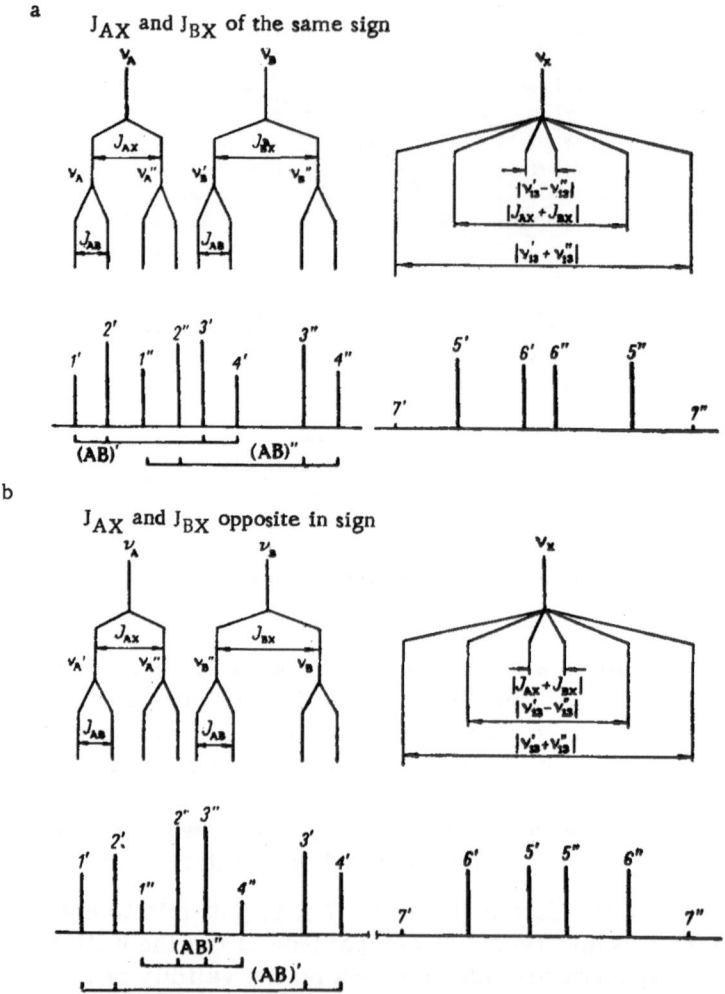

Fig. IV-10. Theoretical spectra of the system ABX with identical (a) and opposite (b) signs of the constants J_{AX} and J_{BX}.

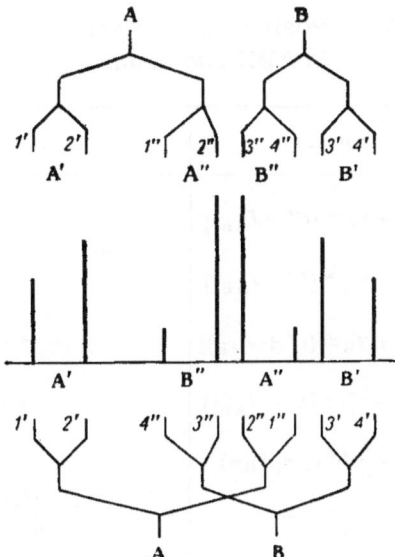

Fig. IV-11. Two variants of the analysis
of the AB part of an ABX spectrum.

The analysis of an actual A_2B spectrum will be given later on using as an example a more complex A_2BX system. It should be noted that in contrast to the AB-type spectra examined above, an A_2B spectrum makes it possible to assign chemical shifts unequivocally to definite nuclei in the molecule. This spectrum is also independent of the sign of the constant J_{AB} and therefore it cannot be used to determine it.

ABX System

Spectra of the ABX type are quite common, particularly among unsaturated and aromatic compounds. In addition to a pure ABX system, which is found in cases where the nucleus X is a different magnetic isotope from the nuclei A and B, the ABX analysis is often used for an approximate calculation of the spectra of an unsymmetrical three-spin system ABC if the chemical shift of one of the nuclei differs appreciably from the other two.

In examining ABX spectra it is convenient to use the term effective shift, the meaning of which is clear from the following discussion. The AB part of the ABX spectrum may be repre-

TABLE IV–3. Position and Intensities of Lines in ABX Spectrum

Line No.	Origin	Frequency	Relative intensity
$1'$, $4'$	AB	$\frac{1}{2}[\nu'_A + \nu'_B \pm (C' + J_{AB})]$	$1 - \dfrac{J_{AB}}{C'}$
$2'$, $3'$	AB	$\frac{1}{2}[\nu'_A + \nu'_B \pm (C' - J_{AB})]$	$1 + \dfrac{J_{AB}}{C'}$
$1''$, $4''$	AB	$\frac{1}{2}[\nu''_A + \nu''_B \pm (C'' + J_{AB})]$	$1 - \dfrac{J_{AB}}{C''}$
$2''$, $3''$	AB	$\frac{1}{2}[\nu''_A + \nu''_B \pm (C'' - J_{AB})]$	$1 + \dfrac{J_{AB}}{C''}$
$5'$ $5''$	X	$\nu_X \pm \frac{1}{2}(J_{AX} + J_{BX})$	1
$6'$,$6''$	X	$\nu_X \pm \frac{1}{2}(C' - C'')$	$\frac{1}{2}\left[1 + \dfrac{\nu_{AB}^2 - \frac{1}{4}(J_{AX} - J_{BX})^2 - J_{AB}^2}{(C' + J_{AB})(C'' + J_{AB})}\right]$
$7'$,$7''$	Combination	$\nu_X \pm \frac{1}{2}(C' + C'')$	$\frac{1}{2}\left[1 - \dfrac{\nu_{AB}^2 - \frac{1}{4}(J_{AX} - J_{BX})^2 - J_{AB}^2}{(C' + J_{AB})(C'' + J_{AB})}\right]$

In this table $\quad C' = \sqrt{\nu'^2_{AB} + J^2_{AB}}$

$$C'' = \sqrt{\nu''^2_{AB} + J^2_{AB}}$$

sented as the result of the superposition of two AB spectra, each of which corresponds to the spin state α or β of the nucleus X. Then depending on whether the spin–spin coupling constants J_{AX} and J_{BX} have the same or opposite signs, two situations may arise (Fig. IV–10).

We will call the nominal values of the resonance frequencies corresponding to the two quartets (AB)′ and (AB)″ in the AB part of the ABX spectrum the e f f e c t i v e s h i f t s. There are four such shifts, namely, ν'_A, ν''_A, ν'_B, and ν''_B. Their relation to the actual shifts of the nuclei A and B and the spin–spin coupling constants J_{AX} and J_{BX} are obvious:

$$\nu'_A = \nu_A + \frac{1}{2}J_{AX}; \qquad \nu''_A = \nu_A - \frac{1}{2}J_{AX}$$

$$\nu'_B = \nu_B + \frac{1}{2}J_{BX}; \qquad \nu''_B = \nu_B - \frac{1}{2}J_{BX} \qquad\qquad (IV\text{–}3)$$

Depending on whether the signs of the constants J_{AX} and J_{BX} are identical or opposite, one of the pairs, for example ν'_B and ν''_B, change places and this leads to a change in the form of the whole spectrum.

It is useful to begin the analysis of the ABX spectrum with the separation of the two quartets in the AB part and the determination of the constant J_{AB} and the effective shifts. It is easy to separate the quartets if we take into account the relative intensity of the lines in them and their symmetry, which obey the rules for AB spectra. However, the AB spectrum does not allow us to assign unequivocally lines connected with the resonance of A and with the resonance of B; therefore, the subsequent analysis of the AB part of the ABX spectrum leads to two equivalent results (Fig. IV-11). It is impossible to make an unequivocal choice between them on the basis of this part of the spectrum alone and therefore it is then necessary to examine the X part of the spectrum.

In the resonance region of the nucleus X there may arise up to 6 lines, arranged symmetrically and two of these, namely, 7' and 7" (see Fig. IV-10) are combination lines and are usually of very low intensity. The two most intense lines, 5' and 5", are separated by a distance of $|J_{AX} + J_{BX}|$. If the spectrum makes it possible to distinguish these two lines from the other two lines 6' and 6", which are somewhat less intense, then by comparing the X and AB parts of the spectrum it is possible to make an unequivocal assignment of the lines in the latter. This is done more readily, the stronger the coupling of the nuclei A and B, i.e., the greater the ratio J_{AB}/ν_{AB}. The distance between the lines 6' and 6" equals the difference in the distances between the lines 1 and 3 in each of the two quartets of the AB part of the spectrum $|\nu_{6', 6''}| = |\nu'_{1.3} - \nu''_{1.3}|$, and it is also useful to take this into account in assigning the bands. In individual cases the bands 7' and 7'', the distance between which equals the sum of the distances 1-3 in the two quartets of the AB part, may also be of help in the assignment.

In the limit when $J_{AB} \ll \nu_{AB}$, the lines 5', 5", 6', and 6" become identical in intensity, while the lines 7' and 7" disappear. In this spectrum, which is a first-order spectrum of the type AMX, it is impossible to make a choice between the two possible solutions. These spectra are usually analyzed by comparing each of the variants with the results obtained for compounds of similar structure. In other cases when the AMX or ABX spectrum is part of a spectrum of a more complex spin system, the assignment is simplified if there is spin—spin coupling of the nucleus A or B with some fourth spin of the molecule.

A theoretical ABX spectrum may be constructed from data in Table IV-3, which was compiled taking into account the normalization conditions so that the total intensity of the resonance signals of each of the nuclei (the combination transition belong to the nucleus X) equals four nominal units. In the limit, when the ABX spectrum changes into an AMX spectrum, the latter contains 12 lines, the intensity of each of which equals 1.

Let us examine two examples of ABX spectra.

The analysis of the spectrum of 4–diethylaminobuten-1-yne-3 (Fig. A-28) is simplified by the fact that the spin—spin coupling of the nuclei B and X is so slight that it does not appear in the spectrum. This makes it possible to assign the lines unequivocally and to carry out a complete analysis, despite the fact that some of the lines corresponding to resonance of the nucleus A are overlapped by an intense signal of the ethyl groups:

$$(CH_3CH_2)_2N \diagdown \quad \diagup C\equiv C-H_X \qquad\qquad (CH_3)_2HC \diagdown \quad \diagup H_B$$
$$\qquad\qquad\qquad C=C \qquad\qquad\qquad\qquad\qquad\qquad C=C$$
$$\qquad H_B \diagup \quad \diagdown H_A \qquad\qquad\qquad\qquad H_A \diagup \quad \diagdown P_X(O)Cl_2$$

The analysis of the spectrum of the diacid chloride of 3–methylbutene-1-phosphinic acid is more complex. No spectrum at the resonance frequency of phosphorus is given and therefore the analysis may be carried out only by examining the AB part of the spectrum. In this case an unequivocal assignment is possible because one of the spins of the ABX system, namely the nucleus A, shows spin—spin coupling with a proton of the isopropyl group with the result that all the bands connected with resonance of this nucleus undergo additional doublet splitting. A comparison of the intensities in the two quartets makes it possible to assign the effective shifts and then it becomes obvious that the constants J_{AX} and J_{BX} are the same in sign.

In conclusion we should note a characteristic of ABX spectra which plays an important part in the analysis of more complex spectra. In this spectrum the 12 lines (apart from the combination lines) are arranged such that there are three distances, each of which is repeated 4 times in the spectrum:

$$1' - 2' = 3' - 4' = 1'' - 2'' = 3'' - 4'' = J_{AB};$$
$$1' - 1'' = 2' - 2'' = 5' - 6' = 5'' - 6'';$$
$$3' - 3'' = 4' - 4'' = 5' - 6' = 5'' - 6' \qquad \text{(IV-4)}$$

In the limit AMX, each of these distances equals the corresponding spin−spin coupling constants, J_{AM}, J_{AX}, and J_{MX}. In the general case, when the system examined consists of three spins with strong coupling, these distances differ to a greater or lesser degree from the corresponding coupling constants, but the equality of the distances always holds. In more complex systems there are also analogous equal distances, which in the limit when the spectrum changes to a spectrum with weak coupling become equal to the corresponding spin−spin coupling constants (or the sum of the constants). This rule is called the rule of equal distances. Together with the rule of the sums of intensities, which was examined in Ch. I, this rule is one of the fundamental propositions of the theory of complex spectra. The origin of the equal distances becomes clear when we examine the energy level diagram (see the following section). We should also note the particular case of an ABX spectrum, when the chemical shifts of the nuclei A and B coincide, while the constants J_{AX} and J_{BX} differ. In this case the spectrum, which now belongs to the AA'X type, consists of a doublet and a triplet with distances between the lines equal to $\frac{1}{2}(J_{AX} + J_{A'X})$. Spectra of this type were examined in Ch. I as spectra with masked spin−spin coupling. It is proposed to calculate independently the position and intensities of the lines in ABX spectra with different ratios between J_{AB} and ν_{AB} and also to carry out the calculation for AA'X spectra.

Strongly Coupled ABC System

Diagram of Energy Levels. The fifteen lines in the ABC spectrum are connected with transitions between eight energy states of the spin system, which are called steady states. The steady states may be obtained directly from the spectrum. On the other hand, they may be calculated with given values of ν_i and J_{ij}, i.e., the chemical shifts of the given nuclei and the constant for coupling between them. If the results obtained by these two methods agree

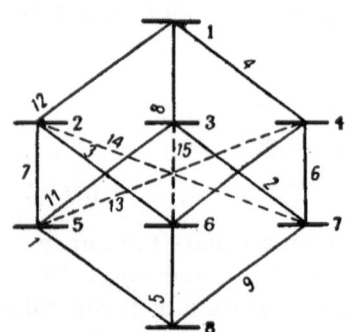

Fig. IV-12. Energy level diagram of the system ABC.

with each other, then it may be concluded that the given set of shifts and the coupling constant correspond to the spectrum. This procedure is the basis for the analysis of systems of strongly coupled spins.

The energy level diagram for a system of three spins is given in Fig. IV-12.

The heavy figures denote the 8 steady states of the system, while the italics denote the 15 possible transitions which produce the 15 lines in the spectrum. The vertical lines denote transitions associated with resonance of the nucleus A; the lines sloping to the left are connected with transitions of the nucleus B and those sloping to the right, with transitions of the nucleus C. The three broken lines correspond to the three combination transitions. The diagram may be represented in the form of a cube, whose points correspond to the steady state and the edges to transitions between them. The steady states, like the transitions, are usually expressed in hertz.

The energy state diagram must satisfty the following conditions.

1. The difference in the energies of two steady states equals the frequency of the line in the spectrum connected with the transition between these states.

2. The sum of the energies of the levels of the diagram equals zero.

The latter condition indicates that the energies of the steady states may be positive or negative.

The steady state 1 corresponds to the highest energy level of the system and the state 8 to the lowest level.

For constructing the energy level diagram of the acrylonitrile system (at 40 MHz; see Fig. IV-1a), the following frequencies of the lines in the system provide the starting data:

A	B	C	Combination transitions
1) 384.7	5) 372.5	9) 353.6	13) —
2) 369.0	6) 361.2	10) 342.3	14) 349.0
3) 382.1	7) 369.0	11) 336.9	15) 330.3
4) 365.3	8) 358.8	12) 325.8	

We assume that lines 2 and 7 in the spectrum overlap and their frequencies cannot be determined accurately.

Initially we assume that $E'_8 = 0$. In this case the levels 5, 6, and 7 are numerically equal to the frequencies of lines 5, 1, and 9, respectively. By summing E'_5 and E'_6 found in this way with ν_3 and ν_{11}, respectively, we find E'_2 and E'_4. As the diagram shows, to find E'_3 it is possible to use two methods and they give practically the same result. E'_1 may be found by three methods. In our case the discrepancy between the results does not exceed 0.3 Hz.

$$E'_1 \ 1084.6; \quad E'_2 \ 754.1; \quad E'_3 \ 714.8; \quad E'_4 \ 725.2; \quad E'_5 \ 372.5;$$
$$E'_6 \ 384.7; \quad E'_7 \ 353.3; \quad E'_8 \ 0$$
$$E_1 \ 535.9; \quad E_2 \ 205.4; \quad E_3 \ 166.1; \quad E_4 \ 176.5; \quad E_5 \ 176.2;$$
$$E_6 \ -164.0; \quad E_7 \ -195.1; \quad E_8 \ -548.7$$

The good agreement indicates, in particular, that the assignment of the lines was carried out correctly.

By using these data it is possible to determine the frequencies of the poorly resolved transitions 2 and 7 more accurately. We find that $\nu_2 = E'_4 - E'_7 = 368.0$ Hz and $\nu_7 = E'_2 - E'_7 = 369.9$ Hz.

So that the diagram of the states obeys condition 2 we find the value T.

The value T is defined as the distance from the reference or the origin adopted for measuring the frequencies to the center of gravity of the spectrum. The same value determines the center of gravity of the energy level diagram.

The energy levels satisfying condition 2 are determined from the formula:

$$T = \frac{1}{8} \sum_{i=1}^{8} E_i' = \frac{1}{12} \sum_{i=1}^{12} v_i \tag{IV-5}$$

($\sum_{i=1}^{8}$ denotes that we sum 8 values from E_1' to E_8').

$$E_i = E_i' - T \tag{IV-6}$$

In our case T = 548.7 Hz.

The energy level diagram is completely determined. The following relations are readily checked:

$$E_1 + E_2 + E_3 + E_4 + E_5 + E_6 + E_7 + E_8 = 0$$
$$E_1 + E_5 + E_6 + E_7 = 0$$
$$E_2 + E_3 + E_4 + E_8 = 0 \tag{IV-7}$$

Let us turn to examining the foundations of the theory relating the energies of the steady state of the spin system to the parameters of the spectrum, namely, the chemical shifts and coupling constants.

Quantum Mechanical Analysis of the Spectrum of a System of Nuclear Spins [4, 6, 7]. A system of nuclear spins is described by the Schrödinger equation, which appears as follows in the operator form:

$$H\psi = E\psi \tag{IV-8}$$

where E is the energy of the steady state, ψ is the wave function of the system corresponding to the given steady state and is called the e i g e n f u n c t i o n, and H is the Hamilton energy operator or h a m i l t o n i a n acting on the wave function.

To a system of nuclei with a spin $\frac{1}{2}$ there correspond 2^p steady states where p is the number of nuclei composing the system. One nucleus may be in two steady states in accordance with the possible values of the spin component relative to the direction of the external field (usually the z component) $+\frac{1}{2}$ and $-\frac{1}{2}$ (Ch. I). To a system of three spins there correspond 8 steady states. In the previous section we showed how to obtain experimentally the values of the energies corresponding to the different states of the system.

To each of the 8 steady states of a system of three spins there corresponds an eigenfunction ψ_i. Finding the complete set of eigenfunctions of the system, which are necessary to calculate the probabilities of transitions and the intensities of the lines in the spectrum associated with them, is one of the last stages in the analysis. The starting set of wave functions of the system, which are called the b a s e f u n c t i o n s , are combined from the spin wave functions of the individual nuclei. To each nucleus there correspond two such functions, i.e., the same number as for the possible steady states. We denote by α the wave function corresponding to $I_z + \frac{1}{2}$, and by β the wave function corresponding to $I_z = -\frac{1}{2}$. By multiplying the functions α and β we obtain the total set of 8 base functions of the system of the type $\alpha\alpha\alpha$, $\alpha\alpha\beta$, etc., in which the position of the factor indicates the nucleus to which it belongs, i.e., the first to the nucleus A, the second to B, and the third to C. To each base function there corresponds a total spin F, whose z-component, which interests us, is found from the relation

$$F_z = \sum_i I_z(i) \tag{IV-9}$$

$F_z = +\frac{1}{2} + \frac{1}{2} + \frac{1}{2} = +\frac{3}{2}$, corresponds to the function $\alpha\alpha\alpha$, $F_z = +\frac{1}{2} + \frac{1}{2} - \frac{1}{2} = +\frac{1}{2}$ to the function $\alpha\alpha\beta$, etc. F_z is c o n n e c t e d w i t h t h e s e l e c t i o n r u l e for transitions between different states of the system: only transitions in which F_z changes by ± 1 are possible. This selection rule determines the 15 possible transitions between the 8 steady states of the ABC system which are shown on the energy level diagram (see Fig. IV-12). Some other selection rules, which reduce the number of possible transitions (for example, to 5 in the system A_2X), will be examined later.

Let us now examine the form of the hamiltonian of the spin system. The general hamiltonian may be represented as the sum of two components

$$H = H^0 + H^1 \tag{IV-10}$$

The term H^0 characterizes the energy of the combination of spins placed in a magnetic field and includes terms connected with the magnetic screening of the nuclei

$$H^0 = \sum_i \nu_i I_z(i)$$
(IV-11)

where ν_i, strictly speaking, denotes the absolute resonance frequency of each of the nuclei examined. However, since only the differences in the frequencies of individual nuclei are important in the analysis spectra, ν_i may be measured relative to any selected point in the spectrum. Apart from specified cases, by ν_i we mean the shift (in Hz) relative to the reference; in this case I_z denotes the z component of the spin in the operator form.

The operator H^1 characterizes the energy of spin−spin coupling between the nuclei of the system

$$H^1 = \sum_{i<j} J_{ij} I(i) I(j)$$
(IV-12)

i.e., it represents the sum of all the possible pairs of products of the spin operators. J_{ij} is a constant characterizing the energy of spin−spin coupling of two nuclei (the i-th and j-th). In the case of a system of three spins, i and j may have 3 values each ($i \neq j$) so that there are three coupling constants: J_{AB}, J_{AC}, and J_{BC} (J_{BA}, J_{CA}, and J_{CB} denote the same).

It is convenient to represent the hamiltonians in the form of square matrices [9], i.e., tables consisting of equal numbers of columns and rows. We will denote the matric elements in the general form by the symbols H_{mn}, where the first subscript denotes the number of the row and the second, the number of the column. Thus, H^1_{23} denotes the matrix element lying in the second row of the third column of the matrix of the operator H^1. The matrix elements H_{mm} lie on the main diagonal of the matrix and are called diagonal elements. The summation of two matrices is achieved by summation of the corresponding elements of these matrices. The order of a square matrix is determined by the number of rows or columns. The sum of the diagonal element is called the trace of the matrix.

The matrix of the hamiltonian of a system of three spins has an order of 8, i.e., the number of base functions and steady states. The hamiltonian of the combination of nuclei H^0 contains only

TABLE IV-4. Diagonal Matrix Elements of the Hamiltonian
of the System ABC

No.	F_z	Base functions	$H_{mm}=H^0_{mm}+H^1_{mm}$
1	$+\dfrac{3}{2}$	$\alpha\alpha\alpha$	$\dfrac{1}{2}(\nu_A+\nu_B+\nu_C)+\dfrac{1}{4}(J_{AB}+J_{BC}+J_{AC})$
2		$\alpha\alpha\beta$	$\dfrac{1}{2}(\nu_A+\nu_B-\nu_C)+\dfrac{1}{4}(J_{AB}-J_{BC}-J_{AC})$
3	$+\dfrac{1}{2}$	$\alpha\beta\alpha$	$\dfrac{1}{2}(\nu_A-\nu_B+\nu_C)+\dfrac{1}{4}(-J_{AB}-J_{BC}+J_{AC})$
4		$\beta\alpha\alpha$	$\dfrac{1}{2}(-\nu_A+\nu_B+\nu_C)+\dfrac{1}{4}(-J_{AB}+J_{BC}\ J_{AC})$
5		$\alpha\beta\beta$	$\dfrac{1}{2}(\nu_A-\nu_B-\nu_C)+\dfrac{1}{4}(-J_{AB}+J_{BC}-J_{AC})$
6	$-\dfrac{1}{2}$	$\beta\alpha\beta$	$\dfrac{1}{2}(-\nu_A+\nu_B-\nu_C)+\dfrac{1}{4}(-J_{AB}-J_{BC}+J_{AC})$
7		$\beta\beta\alpha$	$\dfrac{1}{2}(-\nu_A-\nu_B+\nu_C)+\dfrac{1}{4}(J_{AB}-J_{BC}-J_{AC})$
8	$-\dfrac{3}{2}$	$\beta\beta\beta$	$\dfrac{1}{2}(-\nu_A-\nu_B-\nu_C)+\dfrac{1}{4}(J_{AB}+J_{BC}+J_{AC})$

diagonal elements; all the nondiagonal elements equal zero. The diagonal elements H^0_{mn} are found from formula (IV-11) in which I_z is replaced by $+\frac{1}{2}$ or $-\frac{1}{2}$, depending on whether we are taking the spin function α or β. Thus, for the state $\alpha\alpha\alpha$ the matrix element $H^0_{11}=+\frac{1}{2}\,\nu_A+\frac{1}{2}\,\nu_B+\frac{1}{2}\,\nu_C=+\frac{1}{2}\,(\nu_A+\nu_B+\nu_C)$, while the matrix element H^0_{22}, corresponding to the state $\alpha\alpha\beta$, equals $\frac{1}{2}\,(\nu_A+\nu_B-\nu_C)$.

In the general case the hamiltonian of spin–spin interaction H^1 contains both diagonal and nondiagonal elements which are not equal to zero. The diagonal element may be found from relation (IV-12). Thus, for the state $\alpha\alpha\beta$

$$H^1_{22}=\left(+\tfrac{1}{2}\right)\left(+\tfrac{1}{2}\right)J_{AB}+\left(+\tfrac{1}{2}\right)\left(-\tfrac{1}{2}\right)J_{AC}$$

$$+\left(+\tfrac{1}{2}\right)\left(-\tfrac{1}{2}\right)J_{BC}=\tfrac{1}{4}(J_{AB}-J_{AC}-J_{BC})$$

The nondiagonal matrix elements correspond to the two base functions with a given F_z. Thus, H^1_{23} corresponds to the functions $\alpha\alpha\beta$ and $\alpha\beta\alpha$ with a total spin of $+\frac{1}{2}$. From the selection rules with F_z presented above it follows that H_{mn}^1 does not revert to zero if φ_m and φ_n differ only in the transposition of the spins of the two nuclei. In these cases H_{mn} may be found from formula (IV-13), which follows from (IV-12),

$$H_{mn} = \frac{1}{2} J_{ij} \qquad \text{(IV-13)}$$

with the condition that the base functions corresponding to the states m and n (φ_m and φ_n) differ only in the transposition of the spins of the nuclei i and j. Thus, $H_{23}^1 = \frac{1}{2} J_{BC}$, since the base functions $\varphi_2 = \alpha\alpha\beta$ and $\psi_3 = \alpha\beta\alpha$ differ only in the transpostion of the spins of B and C.

Thus, the full matrix of the hamiltonian of a system of three nuclei has the following form:

	1	2	3	4	5	6	7	8 . . . . No.
	$+\frac{3}{2}$		$+\frac{1}{2}$			$-\frac{1}{2}$		$-\frac{3}{2}$ . . . . F_z
	$\alpha\alpha\alpha$	$\alpha\alpha\beta$	$\alpha\beta\alpha$	$\beta\alpha\alpha$	$\alpha\beta\beta$	$\beta\alpha\beta$	$\beta\beta\alpha$	$\beta\beta\beta$ . . . φ

$$H = \begin{bmatrix} H_{11} & 0 & 0 & 0 & 0 & 0 & 0 & 0 \\ 0 & H_{22} & \frac{1}{2}J_{BC} & \frac{1}{2}J_{AC} & 0 & 0 & 0 & 0 \\ 0 & \frac{1}{2}J_{BC} & H_{33} & \frac{1}{2}J_{AB} & 0 & 0 & 0 & 0 \\ 0 & \frac{1}{2}J_{AC} & \frac{1}{2}J_{AB} & H_{44} & 0 & 0 & 0 & 0 \\ 0 & 0 & 0 & 0 & H_{55} & \frac{1}{2}J_{AB} & \frac{1}{2}J_{AC} & 0 \\ 0 & 0 & 0 & 0 & \frac{1}{2}J_{AB} & H_{66} & \frac{1}{2}J_{BC} & 0 \\ 0 & 0 & 0 & 0 & \frac{1}{2}J_{AC} & \frac{1}{2}J_{BC} & H_{77} & 0 \\ 0 & 0 & 0 & 0 & 0 & 0 & 0 & H_{88} \end{bmatrix}$$

with right-side labels:
$\alpha\alpha\alpha \quad +\frac{3}{2} \quad 1$
$\alpha\alpha\beta \quad 2$
$\left.\begin{matrix} \\ \\ \end{matrix}\right\} +\frac{1}{2}$
$\alpha\beta\alpha \quad 3$
$\beta\alpha\alpha \quad 4$
$\alpha\beta\beta \quad 5$
$\beta\alpha\beta \left.\begin{matrix}\\ \end{matrix}\right\} -\frac{1}{2} \quad 6$
$\beta\beta\alpha \quad 7$
$\beta\beta\beta \quad -\frac{3}{2} \quad 8$

The diagonal matrix elements $H_{mm} = H_{mm}^0 + H_{mm}^1$ are given in Table IV-4.

In the particular case of the system AMX, i.e., a system in which the coupling constants are considerably less than the chemical shifts, the nondiagonal elements (containing only the coupling constants) may be neglected and then the matrix of the hamiltonian becomes diagonal. In this case the diagonal elements represent the steady states of the system, transitions between which determine the frequencies of the spectral lines. In the general case of a system of three spins only H_{11} and H_{88} are steady states

$$H_{11} = E_1$$

$$H_{88} = E_8 \qquad \text{(IV-14)}$$

To find the other steady states it is necessary to transform the matrix of the full hamiltonian into a diagonal matrix, i.e., to find another matrix similar to the given one, but in which all the nondiagonal elements equal zero. The diagonal elements of this matrix are the roots of a characteristic or secular equation of the form

$$\begin{vmatrix} H_{11}-E & H_{12} \ldots H_{1n} \\ H_{21} & H_{22}-E \ldots H_{2n} \\ \cdot \cdot \cdot \cdot & \cdot \cdot \cdot \cdot \cdot \cdot \\ H_{n1} & H_{n2} \ldots H_{nn}-E \end{vmatrix} = 0 \qquad \text{(IV-15)}$$

with the element H_{mn} of this determinant the same as the matrix elements of the hamiltonian. The roots of this equation are the steady states of the system sought.

However, in our case there is no need to calculate the roots of a characteristic equation of the eighth degree since the matrix of the hamiltonian may be broken down into separate submatrices, two of which are separated out by square brackets. Two other submatrices, namely H_{11} and H_{88}, are first-order matrices, which have one element. For them equation (IV-15) is reduced to the form

$$H_{11} - E = 0$$

$$H_{88} - E = 0$$

whence $E_1 = H_{11}$ and $E_8 = H_{88}$. Finding the other six steady states is reduced to the solution of two characteristic equations of the third degree, of which the first

$$
\begin{vmatrix}
H_{22}-E & \frac{1}{2}J_{BC} & \frac{1}{2}J_{AC} \\
\frac{1}{2}J_{BC} & H_{33}-E & \frac{1}{2}J_{AB} \\
\frac{1}{2}J_{AC} & \frac{1}{2}J_{AB} & H_{44}-E
\end{vmatrix} = 0
\qquad\qquad (IV\text{-}16)
$$

gives the steady states E_2, E_3, and E_4, while the other

$$
\begin{vmatrix}
H_{55}-E & \frac{1}{2}J_{AB} & \frac{1}{2}J_{AC} \\
\frac{1}{2}J_{AB} & H_{66}-E & \frac{1}{2}J_{BC} \\
\frac{1}{2}J_{AC} & \frac{1}{2}J_{BC} & H_{77}-E
\end{vmatrix} = 0
\qquad\qquad (IV\text{-}17)
$$

makes it possible to find the states E_5, E_6, and E_7.

Thus, the steady states are connected with the spectral parameters by a strict mathematical dependence.

However, the problem is the reverse in the analysis of NMR spectra: it is necessary to determine the parameters of the spectrum (chemical shifts and coupling constants) from the steady states of the system of nuclear spins found experimentally. There are two routes to the solution of this problem. The first is numerical diagonalization of the submatrices and this is usually carried out by means of electronic computers. If the calculation of the spectrum is carried out without the use of electronic computers, it is more convenient to use the second route, namely, to use the results of algebraic analysis of the hamiltonian of a system of three spins. Below we present a graphical method for accurate analysis of an ABC spectrum, based on the results of [10, 11].

Graphical Method for Analysis of ABC Spectrum. In the accurate analysis of an ABC spectrum it is more convenient to operate with chemical shifts relative to the center of gravity of the spectrum and these are determined from the relations

$$
\nu_a = \nu_A - T; \quad \nu_b = \nu_B - T; \quad \nu_c = \nu_C - T
\qquad\qquad (IV\text{-}18)
$$

where T is the position of the center of gravity of the spectrum relative to some reference point (for example, relative to tetramethylsilane), determined from formula (IV-5). For them the following relation actually holds:

$$\nu_a + \nu_b + \nu_c = 0 \tag{IV-19}$$

Then we introduce the following values which are constant for the given spectrum:

$$M = E_1 - E_8 = E_2 + E_3 + E_4 - E_5 - E_6 - E_7 = E_1' \tag{IV-20}$$

$$\begin{aligned} &e_2 = E_2 - M/6; \quad e_3 = E_3 - M/6; \quad e_4 = E_4 - M/6 \\ &e_5 - E_5 + M/6; \quad e_6 = E_6 + M/6; \quad e_7 = E_7 + M/6 \end{aligned} \right\} \tag{IV-21}$$

(e_1 and e_8 need not be calculated since they are not used subsequently). The values e_n must satisfy the identity

$$e_2 + e_3 + e_4 = e_5 + e_6 + e_7 \tag{IV-22}$$

The following values are derivatives of E_i and e_i:

$$N = -(E_1 + E_2 + E_3 + E_4 + E_5 + E_6 + E_7) \tag{IV-23}$$

$$S = e_2^2 + e_3^2 + e_4^2 + e_5^2 + e_6^2 + e_7^2 \tag{IV-24}$$

$$g = \frac{1}{2}(-e_2^2 - e_3^2 - e_4^2 + e_5^2 + e_6^2 + e_7^2) \tag{IV-25}$$

$$l = e_2 e_3 e_4 \tag{IV-26}$$

$$m = e_5 e_6 e_7 \tag{IV-27}$$

$$t = \sqrt{27 \cdot 8}\left(\frac{N}{16} g - \frac{1}{4} l + \frac{1}{4} m\right) \tag{IV-28}$$

$$h = -\frac{5}{64} N^3 + \frac{1}{8} NS + l + m \tag{IV-29}$$

We also introduce two values which are variables in the given analysis method:

$$r = J_{AB}^2 + J_{BC}^2 + J_{AC}^2 \tag{IV-30}$$

$$u = \nu_a^2 + \nu_b^2 + \nu_c^2 \tag{IV-31}$$

The essence of the method lies in the fact that r depends on u in two ways:

$$r = f_1(u) \text{ and } r = f_2(u) \tag{IV-32}$$

If we draw curves of both relations, then the point of inter-section of the lines is the solution of the problem.

The relation $r = f_1(u)$ is linear

$$r = \frac{1}{12} N^2 + \frac{2}{3} S - \frac{4}{3} u \qquad \text{(IV-33)}$$

The relation $r = f_2(u)$ is more complex and is represented by the following system of equations:

$$\cos \Phi = \frac{t}{\sqrt{u^3}} \qquad \text{(IV-34)}$$

$$\nu_a, \nu_b, \nu_c = \sqrt{\frac{2u}{3}} \cos \left(\frac{\Phi}{3} + n \cdot 120° \right) \qquad \text{(IV-35)}$$

where $n = 0$, 1, and 2.

$$\left.\begin{aligned}
J_{BC} &= \frac{h + g\nu_1 + 2N\nu_2\nu_3}{(\nu_a - \nu_b)(\nu_a - \nu_c)} \\
J_{AB} &= \frac{h + g\nu_2 + 2N\nu_3\nu_1}{(\nu_b - \nu_c)(\nu_b - \nu_a)} \\
J_{AC} &= \frac{h + g\nu_3 + 2N\nu_1\nu_2}{(\nu_c - \nu_a)(\nu_c - \nu_b)}
\end{aligned}\right\} \qquad \text{(IV-36)}$$

The procedure for analysis of the spectrum is as follows:

1. Calculate the constants from formulas (IV-20)-(IV-29).

2. Draw the linear relation $r = f_1(u)$ (IV-33).

3. With various arbitrary values of u, using formulas (IV-34)-(IV-36) draw up a table of the corresponding values of J_{AB}, J_{AC}, and J_{BC} and by means of formula (IV-30), find the corresponding values of r.

4. Use the data obtain in the table to draw the relation $r = f_2(u)$ (IV-32).

5. From the abscissa of the point of intersection of the two lines find the value of u, which is the solution of the system (IV-32). Using this value in formulas (IV-34)-(IV-36) find the true values of ν_a, ν_b, ν_c, J_{AB}, J_{BC}, and J_{AC}. A check on the accuracy of the calculation may be carried out analytically; for this purpose it should be shown that the values obtained satisfy equation (IV-33).

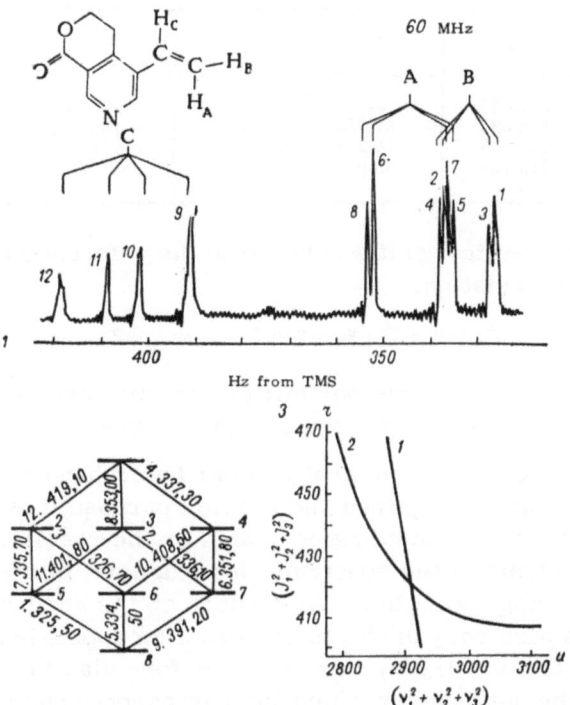

Fig. IV-13. Analysis of the spectrum of the vinyl protons of gentianin. 1) Experimental spectrum of vinyl protons of gentianin (60 MHz); 2) energy level diagram for this spin system; 3) graphical analysis of spectrum.

TABLE IV-5. Signals of Vinyl Protons of Gentianin (Fig. IV-13)

Line No.	Position, Hz	Line No.	Position, Hz
1	325.50	6	351.80
3	326.70	8	353.00
5	334.50	9	391.20
7	335.70	11	401.80
2	336.10	10	408.50
4	337.30	12	419.10

TABLE IV-6. Corrected Energy Levels

n	E_n	e_n	n	E_n	e_n
1	547.425	—	5	−207.375	−27.325
2	128.325	−41.725	6	−198.375	−18.325
3	194.425	14.375	7	−141.675	38.375
4	210.125	30.075	8	−532.875	—

6. The chemical shifts relative to the reference are determined from the relations

$$\nu_A = \nu_a + T; \quad \nu_B = \nu_b + T; \quad \nu_C = \nu_c + T \qquad \text{(IV-37)}$$

and for conversion to parts per million, the frequencies obtained are divided by the working frequency of the spectrometer (in MHz).

In practice it is convenient to limit beforehand the region of the values of u and r required and for this purpose it is possible to use tabular data on coupling constants or the results of an approximate analysis of the spectrum, for example, in the ABX or even the AMX approximation. As a check on the accuracy of the analysis it is also very useful to construct an approximate theoretical spectrum of the system from the formulas in Table IV-3, taking as X the nucleus for which the resonance frequency differs markedly from the resonance frequencies of the other two nuclei so that the following condition holds as far as possible:

$$J_{AX}, J_{BX} \ll |\nu_X - \nu_A|, |\nu_X - \nu_B| \qquad \text{(IV-38)}$$

An example of the analysis of a spectrum by this method is taken from [11]. Figure IV-13 gives the experimental spectrum of the vinyl protons of gentianin (solution in $CDCl_3$). The assignment of the lines was carried out by means of the rule of equal distances and in this case it may be represented by the following equation:

$$1-3=5-7=2-4=6-8$$
$$1-2=3-4=9-11=10-12$$
$$5-6=7-8=9-10=11-12$$

Table IV-5 gives the positions of the individual lines, obtained from the experimental spectrum.

Lines 5 and 2 were not determined accurately from the spectrum, but were found after construction of the energy level diagram. For construction of the latter it was first assumed that $E'_8 = 0$. Then the other energy levels were found and this made it possible to determine the positions of lines 2 and 5. The mean position of the eight energy levels is 532.875 (when $E'_8 = 0$). By subtraction of this value from each value of E' the corrected values were found and these are given in Table IV-6. Then from equations (IV-20) and (IV-21) M and the values from e_2 through e_7 were found (M = 1080.30).

Figure IV-13 gives the relations $r = f_1(u)$ (1) and $r = f_2(u)$ (2), the first of which was constructed in accordance with equation (IV-33) and the second in accordance with (IV-30) using values calculated from equations (IV-22)-(IV-29) and (IV-34)-(IV-36).

The spectral parameters obtained in this way are given below and also the results of calculation in the ABX approximation for comparison.

Method of analysis	ν_A	ν_B	ν_C	J_{AB}	J_{BC}	J_{AC}
ABC	344.95	331.87	403.48	0.89	10.86	17.35
ABX	343.72	331.43	405.15	1.20	10.58	17.32

Unfortunately, the calculation method presented is only suitable for well-resolved spectra with not too strong coupling. Strongly coupled spectra can be analyzed successfully only with electronic computers.

3. SYSTEM OF FOUR MAGNETIC NUCLEI

The spectrum of a symmetrical, strongly coupled system of four spins ABCD in the general case contains 56 lines, produced by different transitions between 16 steady states. The complete analysis of such a system includes the calculation of two characteristic equations of the fourth order, corresponding to $F_z = 1$ and -1 and also one characteristic equation of the sixth order ($F_z = 0$). Moreover, even with the use of electronic computers the complete analysis of such a system is very difficult due to the complexity of the correct assignment of the lines in the spectrum.

We will examine only the particular cases of a four-spin system AA'BB' and ABCX. Spectra of the AA'BB' type are given by

many para-disubstituted and symmetrical ortho-disubstituted derivatives of benzene, γ-substituted six-membered heterocycles,
five-membered heterocycles, symmetrical disubstituted buta-
dienes, and many other compounds. Even with considerable chemical shifts, this case is not trivial due to the presence of crossed
coupling constants. The ABCX spectrum is encountered frequently
among organofluorine and organophosphorus compounds. Of the
other particular cases of a four-spin system, the simplest is the
AB_3 system, but it will not be examined here since only a very limited number of organic compounds give spectra of this type.

AA'BB' Spectra [14]

In the analysis of spectra of the AA'BB' system we will ad-
here to the notation adopted generally:

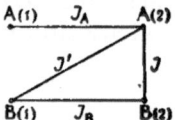

Spectra of the AA'BB' system make it possible to determine
the coupling constants between the chemically equivalent nuclei
A−A' and B−B', which are denoted by J_A and J_B, respectively,
and also the two coupling constants J_{AB}, which are denoted by J
and J'. The origin of the four different constants becomes appar-
ent when we examine, for example, the 1,1-difluoroethylene mole-
cule in which J_{HH} corresponds to J_A and J_{FF} to J_B, while J_{HFcis}
may be denoted by J and $J_{HFtrans}$ by J':

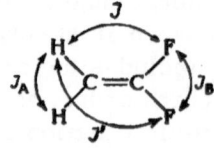

In the analysis of AA'BB' spectra it is convenient to use the
following symbols:

$$K = J_A + J_B; \quad L = J - J'$$
$$M = J_A - J_B; \quad N = J + J \tag{IV-39}$$

The spectrum consists of 24 lines, which are symmetrical relative to the center of the spectrum, and 12 of these are connected with resonance of the nucleus A and 12 with resonance of the nucleus B. Due to the symmetry of the spectrum it is sufficient to examine one part of it, i.e., A or B. Such a considerable reduction in the number of lines in the AA'BB' spectrum in comparison with the general case of a system of four spins is produced by symmetry similar to the way in which the number of lines in an A_2B spectrum is reduced to 9 in comparison with the 15 lines in an ABC spectrum. In an AA'XX' spectrum the number of lines is reduced to 10 in each part of the spectrum (a total of 20 lines). A further reduction in the number of lines is observed when $J = J'$. This spectrum consists of 6 lines, namely, two symmetrical triplets $1 : 2 : 1$, which are connected with the nuclei A and X and may be analyzed by first-order rules. In this case it is impossible to determine J_A and J_X from the spectrum or the relative signs of the coupling constants. In both cases of AA'XX' spectra, each part of the spectrum is symmetrical relative to its center, which corresponds to the resonance frequency ν_A or ν_X. Another interesting case arises when one of the coupling constants, for example, J, is considerably greater than the other three constants. This spectrum consists of a symmetrical quartet which is similar to the quartet of an AB spectrum, whose components are split slightly into five-six peaks due to these three constants. The splitting of the four main lines of this spectrum, which belongs to the $(AB)_2$ or the $(AX)_2$ system, obeys rules similar to those presented for a two-spin system [12, 13]. The different characteristic theoretical spectra of the AA'BB' system are given in Figs. IV-14.

The correct assignment of the transitions is most important for the successful analysis of a spectrum. The same methods are used for this as in a three-spin system.

The complexity of the analysis of an AA'BB' spectrum is connected with the indeterminacy of the parameters due to its symmetry. The spectrum itself offers no possibility of selecting which of the parameters, ν_A, ν_B, J_A, and J_B, actually refer to the nucleus denoted by A and which refer to B. Even if it is possible to assign two of these parameters, for example ν_A and ν_B, from indirect data, the other two remain indeterminate. The same is also true of the constants J and J': it is impossible to determine from the spectrum which of them refers to spin–spin coupling of the nuclei

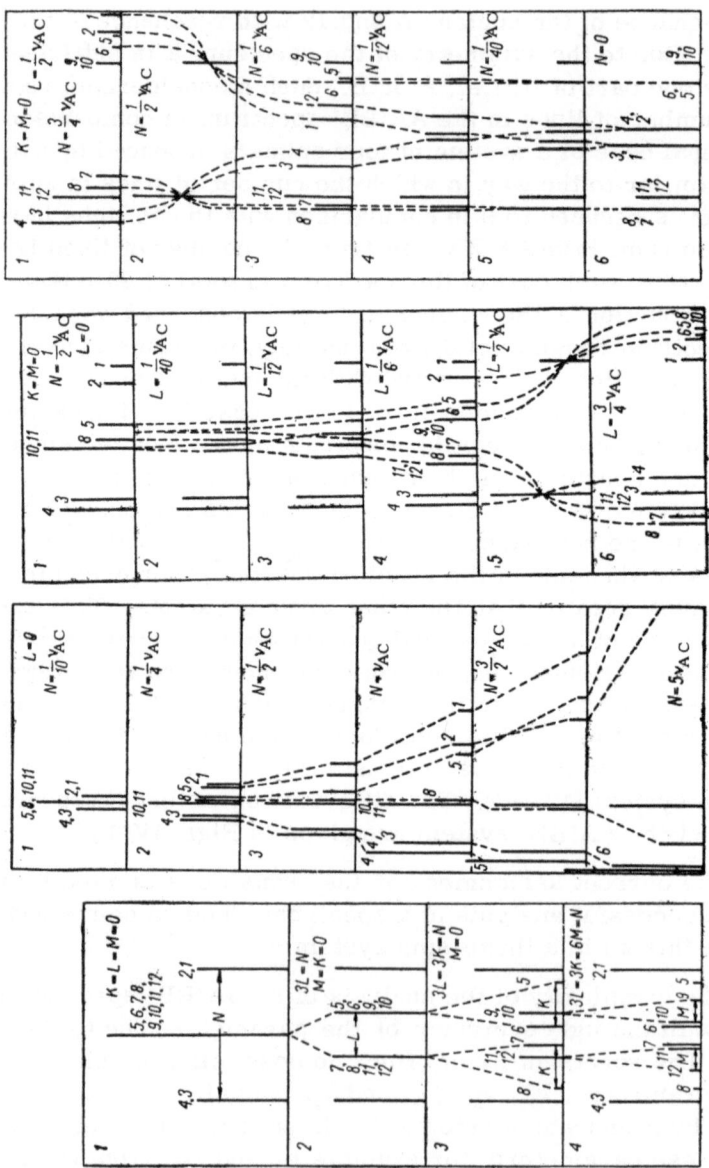

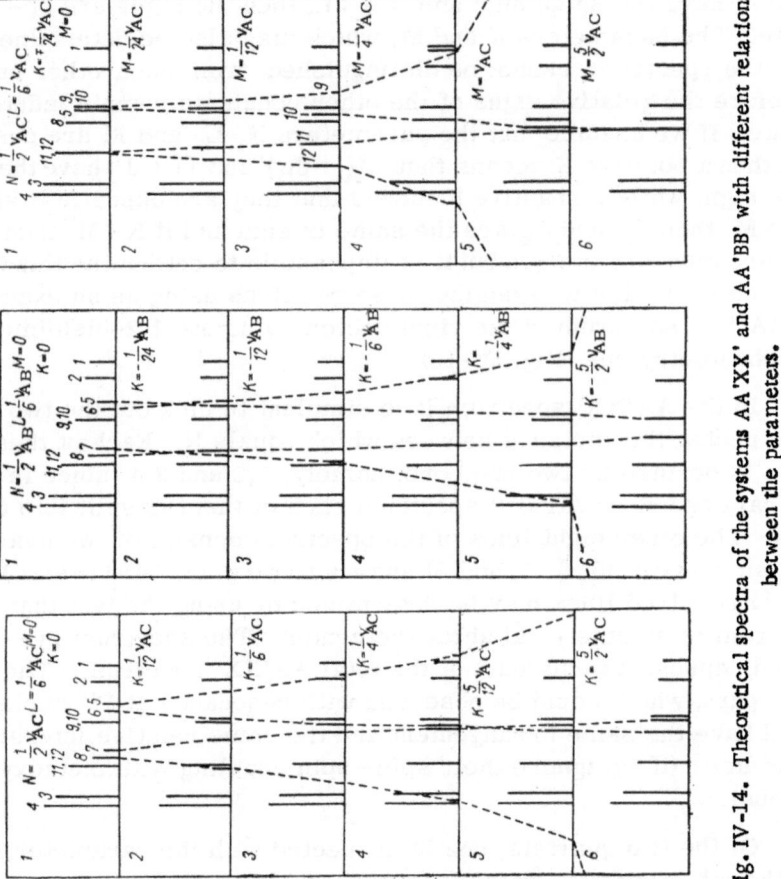

Fig. IV-14. Theoretical spectra of the systems AA'XX' and AA'BB' with different relations between the parameters.

A and B' (or, what is the same, the nuclei B and A') and which to
the pair of nuclei A and B (or, what is the same, A' and B'). Finally,
the spectrum makes it possible only to choose the relative signs
of the constants J and J' due to the fact that it is possible to make
an unequivocal choice between the values N and L. If $N > L$, then
J and J' have the same sign and if $N < L$, then the signs are opposite. The parameters K and M, which may also be determined
from the spectrum, cannot be distinguished from each other and
therefore the relative signs of the other constants remain indeterminate. If we assume that the parameters N, L, and M are positive, then a positive K means that $(J_A + J_B)$ and $(J + J')$ have the
same sign, while a negative K means that they are opposite in sign;
if $K > M$, then J_A and J_B are the same in sign and if $K < M$, then
they are opposite in sign, but it is impossible to determine this from
the spectrum. Let us examine these relations using as an example
the AA'XX' spectrum of the vinyl protons of trans-1,2-bis(dimethyl-
phosphono)ethylene (Fig. IV-15).

In the AA'XX' spectrum it is simplest to pick out the two intense peaks, the distance between which equals N. Each of these
peaks is denoted by two numbers, namely, 1,2 and 3,4, since in the
general case of an AA'BB' spectrum each of them is split into two
lines. The other eight lines of the spectrum consist of two quartets, an outer (5, 6, 7, and 8) and an inner (9, 10, 11, and 12).
Poorly resolved lines may be determined by using the fact that the
spectrum is symmetrical about the center. The spectrum presented represents only part of the total AA'XX' spectrum. The
other part, which could be observed with resonance at P^{19} nuclei,
would have the same arrangement and the same relative intensities
of the lines (if we ignore their spin—spin coupling with methoxyl
protons).

Of the two quartets, one is connected with the parameter K
and the other with M. Since it is impossible to choose between them,
we assume arbitrarily that the outer quartet is connected with the
parameter K and the inner quartet with M. These parameters (K
and M) are determined from the distances between the lines 5-6,
7-8, and 9-10, 11-12, respectively (like the constant J_{AB} in an AB
quartet). In our case $N = 34.6$; $K = 54.6$; $M = 19.9$ Hz.

The parameter L may be determined from both the outer and
the inner quartets. This parameter is determined analogously to

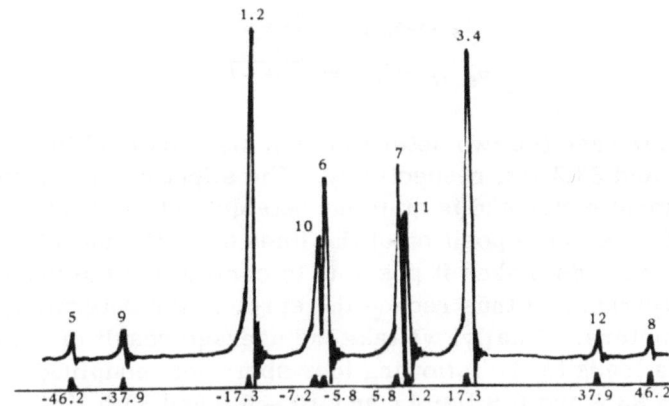

Fig. IV-15. Spectrum of the vinyl protons of trans-1,2-bis(dimethyl-phosphono) ethylene

$$N = 34.6 \qquad J_{HH} = 5.0$$
$$K = 40.4 \qquad J_{PP} = 35.5$$
$$M = 30.5 \qquad J_{HP\ cis} = 0.8$$
$$L = 33.0 \qquad J_{HP\ trans} = 33.8\ Hz$$

TABLE IV-7. Position and Intensities of Lines in Part A of AA'XX' Spectrum

Line No.	Frequency	Relative intensity
1, 2, 3, 4	$\nu_A \pm \frac{1}{2} N$	1
5, 8	$\nu_A \pm \frac{1}{2}(K + \sqrt{K^2 + L^2})$	$\sin^2 \theta_s$
6, 7	$\nu_A \pm \frac{1}{2}(K - \sqrt{K^2 + L^2})$	$\cos^2 \theta_s$
9, 12	$\nu_A \pm \frac{1}{2}(M + \sqrt{M^2 + L^2})$	$\sin^2 \theta_a$
10, 11	$\nu_A \pm \frac{1}{2}(M - \sqrt{M^2 + L^2})$	$\cos^2 \theta_a$

the difference in the chemical shifts ν_A and ν_B in the quartet of an AB two-spin system from the relations:

$$\nu_{6,7} = \nu_{6,8} = \sqrt{K^2 + L^2} \qquad (IV\text{-}40)$$

$$\nu_{9,11} = \nu_{10,12} = \sqrt{M^2 + L^2} \qquad (IV\text{-}41)$$

In our case the two determination methods lead to values of L of 32.7 and 33.2 Hz, respectively. The slight discrepancy is due to instrument error and is quite acceptable. The parameter L is very sensitive to the position of the lines 6, 7, 10, and 11. Finding it by two methods makes it possible to correct for the inaccuracy of the spectrum and thus reduce the error in the determination of its parameters. Finally, we take the average result of 33.0 Hz for L and this leads to the following four spin–spin coupling constants: J and J' − 33.8 and 0.8 Hz; J_A and J_X − 5.0 and 35.5 Hz.

In accordance with the data in Ch. III, of these four values we may assign with certainty only $J_A = J_{HHgem} = 5.0$ Hz and hence it follows that $J_{PPgem} = J_X = 35.5$ Hz. Of the two vicinal constants of coupling of the protons and fluorine, the smaller apparently characterizes a cis-interaction, while the larger characterizes a trans-interaction. Since N > L, then, consequently, these two constants have the same sign. The chemical shift of the two chemically equivalent protons is determined from the position of the center of the spectrum relative to the standard.

Table IV-7 gives the relations of the position and intensity of the lines in an AA'XX' spectrum to its parameters. The values θ_a and θ_s are found from the relations:

$$\left. \begin{array}{ll} \cos 2\theta_a = \dfrac{M}{\sqrt{M^2 + L^2}}; & \sin 2\theta_a = \dfrac{L}{\sqrt{M^2 + L^2}} \\[3mm] \cos 2\theta_s = \dfrac{K}{\sqrt{K^2 + L^2}}; & \sin 2\theta_s = \dfrac{L}{\sqrt{K^2 + L^2}} \end{array} \right\} \qquad (IV\text{-}42)$$

These values are also indeterminate since the choice between the parameters K and M is indeterminate. Figure IV-14, 1 gives the theoretical AA'XX' spectra with different relations between the parameters.

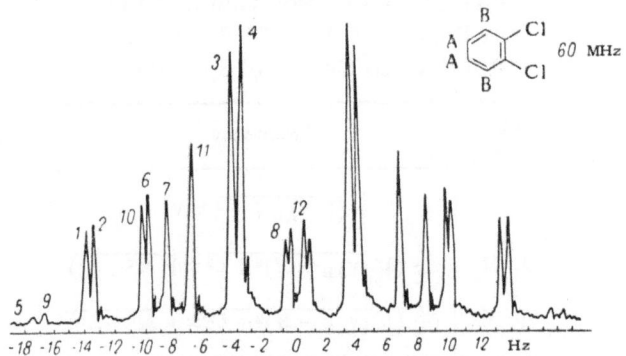

Fig. IV-16. Spectrum of o-dichlorobenzene. Index
H.60.A.AABB.K.

An AA'BB' spectrum with strong coupling between the spins
contains somewhat more information on the spin system because
of the fact that the ratio of the parameters affects the relative intensity of the spectral lines. If this effect is sufficient to make a
choice between a positive or a negative value of the parameter K
(with positive values of N, L, and M), then it is also possible to determine the relative signs of the sums $(J_A + J_B)$ and $(J + J')$. After
finding all the parameters of the spectrum it is usual to construct
two theoretical spectra, one with positive values of N, L, M, and
K and the other with a negative value of K. Comparison of these
spectra with the experimental spectrum makes it possible to choose
the sign of K. However, for the construction of each theoretical
spectrum it is necessary to diagonalize a matrix of the fourth order,
or, what is the same, to solve a secular equation of the fourth degree. This laborious calculation is achieved most simply with an
electronic computer.

Let us examine the course of the analysis of an AA'BB' spectrum without taking into account the relative intensities of the lines,
i.e., without selecting the sign of parameter K.

Analysis of Strongly Coupled AA'BB'

Spectrum [15, 16]

Figure IV-16 gives the spectrum of o-dichlorobenzene at 60
MHz [3], which is typical of an AA'BB' spectrum with strong coupling between the nuclei A and B. The spectrum is symmetrical rela-

TABLE IV-8. Position of Lines
in Part A of AA'BB' Spectrum
Relative to Center of Spectrum

Line No.	Frequency
1, 3	$\frac{1}{2}\left(\sqrt{\nu_{AB}^2 + N^2} \pm N\right)$
9, 11	$\frac{1}{2}\left(\sqrt{(\nu_{AB} + M)^2 + L^2} \pm \sqrt{M^2 + L^2}\right)$
10, 12	$\frac{1}{2}\left(\sqrt{(\nu_{AB} - M)^2 + L^2} \pm \sqrt{M^2 + L^2}\right)$

tive to the center so that each part of it (connected with the resonance of the nucleus A or B — it is impossible to determine this from
the spectrum) may be analyzed separately so as to take the mean
result. It is more convenient to measure the frequencies relative
to the center of the spectrum and to determine the position of the
latter relative to the standard reference.

Each part of the spectrum contains 12 lines, which are usually numbered as shown in Fig. IV-16. The lines 1-4 are connected
with the parameter N, while the 8 remaining lines form two quartets, namely, 5-8, which is connected with the parameter K, and
9-12, which is connected with M. The assignment of the lines may
be carried out using the rule of equal distances, which is expressed
by the following equations in this case:

$$\left.\begin{array}{ll} 5-1=3-8 & 5-6=7-8 \\ 5-3=1-8 & 5-7=6-8 \\ 1-7=6-8 & 9-10=11-12 \\ 1-6=7-3 & 9-11=10-12 \end{array}\right\} \begin{array}{l} \text{Equal to the corresponding} \\ \text{distance in the other} \\ \text{part of the spectrum} \end{array}$$

However, the parameter L cannot be determined by a simple
method as in the case of an AA'XX' spectrum. Moreover, it is necessary to calculate the parameter ν_{AB} since in this case ν_A and
ν_B do not lie at the center of the corresponding parts of the spectrum.

Table IV-8 gives the frequencies of the lines which may be
related to spectral parameters by an explicit dependence.

It is obvious that the distance between line 1 and 3 gives the
value of N, while the distance from the center point between these

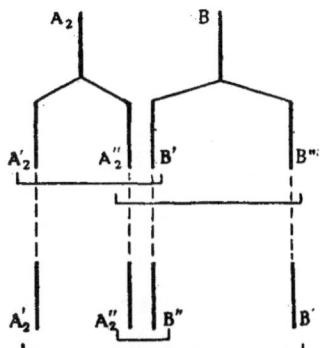

Fig. IV-17. Two variants of the origin of the subspectra in the A_2B part of the A_2BX spectrum in relation to the relative signs of the constants J_{AX} and J_{BX}. Top: J_{AX} and J_{BX} of the same sign ($\pm J_{AX}$, $\pm J_{BX}$); bottom: J_{AX} and J_{BX} opposite in sign ($\pm J_{AX}$, $\mp J_{BX}$).

lines to the center of the spectrum (to the origin) gives the value $\frac{1}{2}\sqrt{\nu_{AB}^2 + N^2}$. Thus, from lines 1 and 3 it is possible to determine N and ν_{AB}.

The parameter K may be obtained by using the following relations:

$$|\nu_4 + \nu_6 - \nu_5 - \nu_2| = |K + N|;$$
$$|\nu_4 - \nu_7 + \nu_8 - \nu_2| = |K + N|;$$

$$\left.\begin{array}{l} \nu_3 + \nu_6 + \nu_9 - \nu_1 - \nu_6 - \nu_4 \\ \nu_3 + \nu_7 + \nu_2 - \nu_1 - \nu_8 - \nu_4 \\ \nu_5 + \nu_7 + \nu_9 - 2\nu_1 - \nu_4 \\ 2\nu_3 + \nu_9 - \nu_6 - \nu_8 - \nu_4 \end{array}\right\} = K \quad \text{(IV-43)}$$

To obtain a more accurate result it is possible to use the mean value. The parameter M is determined from the following equation:

$$M = \frac{\nu_9 \nu_{11} - \nu_{10}\nu_{12}}{\nu_{AB}} \quad \text{(IV-44)}$$

After this the remaining parameter L is readily determined from lines 10 and 12:

$$L = \sqrt{\nu_{10,12}^2 - M^2} \quad \text{(IV-45)}$$

Below we give the frequencies of the lines in the low-field part of the experimental spectrum of o-dichlorobenzene and the parameters of this spectrum (in Hz).

Line No.	Frequency	Line No.	Frequency
1	13.73	7	8.42
2	13.19	8	0.74
3	4.06	9	16.56
4	3.52	10	10.06
5	17.40	11	6.80
6	9.73	12	0.41

ν_{AB} 14.9; N 9.7; J 8.5; K 7.5; J' 1.2; M 6.3; J_A 6.9; L 7.3; J_B 0.6 Hz

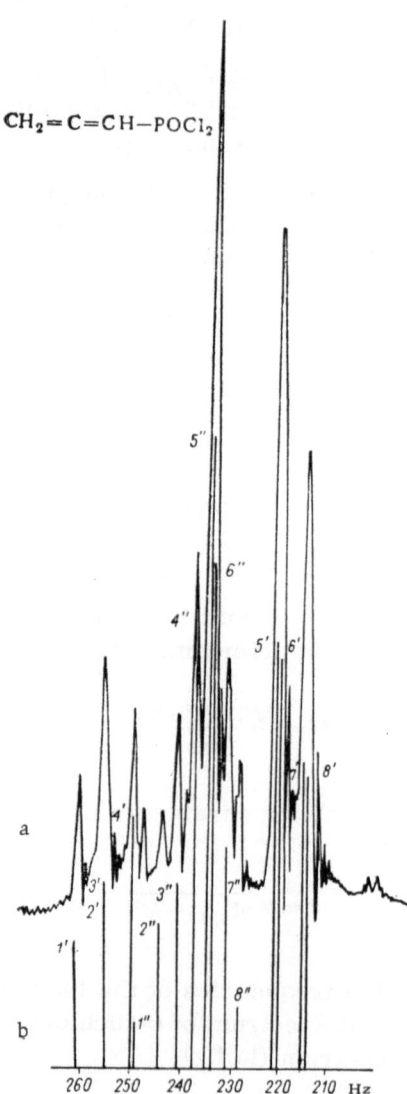

Fig. IV-18. Proton spectrum of the diacid
chloride of propadienephosphinic acid (40
MHz). Index **HP.40.JI = ,Z.A2BX.T.**
a) Experimental spectrum; b) theoretical
spectrum.

The parameters (in Hz) found by complete analysis of the spectrum (taking into account the position of the lines in the high-field part) with the use of a computer are as follows:

$$\nu_{AB} \; 15.23; \quad J \; 8.17; \quad J' \; 1.61; \quad J_A \; 7.44; \quad J_B \; 0.36 \; Hz$$

ABCX System

The analysis of a spectrum of the ABCX type is very similar to the analysis of an ABX spectrum. The ABC part of the ABCX spectrum may be represented as the result of the superposition of of two ABC spectra, each of which corresponds to a spin state of the nucleus $X - \alpha$ or β. In addition to the chemical shifts and relative signs of the constants J_{AB}, J_{AC}, and J_{BC}, this spectrum makes it possible to determine the values and relative signs of the spin-spin coupling constant of the nuclei A, B, and C with the nucleus X. We will examine the principles of analysis of an ABCX spectrum on a simpler example, namely, the proton spectrum of the diacid chloride of propadienephosphinic acid $CH_2 = C = CH - POCl_2$, which belongs to the A_2BX type. This spectrum may be analyzed relatively simply without the use of electronic computers.

Depending on the relative signs of the constants J_{AX} and J_{BX}, there may be two distributions of the lines in the A_2BX spectrum (Fig. IV-17). With identical signs of the constants one would expect the appearance of two subspectra of the A_2B type with approximately identical ratios J_{AB}/ν_{AB}. If the signs of the constants are opposite, then it is more probable that the two subspectra will differ markedly: one of them will be more strongly coupled than the other. The form of each subspectrum is determined by the effective shifts ν'_A, ν'_B, ν''_A, ν''_B, which are determined analogously to the ABX case:

$$\nu'_A = \nu_A + \frac{1}{2} J_{AX}; \quad \nu''_A = \nu_A - \frac{1}{2} J_{AX}$$

$$\nu'_B = \nu_B + \frac{1}{2} J_{BX}; \quad \nu''_B = \nu_B - \frac{1}{2} J_{BX}$$

In the spectrum of the diacid chloride of propadienephosphinic acid, which is given in Fig. IV-18, it is readily seen that the case with opposite signs of the constants J_{AX} and J_{BX} is realized. The effective shifts are determined directly from the spectrum by analysis of the two A_2B subspectra by the rules presented previously. The four effective shifts make it possible to determine the four spectral parameters ν_A, ν_B, J_{AX}, and J_{BX}. The constant J_{AB} may be determined from either of the A_2B subspectra by means of Fig. IV-7, but naturally it is better to use for this purpose the subspectrum in which the ratio J_{AB}/ν_{AB} is lower. To check the accuracy of the assignment of the lines we construct a theoretical spectrum, the position and intensity of the lines of which are determined from the diagrams in Figs. IV-8 and IV-9 or from the formulas in Table IV-2 (p.181).

From this example it is evident that the analysis of A_2BX and ABCX spectra may be simpler than the analysis of A_2B or ABC subspectra despite the greater number of coupled spins. The simplification arises due to the fact that the greater spin−spin coupling constants and the fortunate combination of their relative signs lead to considerable effective shifts and hence reduce in effect the compactness of part of the spectrum. Since the effective shifts depend to a considerable extent on the spin−spin coupling constants with the nucleus X, spectra of this type change little with a change in the working frequency of the spectrometer. The same picture is observed in A_3B_2X spectra, which are examined below.

4. A_3B_2 AND A_3B_2X SYSTEM

The analysis of spectra of this type is simplified due to the fact that all the spin−spin coupling constants between the nuclei A and B are equal, i.e., each of the groups contains only magnetically equivalent nuclei. Typical spectra of the A_3B_2 system are given by protons of the ethyl group and depending on the character of the atom to which the ethyl group is attached, the spectra vary over a wide range from weakly coupled spectra (for example, in nitroethane) to spectra with strong spin−spin coupling, which degenerate into a single line (in ethyl derivatives of silicon). If the ethyl group is attached to an atom with a magnetic moment, then spectra of the A_3B_2X type arise. These spectra (like the case examined in the previous section) consist in the A_3B_2 part of the superposition of two subspectra, namely $(A_3B_2)'$ and $(A_3B_2)''$, whose form is determined by the effective shifts ν'_A, ν'_B, ν''_A, and ν''_B and the constant J_{AB}. The form of the signal of the nucleus X is much more complex and is not examined here.

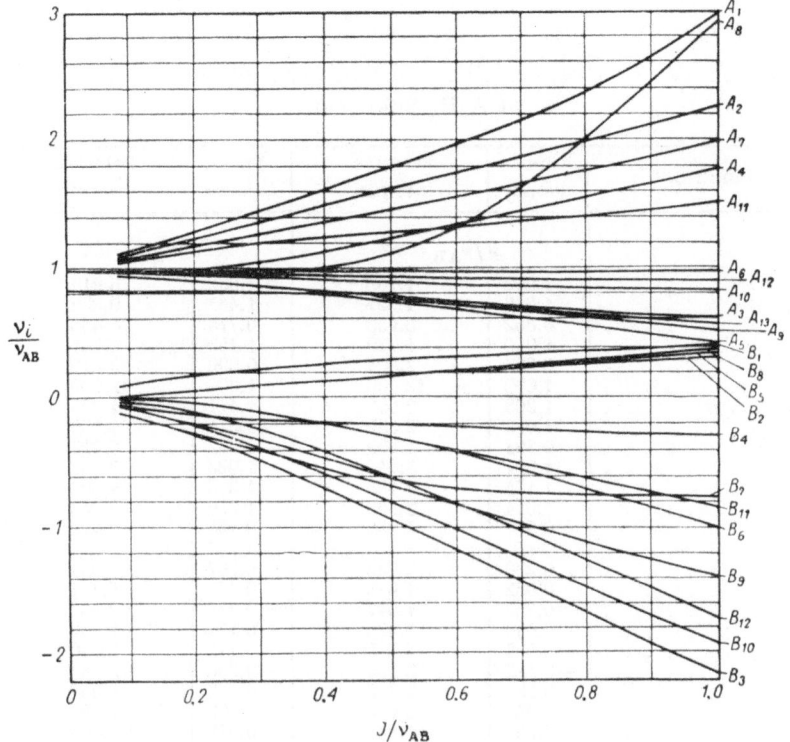

Fig. IV-19. Frequencies of lines in A_3B_2 spectrum (related to the value ν_{AB}).

The form of an A_3B_2 spectrum, as in other cases of two groups of magnetically equivalent nuclei, may be represented as dependent on one parameter, namely, the ratio J_{AB}/ν_{AB}, and therefore it may be represented as a table or graphically. Figure IV-19 shows the relations of the position of the 25 main lines of the A_3B_2 spectrum (ignoring combination transitions) to this ratio and Table IV-9 gives in addition the relative intensities of the same lines [4]. Figure IV-20 gives the theoretical A_3B_2 spectra for different characteristic cases of the ratio J_{AB}/ν_{AB}.

The usual course of the analysis of a spectrum of the A_3B_2 type consists of an approximate first-order analysis with subsequent refinement by selection of the parameters until the best agreement is reached between the theoretical and experimental spectra. The analysis of the A_3B_2X spectrum of the acid bromide of diethylthiophosphinic acid (Fig. IV-21) is simplified because of

TABLE IV-9. Frequencies and Relative Intensities of Lines
of A_3B_2 Spectrum

Line No.	Fre-quency	Relative intensity	Fre-quency	Relative intensity
	$J/\nu_{AB}=0.2$		$J/\nu_{AB}=0.4$	
A_1	1.262	1.768	1.632	0.919
A_2	1.218	1.327	1.463	0.889
A_3	0.852	3.953	0.775	4.436
A_4	1.040	1.779	1.153	1.301
A_5	0.822	2.894	0.690	3.816
A_6	1.000	12.000	1.000	12.000
A_7	1.208	2.695	1.411	1.895
A_8	1.073	2.389	1.374	1.030
A_9	0.798	4.839	0.672	7.024
A_{10}	1.027	3.642	1.023	3.252
A_{11}	1.175	2.193	1.304	1.800
A_{12}	1.014	2.852	1.010	2.753
A_{13}	0.831	5.655	0.738	6.790
B_1	0.238	3.232	0.367	4.081
B_2	0.082	4.673	0.137	5.111
B_3	−0.352	1.047	−0.775	0.563
B_4	−0.096	3.326	−0.173	2.883
B_5	0.096	4.893	0.173	5.810
B_6	−0.122	3.106	−0.290	2.183
B_7	0.281	3.842	0.448	6.047
B_8	0.049	2.070	0.109	1.995
B_9	−0.294	1.193	−0.573	0.803
B_{10}	−0.190	1.298	−0.539	0.645
B_{11}	−0.132	1.502	−0.279	1.096
B_{12}	0.005	1.816	−0.254	0.771
	$J/\nu_{AB}=0.6$		$J/\nu_{AB}=0.8$	
A_1	2.064	0.503	2.525	0.304
A_2	1.726	0.617	2.000	0.444
A_3	0.731	4.666	0.704	4.784
A_4	1.326	0.855	1.540	0.550
A_5	0.600	4.545	0.540	5.030
A_6	1.000	12.000	1.000	12.000
A_7	0.618	1.368	1.839	0.999
A_8	1.835	0.414	2.348	0.209
A_9	0.630	8.049	0.615	8.460
A_{10}	0.972	3.217	0.919	3.336
A_{11}	1.398	1.570	1.466	1.421
A_{12}	0.988	2.747	0.962	2.774
A_{13}	0.692	7.333	0.668	7.590
B_1	0.436	4.496	0.475	4.696
B_2	0.174	5.383	0.200	5.556
B_3	−1.231	0.334	−1.704	0.216

TABLE IV-9 (Continued)

Line No.	Frequency	Relative intensity	Frequency	Relative intensity
B_4	−0.226	2.600	−0.260	2.420
B_5	0.226	6.529	0.260	7.005
B_6	0.500	1.454	−0.740	0.970
B_7	0.511	7.078	0.538	7.484
B_8	0.207	2.100	0.298	2.277
B_9	−0.849	0.577	−1.126	0.432
B_{10}	−0.974	0.333	−1.446	0.193
B_{11}	−0.438	0.799	−0.622	0.579
B_{12}	−0.694	0.291	−1.195	0.139
	$J/\nu_{AB}=1.0$		$J/\nu_{AB}=2.0$	
A_1	3.000	0.200	5.449	0.050
A_2	2.281	0.332	3.732	0.113
A_3	0.686	4.850	0.646	4.957
A_4	1.781	0.363	3.146	0.080
A_5	0.500	5.333	0.414	5.828
A_6	1.000	12.000	1.000	12.000
A_7	2.078	0.739	3.423	0.217
A_8	2.871	0.125	5.449	0.029
A_9	0.609	8.654	0.602	8.914
A_{10}	0.877	3.472	0.782	3.823
A_{11}	1.514	1.320	1.615	1.108
A_{12}	0.939	2.807	0.863	2.914
A_{13}	0.653	7.726	0.652	7.927
B_1	0.500	4.800	0.550	4.949
B_2	0.219	5.668	0.268	5.887
B_3	−2.186	0.150	−4.646	0.043
B_4	−0.281	2.303	−0.318	2.092
B_5	0.281	7.305	0.318	7.807
B_6	−1.000	0.667	−2.414	0.172
B_7	0.552	7.673	0.576	7.920
B_8	0.371	2.441	0.550	2.828
B_9	−1.406	0.334	−2.843	0.126
B_{10}	−1.934	0.124	−4.428	0.030
B_{11}	−0.830	0.420	−2.090	0.108
B_{12}	−1.710	0.080	−4.271	0.017

the great difference in the effective shifts ν'_A and ν'_B and thus the spectrum $(A_3B_2)'$ may be analyzed by first-order rules. The signals B' and B'' are completely superposed and in the first approximation this indicates that the effective signals are equal, from which it follows that $J_{BX} \simeq 0$, $\nu'_B = \nu''_B = \nu_B$. The constant J_{AX} is found directly from the spectrum from the distance between the most intense peaks, which belong to the subspectra $(A_3B_2)'$ and $(A_3B_2)''$, while the constant J_{AB} is found from the distance between the components of the triplet in the spectrum with weaker coupling. In the analysis of more complex spectra of this type it is preferable to use an electronic computer.

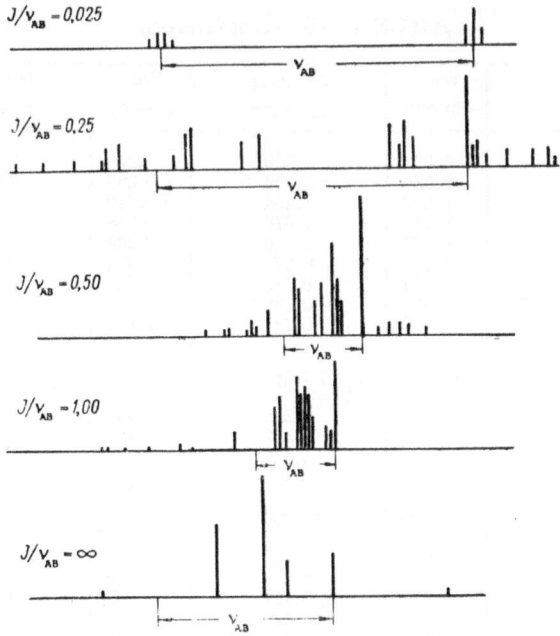

Fig. IV-20. Theoretical spectra of the A_3B_2 system with different ratios of the parameters.

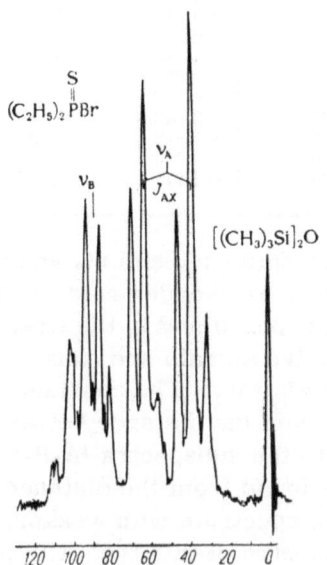

Fig. IV-21. Proton spectrum of the acid bromide of diethylthiophosphinic acid. Index HP.40.JI Z.M.T. $\delta_A 0.90$, $\delta_B 0.22$ ppm.;

$J_{AB} 7.5$, $J_{AX} 24$, $J_{BX} \sim 0$ Hz.

B. NUCLEAR MAGNETIC DOUBLE RESONANCE

Nuclear magnetic double resonance (NMDR) consists of the action on the sample investigated, which is placed in the magnetic field of the spectrometer, of two rf fields used simultaneously in the resonance region of the magnetic nuclei. This comparatively new investigation method has spread rapidly and is used together with single resonance both for structural and physicochemical investigations and also in the study of relaxation processes. The use of NMDR for investigating organic compounds has been described in reviews [17, 57]. In the present section we examine the elementary foundations of the theory of NMDR and the resultant possibilities of using this method and also practical procedures for achieving NMDR and methods of interpreting spectra obtained with the use of double resonance.

In addition to double resonance there is also triple and multi-frequency resonance, where three or more rf fields are used simultaneously. We should also distinguish between homonuclear double resonance, in which both rf fields correspond to the resonance of nuclei of the same magnetic isotope, i.e., they have similar frequencies, and heteronuclear double resonance, in which the two rf fields correspond to resonance frequencies of two different isotopes and hence are quite different.

Let us examine the main effects observed and used in DNMR.

5. ACTION OF STRONG RF FIELD

The action on an individual spectral line of a strong rf field H_2, satisfying the condition $\gamma^2 H_2^2 T_1 T_2^* \gg 1$, leads to saturation of this line and it may be detected no longer by means of the field H_2. The line disappears, but despite the fact that the populations of the corresponding energy levels are comparable [18-21], the interaction of the nuclear spins with the rf field does not cease. This appears as the retention of the dispersion signal and may also be detected by the interaction with a second weak rf field H_1 close to the saturated line. If the angular frequency $\omega_1 = 2\pi f_1$ of the field H_1 is swept, while the angular frequency ω_2 of the field H_2 remains constant, then due to the coherence of motion of the nuclear spins it is possible to observe absorption signals [22-25] in the form of new narrow

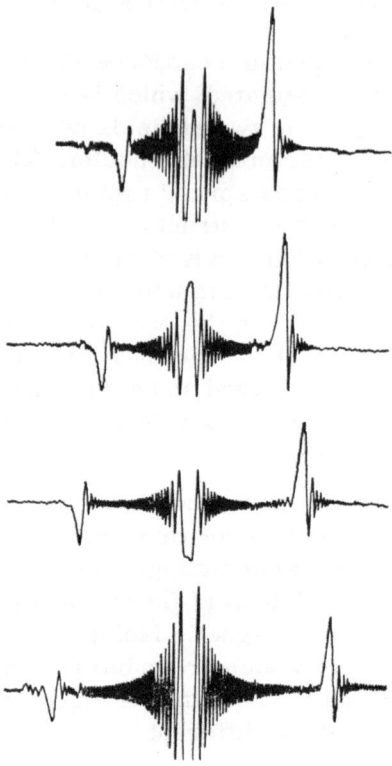

Fig. IV-22. Double resonance of one spin of $\frac{1}{2}$.

spectral lines (Fig. IV-22) [23]. If the frequency ω_2 is tuned accurately to resonance, then the distance between these lines equals double the amplitude $2\gamma H_2$ (rad/sec) of the strong rf field H_2 and depends linearly on the amplitude. The distance between these lines measured in hertz gives the amplitude of the rf field $2\gamma H_2/2\pi$ (Hz) and thus it is possible to measure the strength of the rf perturbing field accurately.

This phenomenon is described conveniently by considering the motion of the perturbed nuclear spins in coordinates rotating about the $+z$ axis with a frequency of $-\omega_2$. In these coordinates their motion is rotation about the direction of the effective magnetic field $\gamma H_{eff} = \pm [(\gamma H_2)^2 + (\gamma H_0 - \omega_2)^2]^{1/2}$, where $\gamma H_0 = \omega_0$ is the resonance frequency of the spins investigated. The

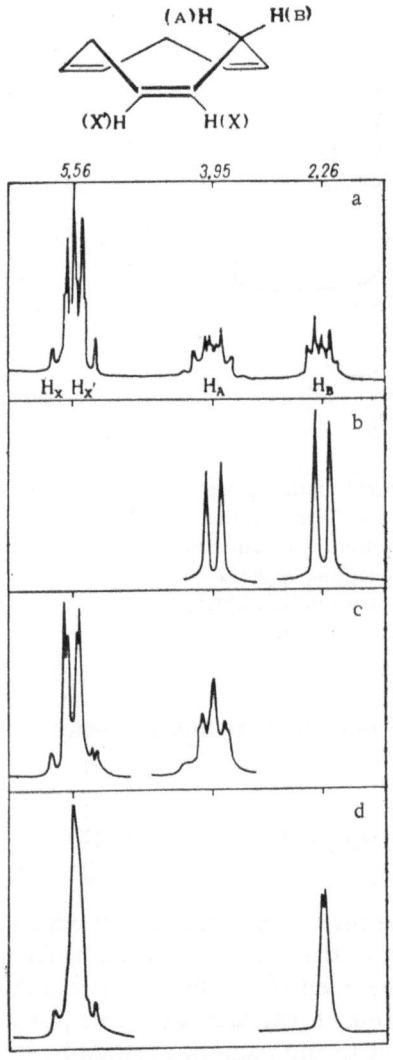

Fig. IV-23. Decoupling of protons in cis,cis,cis-1,4,7-cyclononatriene. a) Single resonance spectrum; b) perturbation of nuclei XX'; c) perturbation of nuclei B; d) perturbation of nuclei A.

rf field γH_2 in these coordinates is as stationary as γH_0 in laboratory coordinates, but is oriented in the x—y plane. If we consider motion in laboratory coordinates, then under these conditions we obtain for the nuclear spins the two resonance frequencies $\omega_1 = \omega_2 \pm [(\gamma H_2)^2 + (\gamma H_0 - \omega_2)^2]^{1/2}$, corresponding to the two spectral lines given above.

From what has been stated it follows that in the case of exact resonance the strong perturbing field γH_2 orients the nuclear spins and the spin magnetization in the x—y plane. Taking into account the fact that the hamiltonian of a system of two spins AX

$$H_0 = -\omega_A I_z(A) - \omega_X I_z(X)$$

$$+ 2\pi J_{AX} I_z(A) I_z(X) \quad (IV-46)$$

it may be assumed that if the strong perturbing field H_2 acts only on the nucleus X and not on the nucleus A, then the magnetizations of the two nuclei will lie in different directions and the directions of quantization (or in the classical treatment, the directions of the axes of precession) of the spin vectors I(A) and I(X) form a right angle, as a result of which the scalar product in the hamiltonian equals zero and the spin—spin coupling and the doublet splitting of the signal of the nucleus A produced by it disappear [19, 26].

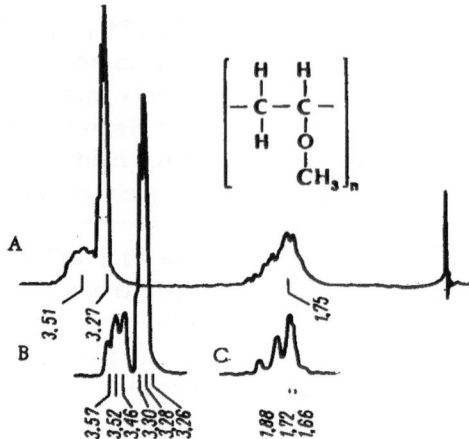

Fig. IV-24. Spectra of poly(vinyl methyl ether):
A) single resonance spectrum (the identification
of most of the lines of the spectrum is practically
impossible). Spectra with decoupling of the α-
(B) and β-protons (C) made it possible to identify
the chemical shifts of the individual stereoregu-
lar triads.

The same result is reached by using rotating coordinates, where
the hamiltonian (IV-46) has the form

$$H_0^R = -(\omega_A - \omega_2) I_s(A) - \gamma H_2 I_x(X) - (\omega_X - \omega_2) I_s(X) + 2\pi J_{AX} I_s(A) \cdot I_s(X)$$

$$\text{(IV-47)}$$

with $\omega_2 = \omega_X$ and $|\gamma H_2| \gg |2\pi J_{AX}|$; the term $\gamma H_2 I_x(X)$ is introduced
to allow for the effect on the spin of X of the strong perturbing field
directed along the axis X. Under these conditions the last two terms
in (IV-47) are small and may be omitted. $I_z(A)$ and $I_x(X)$ are now
quantized independently and in contrast to the case described by
the hamiltonian (IV-46), they may have all the possible values
$m = I, I - 1, ..., -I + 1, -I$ without an intereffect. The decoupling of
the nuclear spins by the perturbing field leads to merging of the
multiplet (doublet in the case when $I = \frac{1}{2}$) of the nucleus A into one
line.

Spin decoupling (collapse) has found wide application in nuclear resonance. Partial or complete collapse makes it possible
to reveal coupling between separate multiplets in complex spectra,

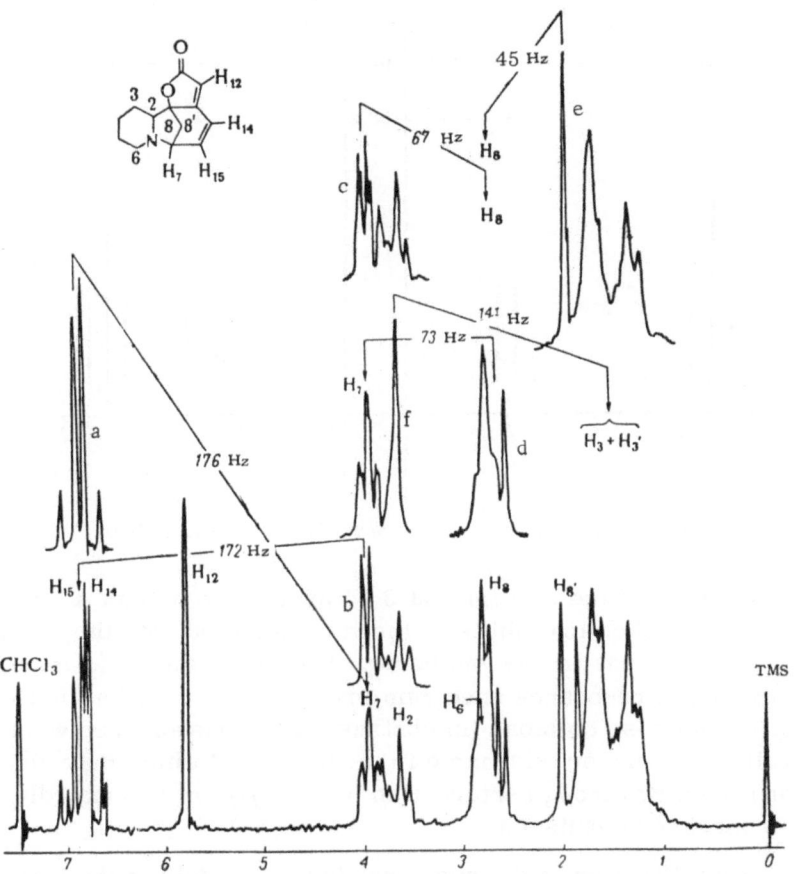

Fig. IV-25. Identification of the multiplets in the proton spectrum of the
alkaloid phyllochrysin.

to determine chemical shifts of nuclei whose signal is not observed
directly, to determine the relative signs of spin−spin coupling constants, and, finally, to simplify spectra. **Figure IV-23** gives the
spectrum of cis,cis,cis-1,4,7-cyclononatriene at −40°C [27]. From
the spectra simplified by double resonance it is clear at first glance
that the compound investigated at this temperature has the structure of a right crown (and not a saddle) and the absolute values of
all the chemical shifts and some of the spin−spin coupling constants may be determined by direct measurement without analysis
of the spectrum. Taking account of the dependence of J on the dihedral angle it is possible to assign the multiplet at 2.26 ppm to

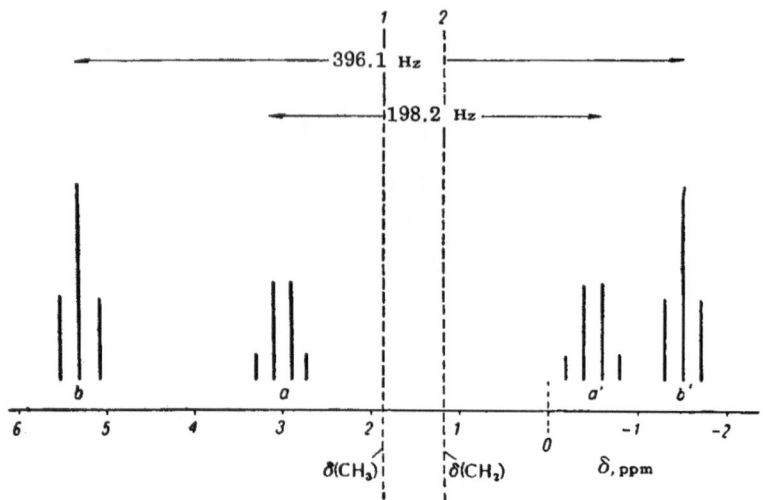

Fig. IV-26. Diagram of proton spectrum of triethylthallium.

the proton B and the multiplet at 3.95 ppm to the proton A. The values of the chemical shifts of these protons indicate the absence of a ring current in the molecule. Obtaining this information from a single resonance spectrum presents great difficulties and thus, in this case collapse (in contrast to the phenomenon which is well-known under this name in medicine and which may also be produced by strong perturbation of the system) undoubtedly gives positive advantages.

Simplification of the spectrum is used widely in the investigation of the stereoregularity of polymers. Figure IV-24 gives the spectrum of poly(vinyl methyl ether) [28]. The chemical shifts of 3.57, 3.52, and 3.46 ppm correspond to syndiotactic (d*l*d or *l*d*l*), heterotactic (d*l**l* or *l*dd) and isotactic (ddd or *l**l**l*) structures of the polymer. These shifts (and the relative contents of the stereo-regular triads) cannot be determined from the single resonance spectrum.

Only the chemical shift $\delta_{H_{12}} = 5.74$ ppm can be determined directly from the spectrum of the solution of the alkaloid phyllo-chrysin in $CDCl_3$ [29] (Fig. IV-25). The use of another solvent, namely, CF_3COOH, made it possible to identify and determine approximately the chemical shifts of the protons H_2, H_6, and H_7 adjacent to the nitrogen. Decoupling the nucleus H_7 leads to the appearance of a quadruplet of the system of protons H_{14} and H_{15} and made it possible to determine $J_{H_{14}H_{15}} = 9.3$ Hz and $|\delta_{H_{15}} - \delta_{H_{14}}|$ $= 0.18$ ppm; at the same time there was merging of the quadruplet of H_8 into a doublet and from this $\delta_{H_8} = 2.64$ ppm and $J_{H_8H_8'} = 9.2$ Hz. Decoupling the nucleus H_8 leads to merging of the doublet of H_8' and the appearance of a quadruplet from the nucleus H_7 and from this it was found that $\delta_{H_8'} = 1.90$ ppm, $\delta_{H_7} = 3.90$ ppm, $J_{H_7H_{14}}$ $= 1.1$ Hz, and $J_{H_7H_{15}} = 4.5$ Hz. Simultaneous irradiation with the perturbing rf field H_2 of the protons H_{14} and H_{15}, H_3 and $H_{3'}$, or H_5 and $H_{5'}$ leads to the appearance of a doublet from the nucleus H_7 and separate lines from the nuclei H_2 and $H_{6,6'}$. Hence $J_{H_7H_8} = 4.5$ Hz. The latter experiments lead to the values $\delta_{H_2} = 3.67$ and $\delta_{H_6} = 2.88$ ppm; the chemical shifts of the overlapping lines of the nuclei H_3 and H_5 are $\delta_{H_3} = 1.32$ and $\delta_{H_5} = 1.80$ ppm, respectively.

Let us examine collapse in a system where the values of the spin–spin coupling constants are very different, for example, in triethylthallium [30]. Thallium produces very large splittings in the spectrum of the ethyl groups and complete decoupling of the methyl and methylene protons is impossible. The proton spectrum of this compound (Fig. IV-26) contains two triplets and a quadruplet each from the methyl and methylene groups, corresponding to the two different values of the spin of thallium ($+\frac{1}{2}$ or $-\frac{1}{2}$, α or β). We assume that the triplet b' corresponds to the spin state of thallium and then strong perturbation of the triplet b' with the field $\gamma H_2 > |2\pi J_{HH}|$ leads to decoupling of the methyl and methylene proton only in molecules where $I_z(Tl) = \alpha$, while the field H_2 has hardly any effect on the other molecules. Now if the two constants J_{TlH} have the same sign, then of the two quadruplets a and a' the spin state of thallium α will correspond to the quadruplet a' and if the signs are opposite, then it will correspond to a. Then on collapse we will observe the quadruplet a' or a, depending on the relative signs of the constants. Experiment showed that collapse of the quadruplet a corresponds to perturbation of the triplet b' while perturbation of the triplet b produces merging of the quadruplet a'. It follows from this that the spin state of thallium α corresponds to unlike spin states of the protons of the methyl and methylene group and consequently, the relative signs of the spin–spin coupling constants are opposite. It is impossible to determine the absolute signs of the constants in this way.

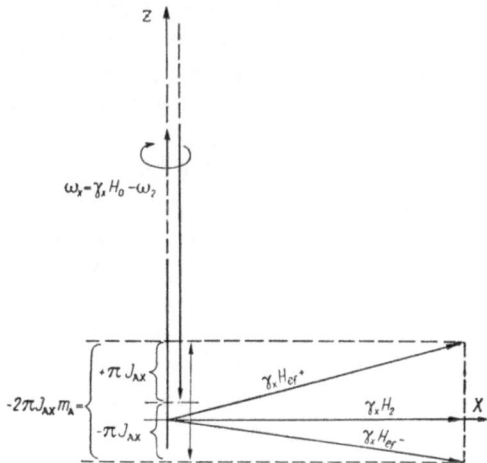

Fig. IV-27. Diagram of the interactions of the spin
I (X) with magnetic fields in rotating coordinates.

An analogous method is applicable in cases where there are
no exceptionally high coupling constants; it is only important that
the spectrum should not be very strongly coupled, but should have
more or less clear separate multiplets for the nuclei whose relative signs of spin−spin coupling with some third nucleus are to be
determined. In contrast to complete decoupling of spins under conditions of complete double resonance, here the perturbing field acts
selectively and we are dealing with selective double resonance.

If a perturbing field acts on a system of two nuclear spins
AX more weakly than in previous cases so that there is not complete orthogonality of I (A) and I (X) and in the hamiltonian (IV-47)
we cannot neglect the last term. At the same time, due to the presence of the term $\gamma H_2 I_X(X)$, the single resonance selection rule
$\Delta m_A = 1$, $\Delta m_X = 0$ becomes invalid so that $[I_z, I_x] \neq 0$ and spin transitions of the nucleus X simultaneously with spin transitions of the
observed nucleus A are possible so that $\Delta m_X = 0$, 1, ..., 2I (X) [22,
31]. The spectrum of the nucleus A is complex in comparison with
its spectrum under single resonance conditions, the number of
lines is increased, and each spectral line of the multiplet is split
into submultiplets [31], whose outer components move further apart
and become weaker as γH_2 is increased. The action of the perturbing field on the spin system is best examined in this case in

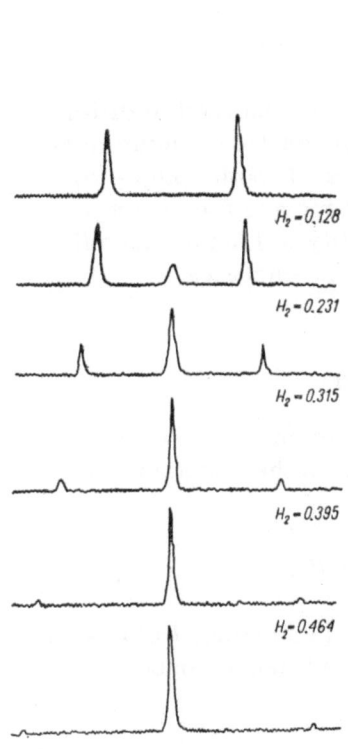

Fig. IV-28. Spectrum of fluorine nucleus in an aqueous solution of Na_2PO_3F with different amplitudes of the perturbing rf field H_2.

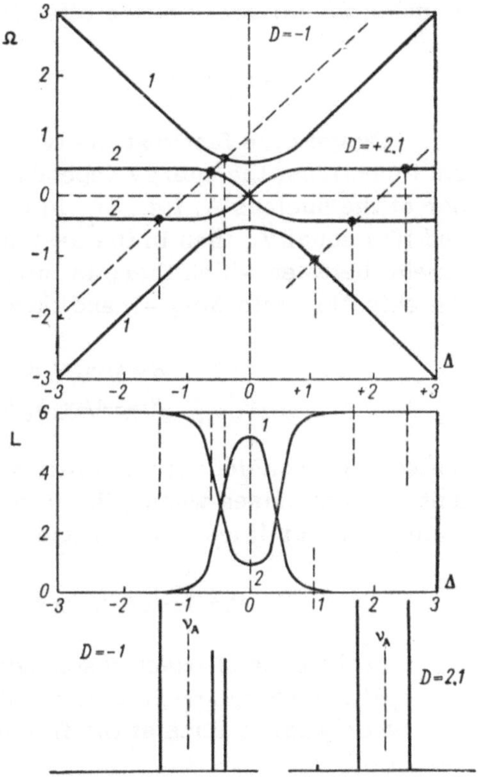

Fig. IV-29. Spectra of the nucleus A of the two-spin system AX, plotted by sweeping the magnetic field H_0 with different degrees of detuning Δ, $\gamma H_2 = 0.4 \pi J$.

rotating coordinates (Fig. IV-27). A stronger perturbing field with an amplitude $2/T_2^* \ll \gamma H_2 \lesssim 2\pi |J|$ and a frequency $\omega_2 \cong \gamma_X H_0$ acts on the nucleus X, but does not affect the nucleus A, whose signal is investigated with a weak measuring field H_1 close to its resonance frequency $\omega_1 \cong \gamma_A H_0$. As in double resonance of a one–spin system, the spin X is quantized along the effective field, but its direction now depends on the magnetic effect $-2\pi J_{AX} m_A$ of the nucleus A, while the direction of quantization of the nucleus A is the z axis as previously and as in single resonance [18, 31, 32]. Figure IV-27 shows that the precession frequency of the nucleus

X under the influence of the effective field is given by

$$\gamma_X H_{eff} = [(\gamma_X H_0 - 2\pi J_{AX} m_A - \omega_2)^2 + \gamma_X^2 H_2^2]^{1/2} \tag{IV-48}$$

Despite the fact that the two nuclei are quantized in different directions, the energy depends on the magnetic quantum number of the nucleus A, i.e., m_A in both cases. If both spins $I(A)$ and $I(X)$ equal $\frac{1}{2}$, then in the system AX there are four energy levels, between which there is the possibility of transitions with the selection rule $\Delta m_A = 1$ and these have the energies

$$E = \frac{1}{2}\gamma_A H_0 \pm \frac{1}{2}\gamma_X H_{eff+}$$
$$E = -\frac{1}{2}\gamma_A H_0 \pm \frac{1}{2}\gamma_X H_{eff} \tag{IV-29}$$

Instead of a doublet with a splitting $2\pi J_{AX}$ in angular frequency units, which arises when $\gamma H_2 = 0$, we obtain in the spectrum of the nucleus A four lines at the frequencies

$$\omega_1 = \gamma_A H_0 \pm (\frac{1}{2}\gamma_X H_{eff}) \pm (\frac{1}{2}\gamma_X H_{eff})$$

In the case of exact resonance of the perturbing field, when $\omega_2 = \gamma_X H_0$, $H_{eff+} = H_{eff-}$, and in the double resonance spectrum there will be only three lines at the frequencies

$$\omega_1 = \gamma_A H_0 \text{ and } \omega_1 = \gamma_A H_0 \pm \gamma_X H_{eff} \tag{IV-50}$$

As the amplitude of γH_2 increases, the fields H_{eff+} and H_{eff-} increase, but as a result of the decrease in the angle between them the probability of simultaneous transitions of the nuclei A and X and the intensities of the corresponding satellites are reduced [18, 32]. When the condition $\gamma H_2 \gg 2\pi |J_{AX}|$ holds, the doublet of the nucleus A merges into one line.

Figure IV-28 gives the spectra of the F^{19} nuclei in an aqueous solution of Na_2PO_3F, plotted with different amplitudes of the perturbing field with the resonance frequency of P^{31} [32]. From the spectra, which were plotted with a constant intensity of the perturbing field, but with different adjustments of the frequency ω_2 relative to ω_0, it follows that the effect of the perturbing field is maximal with exact tuning to resonance.

The effect of a perturbing rf field on NMR spectra of simple spin systems of the type $A_m X_n$ (m, n $\leq$ 3) is described well by Freeman's diagrams [31, 33]. These diagrams make it possible to determine readily the resonance frequencies ω_1 and the relative intensities L of the lines of NMDR spectra in relation to the relative intensity $\gamma H_2/2\pi \mid J \mid$ and the relative degree of detuning $\Omega = (\omega_2 - \omega_X)/2\pi \mid J \mid$ of the frequency of the perturbing rf field H_2. In this case ω_1 is expressed in terms of the relative degree of deturning of the measuring frequency $\Delta = (\omega_1 - \omega_A)/2\pi \mid J \mid$. ω_A and ω_X are the resonance frequencies of the nuclei A and X when there is no spin—spin coupling between them. On these diagrams vertical lines correspond to sweeping of the frequency and sloping lines to sweeping of the magnetic field. In the case of homonuclear double resonance the slope is close to 45° (Fig. IV-29).

6. ACTION OF A WEAK PERTURBING FIELD

The perturbing rf field is regarded as weak if its amplitude is of the order of the width of the spectral lines perturbed and the field H_2 affects only one line in the spectrum:

$$\frac{2}{T_2^*} < \gamma H_2 \ll 2\pi \mid J \mid, \mid (a'-a)-(b'-b) \mid, \mid (a'-a) \mid \qquad \text{(IV-51)}$$

This weak field affects the state of the spin system only if its frequency ω_2 is tuned exactly to some line in the spectrum. The action of the weak field is purely local and as a result there is never a decrease in the number of spectral lines. On the contrary, in addition to saturation of lines, this field may produce splitting of them if the lines are sufficiently narrow such that $2/T_2^* \ll (a'-a)-(b'-b)$. It is also advantageous to use rotating coordinates, but in this case only the conditions of accurate tuning are changed and instead of $\omega_2 = \gamma_X H_0$ we have $\omega_2 = \gamma_X H_0 - 2\pi Jm_A$ and, consequently, from (IV-48) we find that $\gamma_X H_{eff} = \gamma_X H_2$. The energy levels of the nucleus A are split according to equation IV-46 by the value $\pm \frac{1}{2}\gamma_X H_2$ and $\pm \frac{1}{2} 2\pi J_{AX}$, while the spectral lines are split by the value $\gamma_X H_2$. Since the nuclear spin may affect the energy level of another nucleus only with spin—spin coupling between them and quantization of the spin of the nucleus X along the direction of the effective field has a direct effect only on the energy levels of this spin, in the spectrum of the nucleus A the only lines

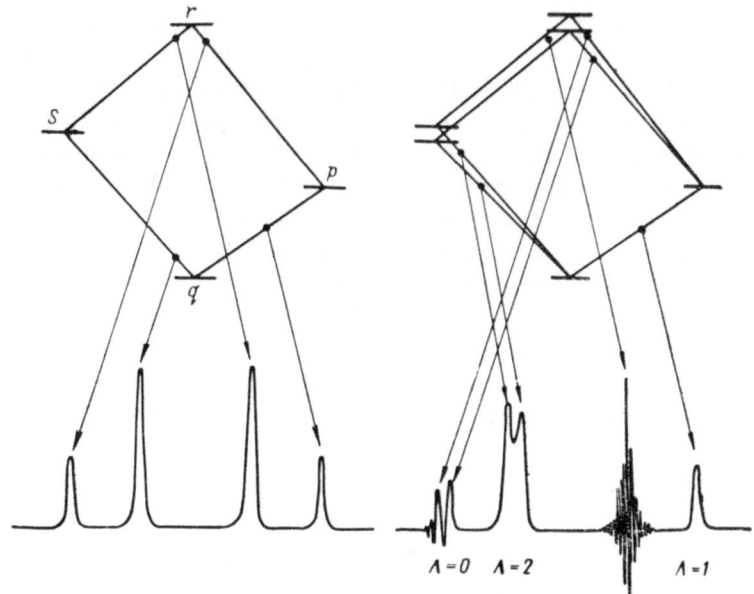

Fig. IV-30. Tickling in an AX two-spin system.

split are those which have common energy levels with the spin X.
The figure IV-30 shows very arbitrarily the appearance of splitting of the spectral lines of the doublet of the nucleus A (tickling)
when γ_A, $\gamma_X > 0$ and $J_{AX} > 0$ [34], starting from the hamiltonian
(IV-46). As the intensity of the perturbing field increases, it will
also act on the other line of the doublet of the nucleus X; then the
inner lines of the four lines of the nucleus A will increase and approach, while the outer lines will decrease and move farther apart.

Similar splitting of the lines is also observed in spectra of
more complex spin systems. Figure IV-31 shows the splitting
in spectra of the type AKM.

From Fig. IV-31 it is clear that local perturbation of only
one line of a multiplet, like the selective decoupling of nuclear
spins (Fig. IV-26), may be used for determining the relative signs
of spin—spin coupling constants. Thus, in styreneimine (Fig. IV-
31a) the signs of the geminal and vicinal constants are the same,
while in styrene sulfide (Fig. IV-31b) they are opposite [35].
Knowledge of the relative signs of the constants facilitates the con-

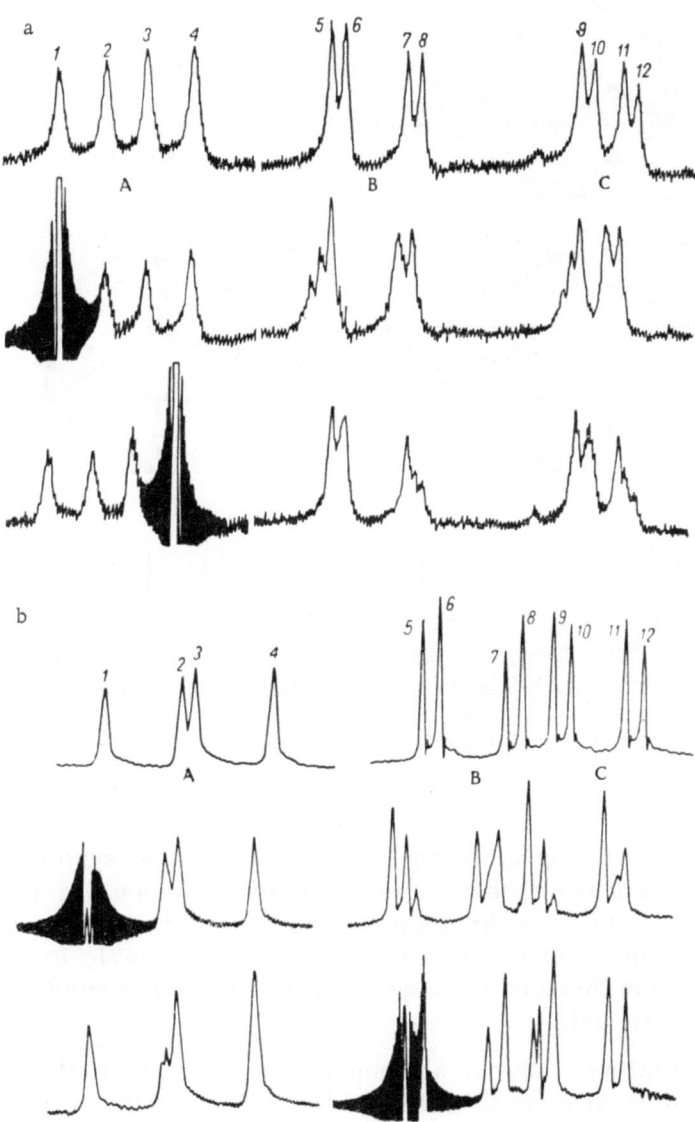

Fig. IV-31. Tickling in proton spectra of the type AKM: a) styrene-imine (J_{gem} – positive); b) styrene sulfide (J_{gem} – negative).

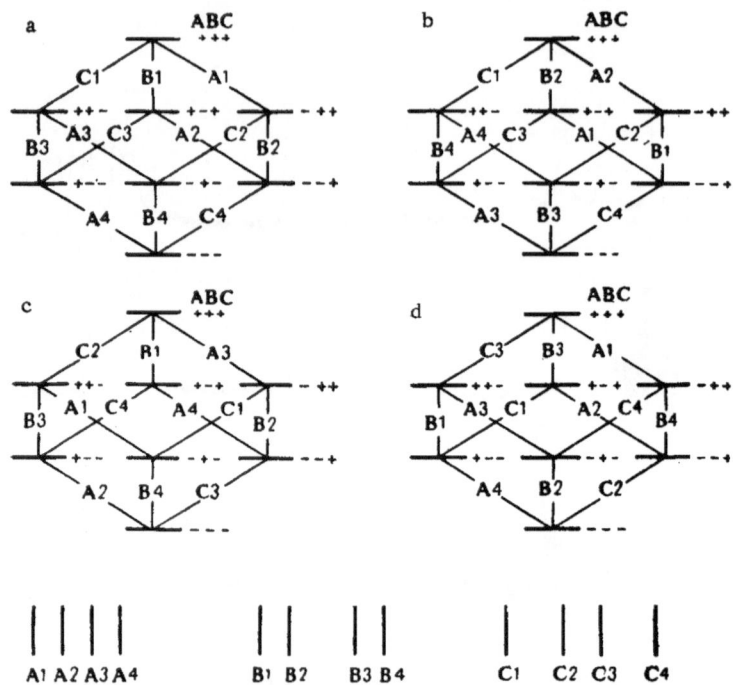

Fig. IV-32. Energy level diagrams of three-spin systems of nuclei with a spin of $1/2$. $\gamma > 0$, but H_0 oriented along the negative direction of the z axis and $|J_{AB}| < |J_{AC}| < |J_{BC}|$. a) All signs of the constants J the same; b) sign of J_{AB} opposite; c) sign of J_{AC} opposite; d) sign of J_{BC} opposite.

struction of a diagram of the energy levels of the system and the transitions between them (spectral lines) [36] (see Fig. IV-32 and p. 189). Tickling is the main method of constructing these diagrams in the interpretation of complex spectra [37]. In degenerate systems there is splitting of lines into a large number of components [31, 37].

Tickling is always accompanied by a change in the intensities of the spectral lines (Fig. IV-33) [38]. The leveling of the populations of the energy levels of the perturbed line leads to a decrease in the difference in populations for the line $\Lambda = 0$ and to an increase for the line $\Lambda = 2$ (Fig. IV-30). This nuclear Overhauser effect accompanies all experiments on nuclear magnetic double resonance with the condition that $\gamma_X^2 H_2^2 T_1 T_2^* \geq 1$. In this respect the data in [32] are inaccurate.

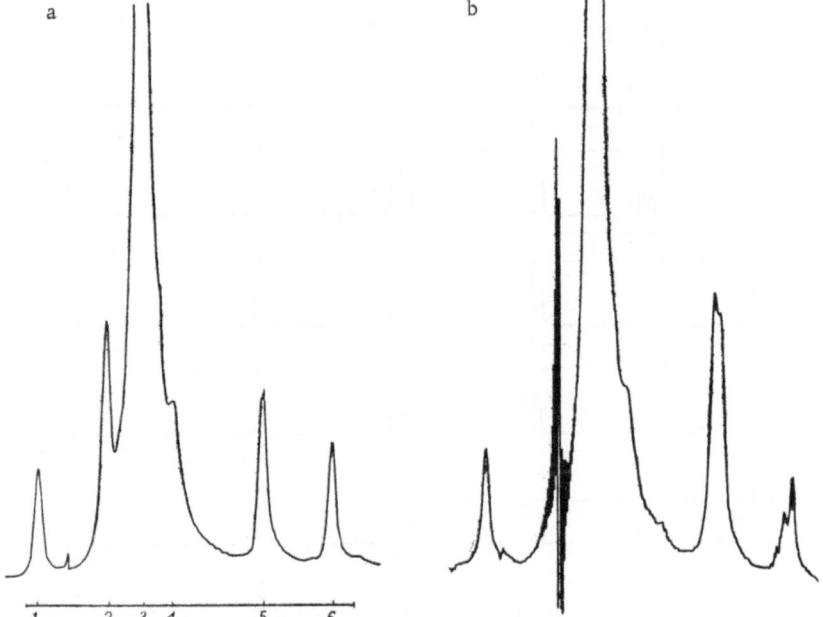

Fig. IV-33. Nuclear Overhauser effect in ethyl cinnamate: a) single resonance spectrum, lines 1, 2, 5, and 6 correspond to the AB system of unsaturated protons; b) perturbation of line 2, the intensity of a line of type $\Lambda = 2$ (5) is increased and that of a line with $\Lambda = 0$ (6) is decreased.

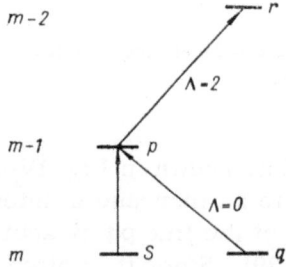

Fig. IV-34. Diagram of energy levels in experiment on transition selective irradiation.

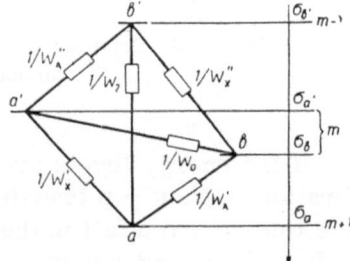

Fig. IV-35. Electrical analog of population of two-spin system.

The intensity of a spectral line $(p \rightarrow r)$ in an NMR spectrum is proportional to the difference in populations $(\sigma_p - \sigma_r)$ of the levels p and r.

$$L_{rp} = \gamma_A H_1 | < p | I_+ (A) | r > |^2 (\sigma_p - \sigma_r) \qquad \text{(IV-52)}$$

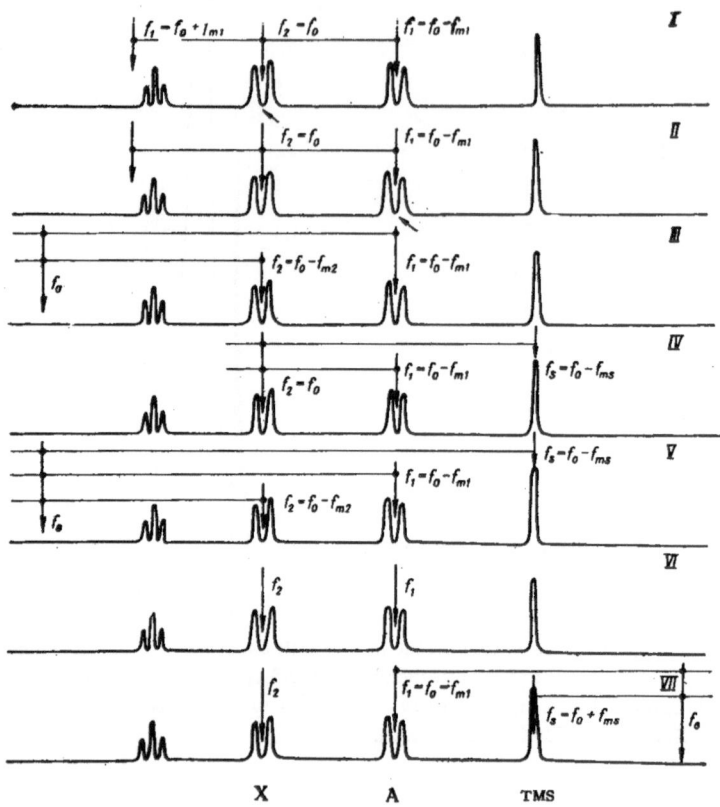

Fig. IV-36. Determination of perturbing and measuring rf fields
for double resonance.

If the energy levels are arranged as shown in Fig. IV-34, then
saturation of the s → p transition leads to an increase in intensity
of the line rp and a fall in the intensity of the line pq by approximately 50% in nondegenerate systems [39]. Since it is often pos-
sible that the conditions $\gamma_X^2 H_2^2 T_1 T_2^* \gg 1$ and $\gamma_X H_2 < 2/T_2^*$ hold simultaneously, then the Overhauser effect may be observed with-
out a substantial change in the wave functions, energy levels, and
frequencies of the spectral lines in the spectrum of the system.
Saturation or inversion of the populations by rapid adiabatic passage of one line in an AX system leads to the appearance of new
populations (different from equilibrium populations). The differences of these populations, which determine the intensities of the

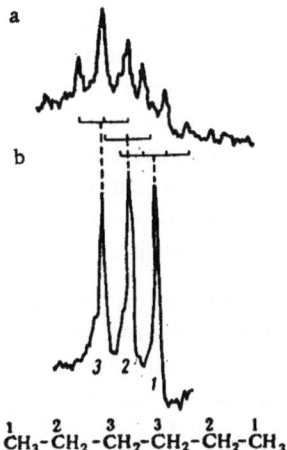

$$\overset{1}{CH_3}-\overset{2}{CH_2}-\overset{3}{CH_2}-\overset{3}{CH_2}-\overset{2}{CH_2}-\overset{1}{CH_3}$$

Fig. IV-37. Spectra of C^{13} nuclei of n-hexane (a) with decoupling of the H^1 nuclei (b).

NMR signals, change practically instantaneously for transitions which have common energy levels with the perturbed line, while the other populations change much more slowly. This difference is the basis of the method of transition selective irradiation (TSI) [40] for determining the relative arrangement of energy levels and the relative signs of spin—spin coupling constants.

New populations in the whole spin system are established as a result of relaxation transitions between levels so that the higher the corresponding probabilities of a transition, the more rapidly is the new equilibrium established. An analogy may be drawn between this phenomenon and the phenomenon of the flow of a current in an electric circuit: the probabilities of transitions correspond to the conductivity and the populations are analogs of potentials. If we assume that the system is linear and the probabilities of relaxation transitions are independent of irradiation, then the two phenomena are completely analogous and the spin system may be described by Kirchhoff's law [22]. Figure IV-35 gives the electrical analog of a two-spin system, which accurately conveys the dependence of the steady values of the populations on perturbation. To simulate the dynamics of the process it is necessary to introduce capacitors [41]. From Fig. IV-35 it follows that the magnitude and sign of the Overhauser effect depends on the relaxation mechanisms, which determine the relative values of $1/W$. From Kirchhoff's laws it follows that since W_0, $W_2 \neq 0$, then the greatest effect will be observed on lines which have energy levels common with the perturbing line and the sign of the effect depends on the character of the transition (on the arrangement of the transition on the energy level diagram).

The Overhauser effect is used for determining the relative arrangement of energy levels [39, 42-44] and also for determining the relative probabilities of transitions [45-47]. An example, which is very special, but is of practical importance, is the interpretation of the spectrum of the thallium derivative of norbornadiene [48] with the aid of four-frequency resonance. The Overhauser effect is used for interpreting the complex splittings produced by long-range spin—spin coupling of different protons with thallium. The Overhauser effect may also be used for determining the steric proximity of protons which are not directly spin—spin coupled [49] and for investigating the microstructure of solutions of organic compounds by the intermolecular double resonance method [50].

The practical application of the double resonance methods listed requires more complex apparatus than for obtaining qualitative single resonance spectra. Collapse may be realized most simply. In the simplest variants the rf field of the spectrometer is used to perturb the multiplet of the nucleus X, while the modulation side band of this field is used as the measuring field. The magnetic field is modulated with a frequency f_{m1}, which equals the difference in the resonance frequency of the perturbed and investigated multiplets, the spectrum is plotted by sweeping the magnetic field, and a low-frequency phase detector is used (Fig. IV-36, II) [17, 51]. This method is used widely, but it has the drawback that another modulation band may appear in the spectrum and interfere with the work. It is better if the measuring and perturbing fields have side modulation bands, using the modulation frequencies f_{m1} and f_{m2} from 1 to 5 kHz [52] (Fig. IV-36, III, V). With this procedure, in addition to sweeping the magnetic field it is also possible to have a frequency sweep if it is possible to change one of the frequencies smoothly, for example, by using a low-frequency oscillator with a motor drive. Both procedures make it possible to achieve collapse and selective collapse, i.e., to carry out experiments with simplification of spectra, to determine which multiplets are connected, which is particularly important in the analysis of natural compounds that usually give poorly resolved spectra, and to determine the relative signs of spin—spin coupling constants [53, 54]. For tickling or measurement of the Overhauser effect it is desirable to use a frequency sweep and internal stabilization (Fig. IV-36, V and VII). The magnetic field H_0 is modulated with a third low frequency f_{ms} so that its side band is in resonance with the line of an internal reference such as tetramethylsilane. The dispersion signal of tetramethylsilane from an appropriate phase detector is fed to a superstabilizer or spin generator for compensation of small changes and drift of the magnetic field so that a long-term stability of $1 \cdot 10^{-9}$ is achieved [31, 37, 55]. In this case the spectrum is usually plotted with a frequency sweep by means of a tunable low-frequency oscillator. For this purpose it is necessary to use a high-quality low-frequency oscillator (for example, GZ-18) so that the drift and instability of the frequency does not exceed 0.1 Hz during the plotting of the spectrum.

The new types of NMR spectrometers are fitted with various forms of spin stabilization (the use of a spin generator for this purpose has an advantage in the suppression of high-frequency interference) and also units for double resonance. These instruments usually use modulation methods to obtain the required frequencies, but since the modulation index depends on the frequency, some quantitative inaccuracies may arise in the spectra. In this respect a synthesis of all the required frequencies is more convenient [51, 56, 57]. Synthesis is also used frequently in instruments for heteronuclear double resonance with perturbation of the nuclei P^{31}, B^1, B^{11}, F^{19}, C^{13}, N^{14}, D^2, etc., for simplifying a proton spectrum, and for indirect determination of the chemical shifts of these nuclei and the relative signs of the spin—spin coupling constants. On the other hand, decoupling of the H^1 nuclei in the investigation of a C^{13} spectrum considerably increases the sensitivity of this method (Fig. IV-37). For simple spin decoupling in heternuclear double resonance it is often sufficient to use a powerful stable rf oscillator with a range of frequencies from 3 to 30 MHz.* Thus, for work involving double resonance methods it is necessary to have an additional one to three very stable low-frequency oscillators, one of them with a motor drive, one or two phase detectors, one rf generator, and a digital frequency meter for accurate measurement of the frequencies used.

Literature Cited

1. A. D. Cohen, N. Sheppard, and J. J. Turner, Proc. Chem. Soc., 1958:118.
2. J. D. Swalen and C. A. Reilly, J. Chem. Phys., 37:21 (1962).

---

*There are also commercial synthesizers of the required frequencies which have high stability, for example, PDG-1 Messelektronik, Berlin, DDR. Universal instruments for homo- and heteronuclear decoupling and tickling with all possible combinations of nuclei are produced in several variants by Nuclear Magnetic Resonance Specialties Co., Inc., Box 145, Greensbury Road, New Kensington, Pennsylvania, USA. The SD-60 is intended for heteronuclear and the PD-60 for homonuclear decoupling.

3.   D. R. Whitman, J. Chem. Phys., 36:2080 (1962).

4.   P. L. Corio, Chem. Rev., 60:363 (1960).

5.   J. I. Musher and E. J. Corey, Tetrahedron, 18:791 (1962).

6.   J. W. Emsley, J. Feeney, and L. H. Sutcliffe, High Resolution Nuclear Magnetic
      Resonance Spectroscopy, Vol. 1, Pergamon, London (1965).

7.   J. Pople, W. Schneider, and H. Bernstein, High-Resolution Nuclear Magnetic
      Resonance, McGraw Hill, New York (1959).

8.   J. D. Roberts, Spin–Spin Splitting in High-Resolution NMR Spectra, Benjamin,
      New York (1961).

9.   D. K. Faddeev and V. N. Faddeeva, Computational Methods of Linear Algebra,
      Fizmatgiz (1963).

10.   S. Castellano and J. S. Waugh, J. Chem. Phys., 34:295 (1961).

11.   Liang Xiao-Tian, Scienta Sinica, 8:589 (1964).

12.   B. Gestblom, R. A. Hoffman, and S. Rodmar, Mol. Phys., 8:425 (1964).

13.   B. Gestblom, R. A. Hoffman, and S. Rodmar, Acta Chem. Scand., 18:1222
      (1964).

14.   D. M. Grant, R. C. Hirst, and H. S. Gutowsky, J. Chem. Phys., 38:470 (1963).

14a. J. D. Roberts, Nuclear Magnetic Resonance: Applications to Chemistry, McGraw,
      New York (1959).

15.   P. F. Cox, J. Am. Chem. Soc., 85:380 (1963).

16.   B. Dischler and W. Maier, Z. Naturforschung, 16a:318 (1961).

17.   J. D. Baldeschwieler and E. W. Randall, Chem. Rev., 63:81 (1963).

18.   A. Abragam, Principles of Nuclear Magnetism, Oxford Univ. Press, New York
      (1961).

19.   A. Leshe, Nuclear Induction [Russian translation], IL, 1963.

20.   C. P. Slichter, Principles of Magnetic Resonance, Harper and Row, New York
      (1963).

21.   E. Andrew, Nuclear Magnetic Resonance, Cambridge Univ. Press (1956).

22.   F. Bloch, Phys. Rev., 102:104 (1956).

23.   J. D. Baldeschwieler, J. Chem. Phys., 40:459 (1964).

24.   W. A. Anderson, Phys. Rev., 102:151 (1956).

25.   S. L. Gordon, J. Chem. Phys., 45:1145 (1966).

26.   J. G. Powles, Rep. Progr. Phys., 22:433 (1959).

27.   K. G. Untch and R. J. Kurland, J. Mol. Spectr., 14:156 (1964).

28.   K. C. Ramey, N. D. Field, and I. Hasegawa, J. Polymer Sci., B2:865 (1964).

29.   J. Parello, A. Melera, and R. Goutarel, Bull. Soc. Chim. France, 1963:898.

30.   J. P. Maher and D. F. Evans, Proc. Chem. Soc., 1961:208.

31.   W. A. Anderson and R. Freeman, J. Chem. Phys., 37:85 (1962).

32.   A. L. Bloom and J. N. Shoolery, Phys. Rev., 97:1261 (1955).

33.   R. Freeman, Mol. Phys., 5:499 (1962).

34.   H. Shimizu and S. Fujiwara, J. Chem. Phys., 34:1501 (1961).

35.   S. L. Manatt, D. D. Elleman, and S. J. Brois, J. Am. Chem. Soc., 87:2220(1965).

36.   R. A. Hoffman, B. Gestblom, and S. Forsen, J. Mol. Spectr., 13:221 (1964).

37.   R. Freeman and W. A. Anderson, J. Chem. Phys., 37:2053 (1962).

38.   É. Lippmaa, Yu. Puskar, M. Alla, and A. Syugis, Izv. Akad. Nauk ÉstSSR, ser.
      fiz.-mat. i tekhn. nauk, 14:306 (1965).

39.   R. A. Hoffman, B. Gestblom, and S. Forsen, J. Chem. Phys., 40:3737 (1964).

40.   R. A. Hoffman, B. Gestblom, and S. Forsen, J. Chem. Phys., 39:486 (1963).

41.   P. R. Solomon, Rev. Sci. Instr., 36:1130 (1965).

42.   V. J. Kowalewski, D. G. de Kowalewski, and E. C. Ferra, J. Mol. Spectr., 20:203 (1966).

43.   R. Kaiser, J. Chem. Phys., 39:2435 (1963).

44.   K. K. Kuhlmann and J. D. Baldeschwieler, J. Am. Chem. Soc., 85:1010 (1963).

45.   J. H. Noggle, J. Chem. Phys., 43:3304 (1965).

46.   V. Sinivee and É. Lippmaa, Izv. Akad. Nauk ÉstSSR, ser. fiz.-mat. i tekhn. nauk, 14:258, 564 (1965); 15:64 (1966).

47.   V. Sinivee, Izv. Akad. Nauk ÉstSSR, ser. fiz.-mat. i tekhn. nauk, 15:182 (1966).

48.   F. A. Anet, Tetrahedron Letters, 1964:3399.

49.   M. C. Woods, J. Miura, Y. Nakadaira, A. Terahara, M. Maruyama, and K. Nakanishi, Tetrahedron Letters, 1967:321.

50.   É. Lippmaa and M. Alla, Izv. Akad. Nauk ÉstSSR, ser. fiz.-mat. i tekhn. nauk, 15:476 (1966).

51.   R. Freeman and D. H. Whiffen, Proc. Phys. Soc., 79:794 (1962).

52.   R. Freeman, Mol. Phys., 3:435 (1960).

53.   J. Parello, Bull. Soc. Chim. France, 1964:2033.

54.   E. B. Whipple and Y. Chiang, J. Chem. Phys., 40:713 (1964).

55.   J. H. Noggle, Rev. Sci. Instr., 35:1166 (1964).

56.   A. Syugis and É. Lippmaa, Izv. Akad. Nauk ÉstSSR, ser. fiz.-mat. i tekhn. nauk, 16:81 (1967).

57.   É. Lippmaa, Zh. Strukt. Khimii, 8:731 (1967).

Chapter V

NMR Spectra and the Structure
of Organic Molecules

In previous chapters we examined the rules of the changes in the main NMR parameters, namely, the chemical shifts of magnetic nuclei and the spin–spin coupling constants between them, in relation to various structural factors of the molecule and the external conditions under which the spectra were plotted, and we also gave accounts of methods for analyzing the spectra to extract these parameters from the spectral data. The aim of the present chapter is to examine actual cases of the use of these methods and rules in studying various classes of organic compounds by means of NMR spectroscopy.

Among the proton-containing groups examined in the diagrams in Ch. I the most important role in the elucidation of the structure of molecules is played by methyl and methylene groups since they have characteristic intense signals and are present in an overwhelming majority of organic compounds. Therefore, before turning to the examination of the spectra of actual compounds, it is worthwhile to define more precisely the chemical shifts of these groups in relation to the immediate environment. To facilitate structural assignments, diagram V-1 shows the chemical shifts of methyl groups in the form of several sections, each of which is subdivided further, depending on the more remote substituents. In the case of methylene groups, which bear two substituents each, the number of variants is increased markedly. In the diagram V-2 the characteristic groups are given in order of the shifts of the methylene group signal to lower fields [1].

Chemical shifts of the methyl group

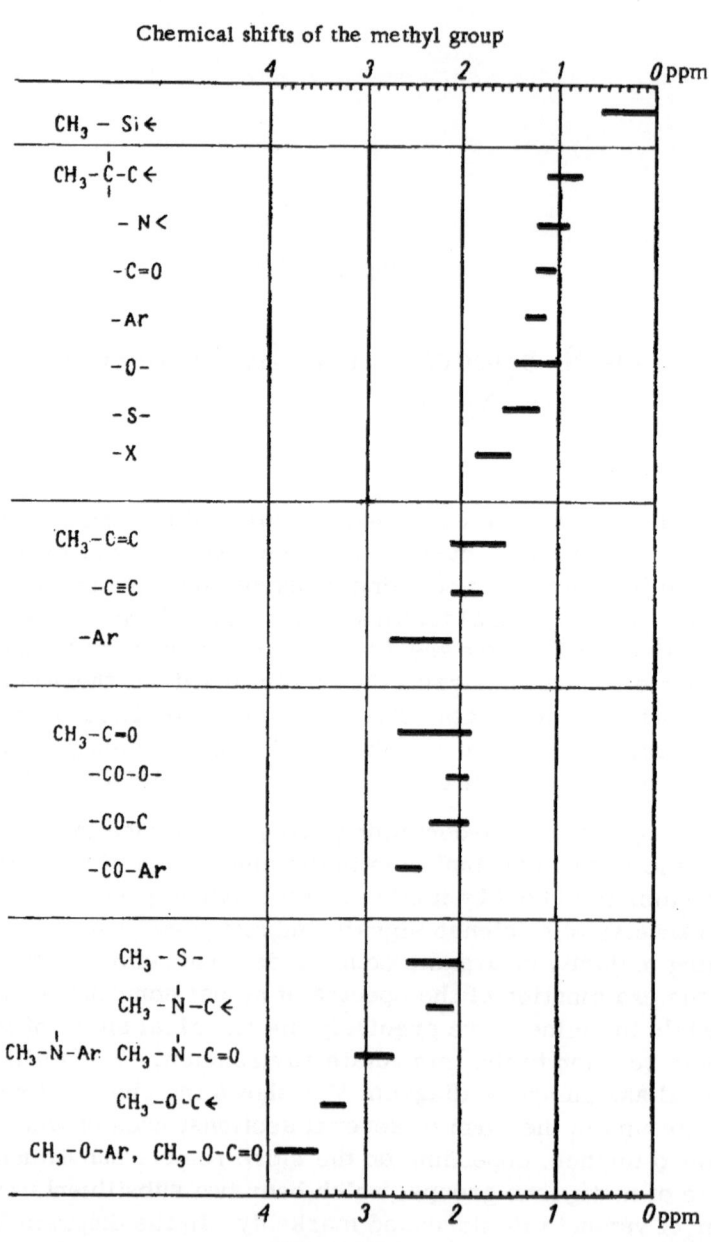

Diagram V-1.

Chemical shifts of the methylene group

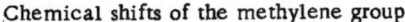

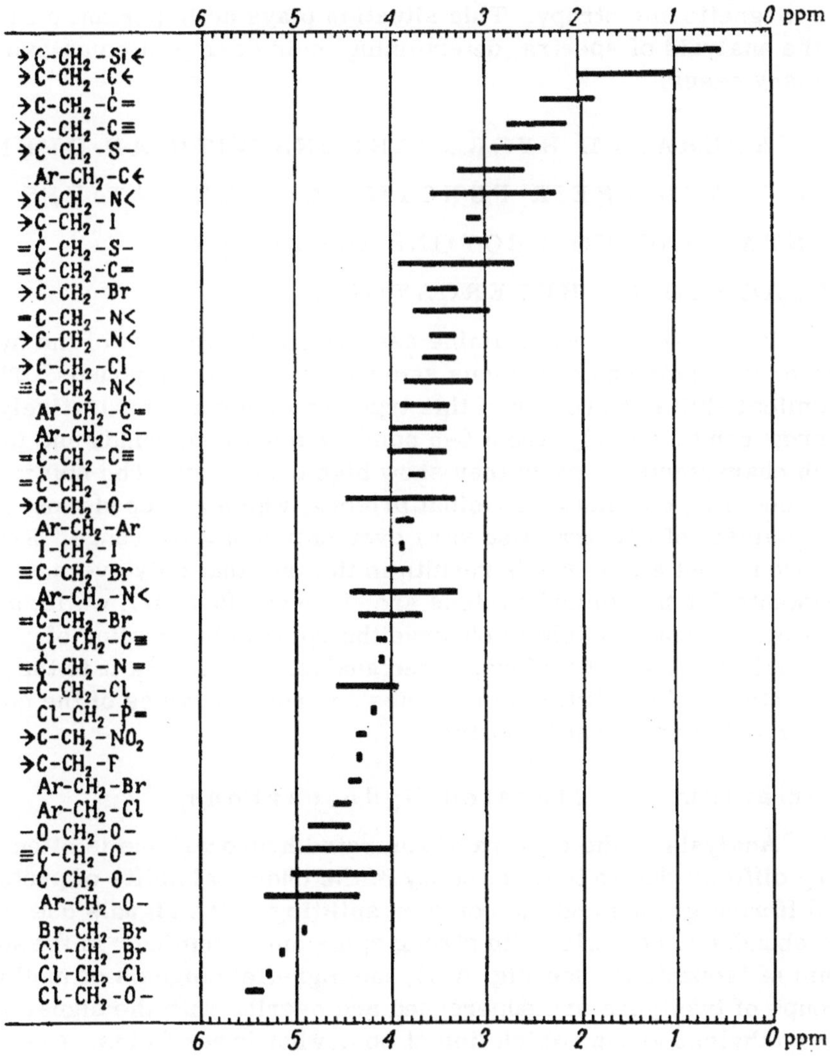

Diagram V-2.

As the diagrams presented show, the regions of the chemical shifts of methyl and methylene groups overlap to a considerable extent. The chemical shifts of these groups change appreciably when they are attached to a heteroatom or a group which has magnetic anisotropy. This situation plays an important part in the analysis of spectra, determining their degree of complexity in many cases.

1. SATURATED HYDROCARBONS WITH AN OPEN CHAIN AND THEIR FUNCTIONAL DERIVATIVES CONTAINING NO PROTONS DIRECTLY ATTACHED TO HETEROATOMS

In this section we examine essentially the spectra of the hydrocarbon skeletons of various saturated organic compounds. The chemical shifts of protons of this type vary over a comparatively narrow range, namely, from 0-5 ppm for most compounds and for each characteristic group they show high constancy. The spin—spin coupling constants of vicinal protons, which are used most in the analysis of spectra, also vary over only a narrow range. Free rotation about single bonds results in the fact that only average constants for the vicinal protons are observed (6-8 Hz), while, as a rule, it is not possible to observe the spin—spin coupling of geminal protons. Therefore, in the analysis of spectra it is the character of the splitting of the lines and not the values of the constants which are most important.

Isomerism of Saturated Hydrocarbons

Analysis of the spectra of saturated hydrocarbons is often very difficult due to the similarity in the chemical shifts of protons of different groups and the complex splitting of the signals due to the abundance of nuclei with strong spin—spin coupling. In the spectrum of isopentane (see Fig. A-1), the high-field signals of methyl groups of two types are superposed and overlap with the signal of the methylene group, which lies at somewhat lower fields. Although the signal of the tertiary proton is at still lower fields, it is strongly split and overlaps the signal of methylene protons.

The presence of a definite symmetry in hydrocarbon molecules substantially simplifies their spectra. Thus, in the spectra

of unbranched paraffins (see Fig. A-2) it is possible to pick out
signals of two types, namely those of methyl groups lying at the
ends of the molecule and those of methylene groups. The signals
of the methyl groups in such compounds do not give clear triplets
due to the high value of the ratio $J/\Delta\nu$ of the methyl and methylene
groups. The signal of the methylene groups usually consists of a
broad band, whose intensity in comparison with the intensity of the
terminal methyl groups may be used to estimate the chain length
of normal alkanes. For the simplest compound of this type, namely,
propane, it was possible to carry out a complete analysis of the
spectrum ($A_3A'_3B_2$ type) using a computer [2] and this gave the following results: δ (CH_3) 0.912 ± 0.005; δ (CH_2) 1.350 ± 0.005 ppm;
J_{CH_3, CH_2} 7.35 ± 0.02 Hz.

If the molecule gives no other peaks in the same region, the
isopropyl radicals in the groups $(CH_3)_2CH-C$ give characteristic
doublets with coupling constants of ~5.5 Hz with the low-field peaks
of the doublets usually more intense due to the relatively small
difference in the chemical shifts of CH_3 and CH groups (the ratio
$J/\Delta\nu$ at 40 MHz is ~0.25). The spectra of hydrocarbon skeletons
are simplest when tert-butyl radicals are present as these give
intense single peaks in the high-field region at ~0.8-0.9 ppm.

The elucidation of the isomerism of saturated hydrocarbons
of complex structure is simplified if resonance at C^{13} nuclei is
used with the natural content of the isotope with suppression of
spin-spin coupling with protons (Ch. II and IV) since the chemical shifts of C^{13} depend strongly on the number and type of the substituents (see Fig. II-17). However, investigations of this type essentially are still in the experimental stage.

Derivatives of Saturated Hydrocarbons
Containing Heteroatoms

The introduction into a hydrocarbon chain of heteroatoms
which are more electronegative than carbon leads to a substantial
shift in the signal of a neighboring proton group to lower fields,
considerably simplifying the analysis of the spectra. As a characteristic example we will examine the spectrum of 1-bromo-3-
chloropropane (see Fig. A-3). At lower fields there are the signals
of methylene groups attached to halogens and each of them is split
into a triplet due to spin-spin coupling with protons of the central

TABLE V-1. NMR Spectral Parameters of Ethyl Groups

Compound	δ_{CH_2}	δ_{CH_3}	J	$J/\nu_{CH_2} - \nu_{CH_3}$	
				at 40 MHz	at 60 MHz
$(C_2H_5)_3Al$	—	—	8.01	—0.278	—0.186
$(C_2H_5)_3AlCl$	—	—	8.12	—0.247	—0.166
$(C_2H_5)_2AlBr$	—	—	8.07	—0.277	—0.184
$(C_2H_5)_2AlI$	—	—	8.11	—0.338	—0.225
$(C_2H_5)_2C(CH_3)_2$	1.206	0.798	7.52	0.461	0.308
$(C_2H_5)_4C$	1.171	0.728	7.53	0.426	0.284
$C_2H_5C\equiv N$	2.311	1.289	7.60	0.187	0.124
$C_2H_5C_6H_5$	2.606	1.224	7.62	0.139	0.092
$C_2H_5COCH=CH_2$ . . .	2.541	1.067	6.97	0.078	0.052
$C_2H_5OC(O)CH_3$	4.046	1.232	7.12	0.061	0.041
$(C_2H_5)_2O$	2.425	0.970	7.18	0.123	0.082
C_2H_5I	3.155	1.059	7.45	0.089	0.059
C_2H_5Br	3.366	1.677	7.33	0.108	0.072
C_2H_5Cl	3.505	1.488	7.23	0.089	0.060

methylene group. Despite the fact that these triplets overlap, no
distortion of the form of the signal occurs since the spin–spin
coupling of protons of these groups, which are separated by four
ordinary bonds, is negligibly small. According to an estimate of
the chemical shifts from the Shoolery constants (see Table II-4),
the triplet which lies at lower fields corresponds to the methylene
group attached to chlorine atoms. The quintuplet in the high-field
region (δ 2.28 ppm) corresponds to the central methylene group;
the nature of the splitting is due to the fact that the spin–spin
coupling constants with the other two methylene groups are equal.

The introduction of heteroatoms which have a positive inductive effect shifts the signal of a neighboring proton group to higher
fields. In most cases this leads to considerable complication of
the spectra due to the overlapping of the signals of methylene
groups and the increase in multiplicity due to the increase in the
ratio $J/\Delta\nu$. In this respect the form of the signal of protons of
ethyl groups attached to groups of different character is very significant. The NMR spectral parameters given in Table V-1 for
ethyl groups were obtained by computer analysis of A_3B_2 spectra
for most of the compounds.

A theoretical examination of A_3B_2 spectra (see Ch. IV) leads
to the conclusion that only when $J/\Delta\nu$ does not exceed 0.25 is it
possible to avoid a complete analysis of the spectrum and separate
the resonance regions of the methyl and methylene parts of the

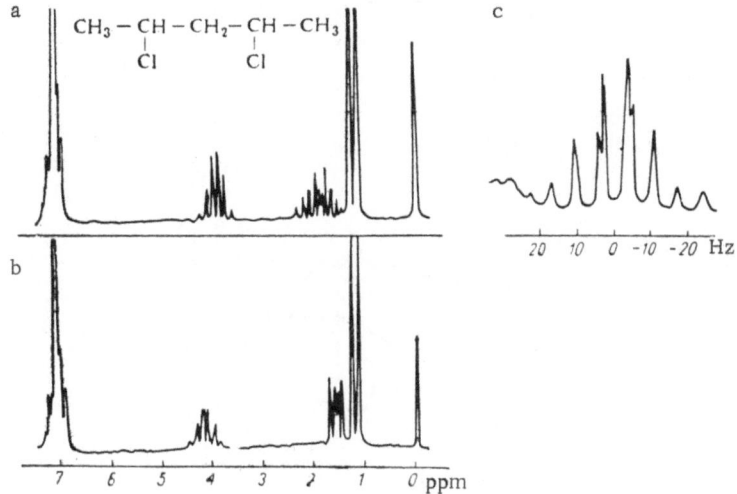

Fig. V-1. Spectra of optical isomers of 2,4-dichloropentane: a) d*l*-2,4-dichloropentane; b) meso-2,4-dichloropentane; c) methylene protons with increased resolution.

ethyl radicals. Among the substances listed in Table V-1, the spectra of compounds lying below propionitrile may be analyzed by first-order rules, while substances lying above it require quantum mechanical analysis of the spin system. Propionitrile itself is essentially an intermediate compound. While the spectrum at 60 MHz may be assigned approximately to first-order spectra, the' spectrum at 40 MHz requires complete analysis.

In conclusion we will examine the spectra of optical isomers of 2,4-dichloropentane (Fig. V-1) [3]. The spectra of both isomers contain intense doublets with δ = 1.54 ppm and J = 6.7 Hz, which correspond to methyl groups, multiplets of methylene groups at ~2-2.3 ppm, and signals of CH groups at ~4.2 ppm. Examination of the steric structure of these compounds shows that in both stable conformations of the d*l*-isomer the protons of the methylene group are completely equivalent to each other. This fact and also the high value of the chemical shifts of the groups of protons in comparison with the coupling constants and the absence of appreciable spin—spin coupling between protons of the methyl and methylene groups make it possible to treat the signal of the methylene protons as part of an AA'XX' spectrum. Another part of this spectrum, namely,

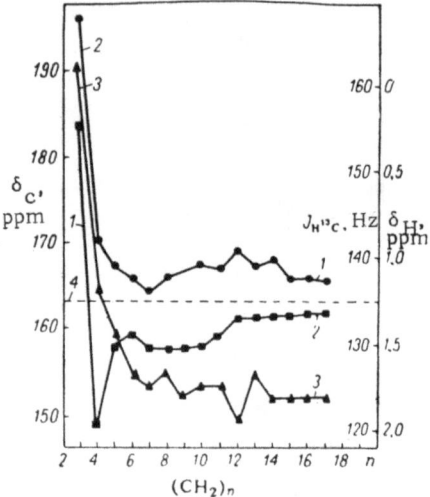

Fig. V-2. Chemical shifts of C^{13} and $H-C^{13}$
spin—spin coupling constants J_{HC} in cycloparaffins: 1) chemical shifts of H^1 (on δ scale);
2) chemical shifts of C^{13} relative to the external
standard CS_2; 3) constants J_{HC} (in Hz); 4) chemical shifts of H^1 and C^{13} of methylene groups in
higher normal paraffins.

the signal of CH protons, is complicated due to additional spin—
spin coupling with methyl protons. In both stable conformers of
meso-form (these are mirror images of each other) the two methylene protons are in different positions relative to the CH protons.
Strictly speaking, the four protons should form a system of four
nonequivalent protons ABXY, but the energy equivalence of the two
conformers and the absence of hindrance to their interconversions
simplifies the spectrum to the ABX_2 type, whose AB part consists
of the protons of the methylene group. Analysis of the signal of the
CH protons in this case is naturally difficult due to spin—spin
coupling with the three methyl protons. The difference between the
two types of spectra makes it possible to distinguish the two optical isomers of dichloropentane.

Even a cursory glance shows that the signal of the methylene
group in spectrum 1 (dl-isomer) has the typical form of half an
AA'XX' spectrum (see Fig. V-1). More accurate analysis of the
two spectra, taking into account the fact that the chemical shifts

and coupling constants are still commensurate values and comparing the experimental and theoretical spectra lead to the following parameters:

dl-2,4-Dichloropentane: ν_{AX} 141.3 Hz; J_{AX} 2.1-2.8 Hz; $J'_{AX'}$ 10.6-9.9 Hz. meso-2,4-Dichloropentane (20% solution in CCl_4): ν_{CH_2CH} 119.1 Hz; ν_{AB} 16.7 Hz; J_{AB} 14.3 Hz; $\frac{1}{2}$ (J_{AX} + J_{BX}) 7.0 Hz.

The investigation of the stereochemistry of organic compounds is an important problem in NMR spectroscopy. The great value of this method is revealed particularly clearly in the study of the conformations of cyclic compounds.

2. SATURATED CARBO-
AND HETEROCYCLIC COMPOUNDS

The NMR spectra of cyclic compounds are very similar to the spectra of hydrocarbons and their derivatives with an open chain. However, the presence of the closed ring superposes certain peculiarities, which appear even on comparing the NMR spectra of the methylene groups in cycloparaffins (Fig. V-2) [4]. The presence of high strain in the three-membered ring of cyclopropane is accompanied by an increase in the p-character of the C−C

TABLE V-2. Spectral Parameters of Methylene Groups and C^{13} Satellites of Three-Membered Rings

Compound	Chemical shift, ppm	Spin−spin coupling constants, Hz			
		J_{cis}	J_{trans}	J_{gem}	J_{HC}
H_2C——CH_2 $\diagdown$ $\diagup$ CH_2	0,3	9,2	6,2	−5,6	161
H_2C——CH_2 $\diagdown$ $\diagup$ O	2,54	4,45	3,1	—	175,8
H_2C——CH_2 $\diagdown$ $\diagup$ S	2,27	7,15	5,65	—	170,5
H_2C——CH_2 $\diagdown$ $\diagup$ N $\mid$ H	1,36	6,3	3,8	—	168,1

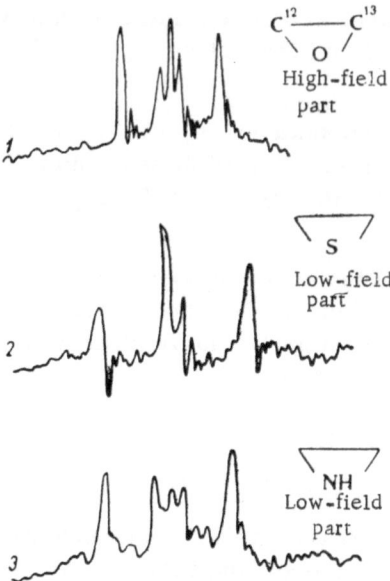

Fig. V-3. Spectra of the C^{13} satellites of three-membered heterocycles: 1) ethylene oxide (high-field part of spectrum); 2) ethylene sulfide (low-field part of spectrum); 3) ethyleneimine (low-field part of spectrum).

bond and consequently an increase in the s-character of the C−H bond. The latter leads to such an increase in the constant J_{CH} that in cyclopropane it exceeds J_{CH} of a benzene ring. The increased double-bond character of the C−C bond results in the appearance of effective ring currents and this leads to appearance of considerable magnetic anisotropy of the three-membered ring so that in this case there is increased screening of the protons and C^{13} nuclei. An increase in the size of the ring leads to a smoothing out of these differences and the spectral parameters of higher cycloparaffins do not differ substantially from those of compounds with an open chain.

NMR Spectra of Three-Membered Rings

The signals of the methylene protons of three-membered rings lie at considerably higher fields than those of the corresponding protons in open chains. The rigid structure of these compounds

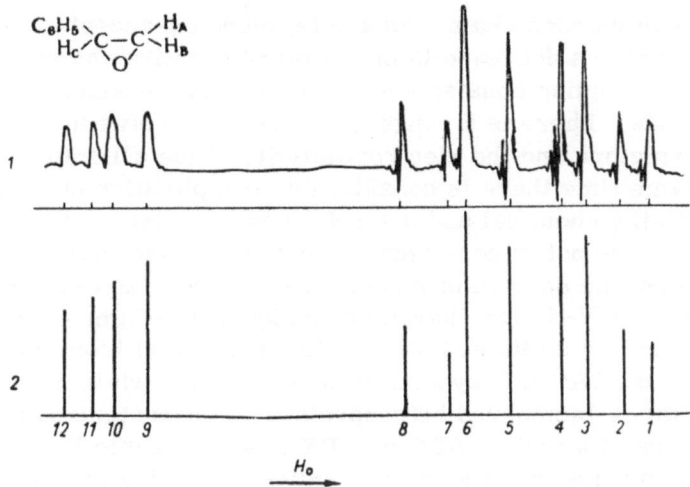

Fig. V-4. Spectrum of the protons of the epoxide ring in styrene oxide:
1) experimental spectrum; 2) theoretical spectrum calculated from the
following parameters: δ_A 2.51; δ_B 2.81; δ_C 3.51 ppm; J_{AB} 5.65; J_{AC}
2.42; J_{BC} 4.10 Hz; all the spin—spin coupling constants have the same
sign.

Frequencies and Intensities of Transitions in the Spectrum of Styrene Oxide

Line No.	Observed		Calculated (ABC)	
	frequency	intensity	frequency	intensity
A_1	95.60	0.134	95.66	0.133
A_2	98.19	0.162	98.16	0.153
A_3	101.37	0.331	101.31	0.339
A_4	103.83	0.364	103.82	0.376
B_5	108.05	0.132	108.06	0.320
B_6	112.11	0.380	112.07	0.395
B_7	113.70	0.141	113.72	0.118
B_8	117.70	0.174	117.73	0.168
C_9	141.20	0.285	141.21	0.298
C_{10}	143.74	0.272	143.71	0.264
C_{11}	145.23	0.239	145.22	0.231
C_{12}	147.68	0.203	147.73	0.207

leads to the appearance of three spin—spin coupling constants with
different values. In symmetrical molecules these were determined
from the spectra of the C^{13} satellites, which are spectra of the
AA'XX' type (Fig. V-3) [5].

The introduction of substituents produces a considerable
change in the parameters, which obeys the same rules as in com-

pounds with an open chain. As a rule, electron-acceptor substituents produce a decrease (a shift toward negative values) in the spin-spin coupling constants and a shift in the proton signals to lower fields. There is a rough linear relation between the spectral parameters and the electronegativity of the substituents [6, 7]. At the same time there is considerable complication of the spectra due to the chemical and magnetic nonequivalence of the protons which are not in equivalent positions relative to the substituents. Thus, the spectra of monosubstituted derivatives of cyclopropanes (see V-4) are characterized by at least nine parameters (three chemical shifts and six coupling constants) from the partly symmetrical AA'BB'C system of nuclear spins, while monosubstituted three-membered heterocycles are characterized by six parameters of a typical ABC or ABX system. Table V-2 gives the spectral parameters of some molecules containing three-membered rings.

The spectrum of styrene oxide is a classical example of an ABC spectrum (Fig. V-4) [8]. The spectrum contains 12 basic lines, which are grouped in three quartets.

It is possible to assign the lines after assessing the character of the substituents and by comparison of the spectra of the substances of similar structure.

Strained Rings with More Than Three Atoms

With a change to cyclic compounds with a large number of atoms in the ring the signals of the methylene protons are shifted sharply toward lower fields (see Fig. V-2) and the chemical shifts approach the normal values which are characteristic of compounds with an open chain. Thus, even in the peculiar derivative of the cyclic hydrocarbon cubane I, the protons at the points of the trihedral angle give a single peak, which is characteristic of a tertiary proton in compounds with an open chain [9]. In complex molecules the signal of these protons overlap, hampering the analysis of the spectra.

I

For cyclic compounds of this type the spin—spin coupling constant for protons and C^{13} are of great importance and they have certain characteristic peculiarities [10].

Constants J_{HC}, Hz

As is well-known, the high value of the constants J_{HC} indicates the high s-character of this bond. Though this criterion is not definitely valid for all cases, it is used for qualitative estimation of the character of an orbital in substances of similar structure. It is characteristic that high values of J_{HC} correspond to protons attached to carbon atoms which lie at the points of the trihedral angles, where the increased s-character of the C—H bond is indicated by other considerations.

It is interesting to note the substantial difference in the constant J_{HC} for exo- and endo-methylene protons of bicyclobutane (170 and 152 Hz). The low value of J_{HC} in the second case may be explained by the bond strain, which is produced by the repulsion of the two end-protons due to their close steric arrangement. The result may be disruption of the normal bond angles and a consequent redistribution of the character of the orbitals, which is reflected in the values of J_{HC}.

Constants J_{HH}, Hz

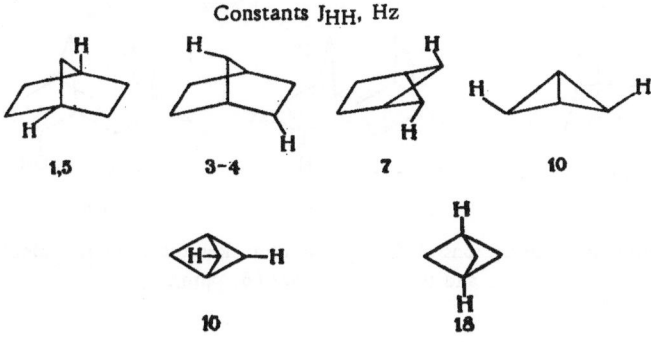

A peculiarity of the proton—proton coupling constants is the fact that high values are observed in cases where the two protons are in the exo-position so that more effective overlapping of small parts of their hybrid C—H orbitals is possible (see Ch. III).

The disruption of the normal bond angles or the s t e r i c c o m p r e s s i o n effect leads in individual cases to considerable deviations from the normal values of the chemical shifts and spin— spin coupling constants of protons (II-X). In this respect the chemical shifts of protons attached to carbon atoms in the skeleton of compounds with half-cage or endo-, endo-bridge-cage structures are very significant [11].

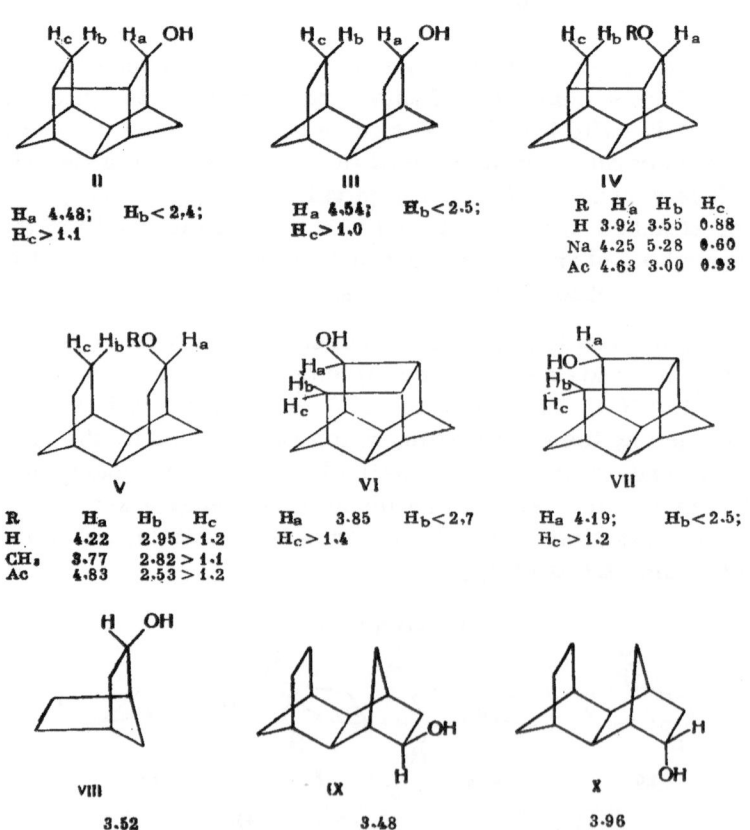

II

H$_a$ 4.48;   H$_b$ < 2.4;
H$_c$ > 1.1

III

H$_a$ 4.54;   H$_b$ < 2.5;
H$_c$ > 1.0

IV

R	H$_a$	H$_b$	H$_c$
H	3.92	3.55	0.88
Na	4.25	5.28	0.60
Ac	4.63	3.00	0.93

V

R	H$_a$	H$_b$	H$_c$
H	4.22	2.95 > 1.2	
CH$_3$	3.77	2.82 > 1.1	
Ac	4.83	2.53 > 1.2	

VI

H$_a$ 3.85   H$_b$ < 2.7
H$_c$ > 1.4

VII

H$_a$ 4.19;   H$_b$ < 2.5;
H$_c$ > 1.2

VIII

3.52

IX

3.48

X

3.96

Chemical shifts of protons of the hydrocarbon skeleton in cyclic alcohols and their derivatives (δ, ppm).

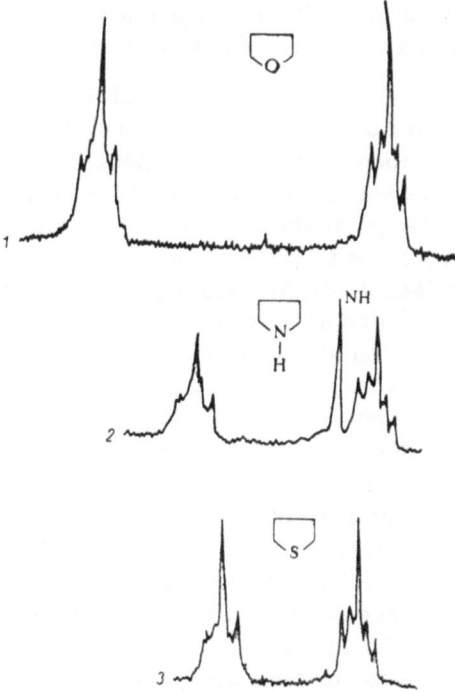

Fig. V-5. Spectra of five-membered saturated
heterocycles: 1) tetrahydrofuran; 2) tetrahydropyrrole; 3) tetrahydrothiophene.

In all cases the compression effect leads to a shift of the
signal toward lower fields in comparison with substances where
this effect is insignificant (for example, VIII-X). In the case of
compounds III and V, which do not have a cage structure, compression has less effect since there is the possibility of partial
compensation for this effect by distortion of other bond angles.
It is characteristic that the strain is also reflected in the change
in the chemical shifts of the protons H_C in the exo position, possibly due to a deviation from the normal position of the whole system of bonds H_C-C-H_b. Another surprising peculiarity is the
fact that the greatest effect is produced by the group O^- (compounds
IV-Na), which would seem at first glance to be smaller in volume
than, for example, an acetoxyl group. This result indicates the importance of electronic interactions which are responsible for steric
effects in comparison with purely m e c h a n i c a l repulsion, as a

result of which the steric effect of a negative alkoxyl ion with three unshared electron pairs at the oxygen is greater.

The parameters of spectra of 4- and 5-membered saturated heterocycles are similar to the parameters of the corresponding compounds with an open chain. In the spectrum of trimethylene oxide (see A-13) the signal at lower fields corresponds to protons of the methylene groups at the oxygen atom. The chemical shift of these protons is close to the chemical shift of the α-methylene protons in ethers (see A-4), but has a greater multiplicity due to the inequality of the constants of the cis- and trans-protons since there is no free rotation about the $C-C$ bonds. A similar effect is observed in the spectra of tetrahydrofuran, tetrahydropyrrole, and tetrahydrothiophene (Fig. V-5). The complete analysis of these spectra is difficult due to the abundance of magnetically nonequivalent protons with unequal chemical shifts. In contrast to these compounds, in the spectrum of cyclopentane there is a narrow single peak (1.51 ppm), which is due to the chemical equivalence of the ten ring protons. This situation and also the chemical inertness and the absence of appreciable magnetic anisotropy of the cyclopentane ring make this compound a convenient internal reference in the investigation of substances which do not resonate in this region.

Six-Membered Carbocyclic and Heterocyclic

Saturated Compounds

The study of the spectra of six-membered rings plays an important part in the investigation of many natural compounds (sugars, steroids, terpenes, etc.), containing cyclic structural elements and also ring compounds with a larger number of atoms in the ring. A peculiarity of many six-membered rings is the fact that they undergo rapid conformational conversion (at normal temperatures), leading to averaging of the parameters of the spectra belonging to the individual conformers.

Six-membered rings contain magnetically nonequivalent protons in axial and equatorial positions and this leads to different spin−spin coupling constants for the vicinal protons, namely, J_{aa} (axial−axial), J_{ee} (equatorial−equatorial), and J_{ae} or J_{ea} (axial−equatorial or vice versa). Naturally, in cyclohexane itself at room temperature there is only a singlet because the transitions between

the two equivalent chair conformations leads to transitions of the
axial protons to the equatorial position and vice versa (XI).

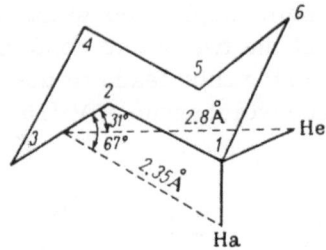

XI

However, even in the simplest derivatives of cyclohexane it
is possible to observe multiplicity due to this nonequivalence.
Analysis of the spectra of a large number of six-membered cyclic
compounds indicates the following main rules.

1. In most cyclic compounds the signals of the axial protons
occur at a higher field than the corresponding equatorial protons.

This property is connected with the magnetic anisotropy of
carbon-carbon bonds (Ch. II), whose magnitude is of the order of
$\Delta \chi - 5 \cdot 10^{-30}$ cm^3 per molecule. While the contribution due to the
bonds 1-2 and 1-6 are the same for axial and equatorial protons,
the bonds 2-3 and 5-6 screen the axial protons H_a and descreen
the equatorial proton [12]. Calculation by McConnell's equation
(Ch. III), using the given values for the anisotropy and the geometric parameters leads to a difference in screening due to the contribution from the two bonds of ~0.20 ppm, which agrees well with
experimental data.

Geometric parameters of the cyclohexane ring

The rule given is broken when the molecule contains a substitutent which has a considerable effect in the opposite direction.

2. The signals of axial protons attached to carbon atoms adjacent to a carbonyl group (in compounds such as cyclohexanone or

acetoxycyclohexane) may appear at lower fields than the equatorial protons [13]. This deviation is readily explained from the point of view of the theory of the reaction field (Ch. II).

The component of the electrostatic field of the dipole of the molecule oriented along the C$-$H bond is in the opposite direction in trans-XII and cis-2-bromocyclohexanone XIII and this leads to screening of the equatorial proton H_e and descreening of the axial proton H_a.

Screening of protons in isomeric 2-bromocyclohexanes

3. The spin$-$spin coupling constants of vicinal protons in fixed cyclohexane systems lie in the following range (in Hz) [14]:

$$J_{aa} = 10-12.5; \quad J_{ee} = 3.5-4.5; \quad J_{ae} \approx 2.7$$

4. In mobile cyclohexane systems it is possible to observe only the average spin$-$spin coupling constants of vicinal protons. With equal populations of the two stable conformations of the chair-form of the cyclohexane ring this leads to the following results (using trans-1,3-dibromocyclohexane XIV [15] as an example):

$$J_{14} \text{ (av)} = J_{23} \text{ (av)} = \frac{1}{2}(J_{aa}' + J_{ee}) = J = 6.1 \text{Hz (6--7Hz)}$$

$$J_{13} \text{ (av)} = J_{24} \text{ (av)} = \frac{1}{2}(J_{ae} + J_{ea}) = J_{ae} = J' = 3.7 \text{Hz (3.5--4.5 Hz)}$$

The signal of protons 3 and 4 is usually resolved inadequately. The width of the signal at $^1/_4$ the height is determined by the equation

$$W_{1/_4} = J_{aa} + 2J_{ae} + J_{ee} = 19.6 \text{ Hz } (19-22 \text{ Hz})$$

The spectral parameters change when the populations of the conformations are different and this may be used for conformational analysis.

In substituted six-membered rings the spin—spin coupling constants of the vicinal protons vary in relation to the electronegativity of the substituent. These changes are also determined by the orientation of the substituent relative to the protons examined [16] and there appears the so-called effect of "trans-coplanarity" of the substituent (Booth's rule), which results in the effect of the substituent with the steric arrangement XV being greater than in XVI.

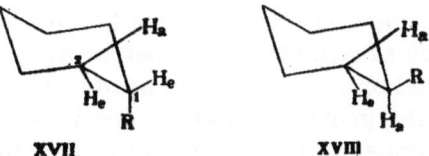

5. If R is an electronegative substituent attached directly to the cyclohexane ring, the effect of the electronegativity on J_{HHvic} is maximal in structure XVII and not XVIII. In other words J_{e1a2} in XVII is less than in XVIII (under otherwise equal conditions).

XVII   XVIII

6. If an electronegative atom X (for example, O or N) is part of a saturated six-membered heterocycle in the chair conformation (XIX) then the effect of the electronegativity of this atom on J_{HHvic} appears at the proton H_e in position 3, i.e., J_{a2e3} is less than J_{e2a3}.

XIX

Analysis of the spectra of cyclohexane derivatives is simpler with the presence of substituents, which reduce the number of protons in the molecule and lead to unequal chemical shifts for the remaining protons. Determination of the stereochemistry of isomeric trans-2-(p-chlorophenyl)-cis-3-nitro-cis-1,4-cyclohexanedicarboxylic acid (Fig. V-6a) and cis-2-(p-chlorophenyl)-trans-3-nitro-cis-1,4-cyclohexanedicarboxylic acid (Fig. V-6b) was possible by first-order analysis of the spectra of these compounds determined at 60 MHz [17].

In the high-field region (2.2 ppm) there are the signals of the protons at carbon atoms 5 and 6, which do not bear substituents. Analysis of this part of the spectrum is difficult due to the similarity in the chemical shifts of these four protons. In the low-field region lie the signals of the phenyl protons (7.4 ppm) and carboxyl protons (not given in the spectra). The remaining protons form comparatively simple spectra, which make it possible to establish the structure of the stereoisomers unequivocally.

The lines in the spectra may be assigned to definite nuclei on the basis of the following arguments. The broad signals at higher fields are produced by protons at the carboxyl groups (1 and 4). The broadening of these signals is due to the presence in the vicinity of each proton of three other protons, which produce complex unresolved splitting of these bands. In the low-field region lie the signals of the protons at carbon atoms 3, which are attached to nitro groups (5.28 and 6.10 ppm) (see additive components of chemical shift, Ch. I and II). In the region of 4.11 and 3.58 ppm lie the multiplets of the protons in position 2.

It is characteristic that in the spectrum (Fig. V-6a) the signal of the proton H_3 appears at higher fields and that of H_2 at lower fields than the signals of the corresponding protons in the second spectrum (Fig. V-6b). This may be explained by the effect of the reaction field of the carboxyl groups and, consequently, in the first case the proton H_2 lies cis to the carboxyl group, while spectrum

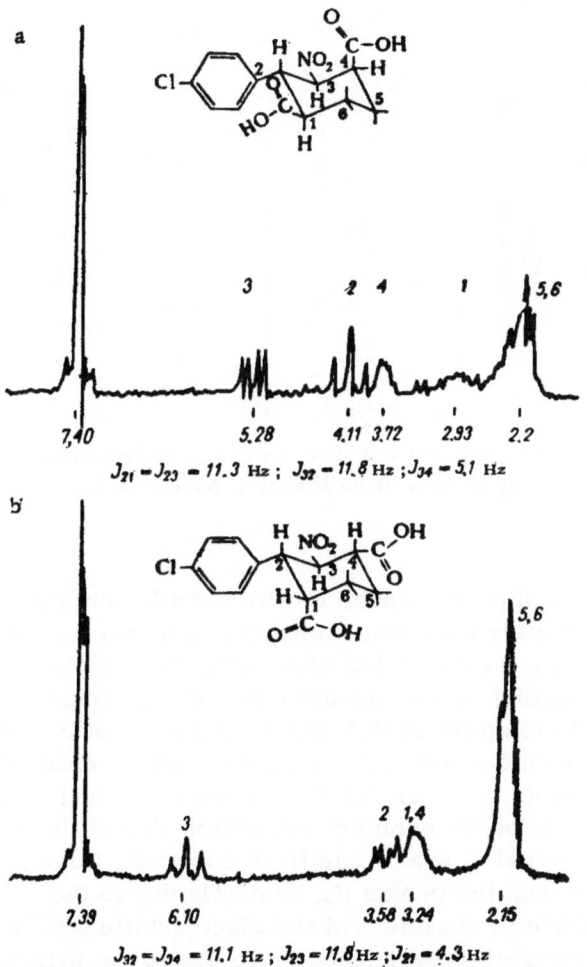

Fig. V-6. Spectra of derivatives of cyclohexanedicarboxylic acid: a) trans-2-(p-chlorophenyl)-cis-3-nitro-cis-1,4-cyclohexanedicarboxylic acid; b) cis-2-(p-chlorophenyl)-trans-3-nitro-cis-1,4-cyclohexanedicarboxylic acid.

(b) indicates the cis arrangement of the carboxyl groups and the proton H_3. Confirmation for this may be found by examining spin—spin splitting. In the spectrum (a) the signal of H_2 is a triplet with a splitting of 11.3 Hz, which is characteristic of protons in an axial position; the form of the signal is due to the two approximately equal

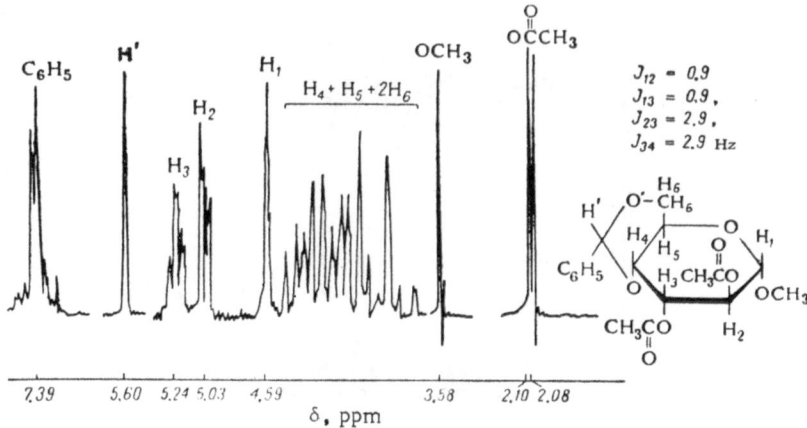

Fig. V-7. Spectrum of methyl-2,3-di-O-acetyl-4,6-O-benzylidene-α-D- altro-
pyranoside. Index H.60.Ц + A Z.I + M. B.

constants for spin—spin coupling with the adjacent protons H_3 and
H_4. Approximately the same splitting is observed in the spectrum
of the proton H_3 (11.8 Hz), but each component of this signal is also
split into a doublet due to the presence of the proton H_4 and the
magnitude of the splitting (5.1 Hz) is characteristic of axial—equatorial spin—spin coupling of vicinal protons. In exactly the same
way the isomeric structure of the second compound agrees with
the triplet form of the signal of the proton H_3 and the quartet of H_4.
Finally, the signal of the proton H_1 in spectrum (a) appears at
higher fields than the proton H_4, evidently due to the fact that it is
less susceptible to the effect of the electrostatic field of the nitro
group. This signal is broadened more due to the high value of J_{12}
in comparison with the corresponding constant J_{34} and this explains
the smaller broadening of the band of H_4. In the spectrum of the
second isomer the two protons at the atoms bearing the carboxyl
groups are equivalent.

Many carbohydrates of the pyranose type also bear protons,
which lead to the appearance of first-order NMR spectra. The
spectra of methyl-2,3-di-O-acetyl-4,6-O-benzylidene-α-D-altro-
pyranoside (Fig. V-7) and methyl-3-O-acetyl-4,6-O-benzylidene-
2-deoxy-α-D-arabinohexopyranoside (Fig. V-8) at 60 MHz provide
examples of well-resolved spectra, which make it possible to es-

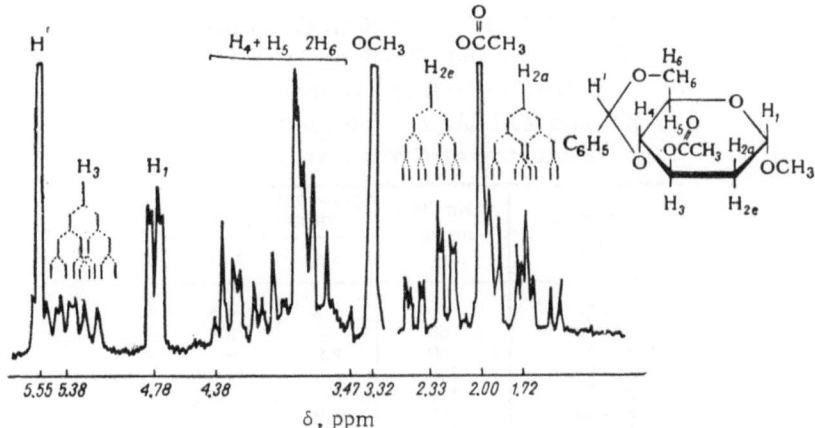

Fig. V-8. Spectrum of methyl-3-O-acetyl-4,6-O-benzylidene-2-deoxy-α-D-arabinohexopyranoside. Index H.60.Ц + A Z.I ✦ M.B.

tablish the chair conformations **XX** and **XXI** for these compounds [18].

Both spectra contain intense signals due to the protons of the benzene ring (not given in the spectrum in Fig. V-8) and methyl and acetyl groups. The signals of the protons at carbon atoms 4, 5, and 6 are poorly resolved and are not used in the analysis of the conformations of these substances. At the same time, the signals of the protons H_1, H_2, and H_3 form simple first-order spectra making it possible to determine the spin—spin coupling constants between them with high accuracy directly from the spectrum without resorting to complex analysis. The assignment of the bands shown in the spectrum was derived from a comparison of a large number

TABLE V-3. Spin—Spin Coupling
Constants and Dihedral Angles
of H−C−C−H in Compounds
XX and XXI (Assuming Idealized
Geometry of Cyclohexane)

Position of nuclei	Dihedral angle, deg	Spin–spin coupling constants, Hz	
		XX	XXI
1e—2e	60	0.9	1.1
1e—2a	60	—	3.8
2e—·3e	60	2.9	—
2a—3a	180	—	11.5
2e—3a	120	—	5.8
3e—4a	60	2.9	—
3a—4a	180	—	9.4

of spectra of related compounds. For most substances of this type the signal of the proton H_3 is in a lower field region than the equatorial proton H_1. The signals of the protons H_2 in the spectrum of XXI (Fig. V-8) are at much higher fields due to the absence of an electronegative oxygen atom at the second carbon atom of this compound.

An important result is given by an examination of the spin—spin coupling of these three protons. In the spectrum of XX (Fig. V-7) the signal of H_1 is only very weakly split into a triplet (J < 0.9 Hz). The triplet form of the signal may be due to the superposition of doublets with approximately equal splitting, produced by spin—spin coupling of the proton H_1 with H_2 and H_3 (long-range coupling). An examination of the signals ot H_2 and H_3 confirms this conclusion. The spin—spin coupling constant of the protons H_2 and H_3 is split into a triplet due to spin—spin coupling with the proton H_4 with the same constant. Spin—spin coupling constants of similar magnitude are also observed for the same three protons in the spectrum of XXI (Fig. V-8); moreover, additional splitting is observed here and this is produced by the presence of the proton at the second carbon atom in the axial position. A comparison of the magnitudes of the constants and the dihedral angles of H−C−C−H for different conformations leads to the conclusion that the conformations of XX and XXI agree more with the spectra (Table V-3).

As the table shows, the low values of the constants (0.9–3.8 Hz) correspond to protons with a g a u c h e orientation, while the maximal constants (9.4–11.5 Hz) arise with a t r a n s o i d orientation, in complete agreement with Karplus's theory (Ch. III, p. 132). Small deviations may be caused by strain in the rings produced by the presence of bulky substituents and also by the direct effect of electronegative substituents and their orientation on the values of spin—spin coupling constants of vicinal protons. The spin—spin coupling constants of the geminal protons $H_{2a} - H_{2e}$ in compound XXI is of the usual order.

3. OLEFINIC AND ACETYLENIC COMPOUNDS

The identification of the signals of olefinic and acetylenic protons is important in the NMR spectroscopy of organic compounds because the resonance of these protons obeys quite strict rules and may give valuable information both on the structure of the molecules and on their electron distribution. The signals of olefinic protons lie in the region of 4.5–8 ppm. The spectra of unsaturated compounds vary in accordance with the main rules given below.

1. Substituents with an electronegative atom which have a positive conjugation effect (for example, hydroxyl and amino groups) cause the appearance of the signal of olefinic α-protons at much lower fields than the β-protons. Electronegative substituents with a negative conjugation effect (nitro group) may give the opposite effect. The chemical shift of olefinic β-protons correlates roughly with the Hammett constant of the substituent in the α-position.

2. The descreening effect of anisotropic substituents has a much stronger effect on the chemical shift of cis-protons than on trans-protons. As a rule, the signals of olefinic protons in the cis-position to an electronegative substituent appear at lower fields than those of protons in the trans-position.

3. The spin—spin coupling constants of olefinic protons in different positions differ substantially in magnitude and are characterized by great constancy. For most olefinic compounds, the spin—spin coupling constants vary over the following ranges (in Hz):

$$J_{cis} = 8 - 13; \quad J_{trans.} = 14 - 18; \quad J_{gem} = 0 - 2$$

All three constants are positive (absolute sign).

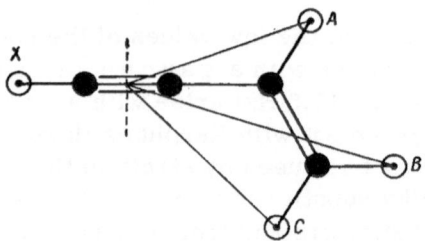

Fig. V-9. Geometric structure of vinylacetylene. Bond lengths and angles in vinylacetylene:
A—C ≡C 1.20; C =C 1.34; ≡C —C = 1.42; =C —H
1.086; ≡C —H 1.055; =C angle —120°.

4. Electronegative substituents produce a decrease in the
spin—spin coupling constants, and for geminal protons the constant
may pass into the region of negative values. The effect of a substituent on a constant is higher with a trans arrangement than with
a cis arrangement (Booth's rule).

Disubstituted olefins form comparatively simple spectra of
the AB type, whose interpretation presents no difficulty, particularly if the substituents differ markedly in electronegativity. The
NMR spectra of olevinic protons provide the most convenient method
of determining the cis-trans configuration of olefins. A typical example is the spectrum of a mixture of cis- and trans-1-dimethyl-
amino-5-ethylhepten-1-yn-3-ol-5 [19] (see Fig. A-27). The spectrum contains two typical AB quartets, which are due to the ethylene protons in the cis- and trans-products. The relative intensity
of these signals and also the signals of the protons of the dimethylamino groups make it possible to estimate the composition of the
mixture of isomers.

The low-field olefinic proton doublet of the cis-compound appears at higher field than the corresponding doublet in the trans-
compound. This is evidently due to the anisotropy of the triple
bond, which is stronger with the trans-arrangement.

Disubstituted olefins containing a terminal =CH$_2$ group may
also form a spectrum of the AB type, but with a much lower spin—
spin coupling constant. A characteristic spectrum of this type is
given by the terminal protons of α-deuterostyrene (Ch. IV). Here
also the form of the spectrum is determined to a considerable ex-

tent by the magnetic anisotropy of the benzene ring, which leads to greater deshielding of the cis-proton than the proton in the trans-position relative to the benzene ring, producing an appreciable difference in the chemical shifts of these protons. In many other cases the terminal vinyl protons give only a single peak.

Unsaturated compounds with a monosubstituted vinyl group give typical spectra of the ABC type. Although the complete analysis of such spectra presents considerable difficulties and generally requires the use of computer techniques, for ethylene compounds it is substantially facilitated by the possibility of estimating the spectral parameters (chemical shifts and spin−spin coupling constants) quite accurately beforehand by using the rules presented above. When one of the protons of ethylene is replaced by a group with an electronegative atom, as a rule the spectra have a simple form and may be analyzed by the ABX approximation.

The vinyl group in vinylacetylene derivatives forms a complex ABC spectrum, which is distinguished by small chemical shifts of the protons. A preliminary estimate of the spectral parameters of these compounds may be made starting from the effect of the anisotropy of the triple bond. Below we given an example of such a calculation using the data in Table II-3 and the usual geometric parameters of such molecules with the aid of McConnell's point dipole approximation (Ch. II). The calculation data may be obtained directly by a graphical method (Fig. V-9).

Geometric parameters for calculating the values of σ_{an} of vinylacetylene

Nucleus	r, Å	$r^3 \cdot 10^{24}$, cm	$\cos \Theta$	σ_{an}
H_A	2.72	20.12	0.0960	+0.76
H_B	3.96	62.10	0.953	+0.25
H_C	2.98	26 46	0.711	+0.17
H_X	1:65	4 19	1.000	+4.00 (starting value)

However, the results obtained only make it possible to estimate the relative chemical shift for the protons H_B and H_C, while the signal of the proton H_A appears at lower fields than the signals of the other two vinyl protons due to the effect of the electronegativity of the nearest sp-hybridized carbon atom. Complete analysis of the spectrum of vinylacetylene and α-deuterovinylacetylene led to the following parameters [20, 21]:

$$\delta_A \ 4.25; \quad \delta_B \ 4.02; \quad \delta_C \ 4.20 \text{ ppm} \quad J_{AB} \ 11.5; \quad J_{AC} \ 17.3; \quad J_{BC} \ 2.0; \quad J_{AX} \ -2.1;$$
$$J_{BX} \ 0.8-0.9; \quad J_{CX} \ 0.7 \text{ Hz}.$$

It is characteristic that the difference in the chemical shifts of the protons H_B and H_C corresponds exactly to the difference in the contributions from the anisotropy of the triple bond (0.18 ppm). This indicates that the electronic effects on the chemical shifts of the cis- and trans-protons are equal.

The long-range spin–spin coupling of the acetylene and vinyl protons, which is observed in vinylacetylene, usually appears in the spectra of vinylacetylene derivative substituted in the vinyl group. In the spectrum of cis-1-diethylaminobuten-1-yne-3 (see Fig. A-28), the long-range coupling constant J_{AX} is 2.5 Hz. In addition to the typical signal of an ethyl group (a quartet and a triplet), the spectrum contains a characteristic AB quartet of vinyl protons. The low-field part of this quartet belongs to the proton in the vicinity of the electronegative nitrogen atom. The components of the higher-field part are also split into a doublet due to long-range coupling with the acetylene proton, whose signal is also split. In the spectrum it is possible to observe only one of the doublets of the high-field part of the AB quartet, while the other one overlaps the quartet of the CH_2 groups. However, these data are already sufficient for a complete analysis of the spectrum. Essentially first-order analysis is permissible for this spectrum.

If there is a methylene group adjacent to a monosubstituted vinyl group as, for example, in allyl compounds, the analysis of this spectrum, which belongs to the $ABCD_2$ type, is much more complicated.

The results of investigating the NMR spectra of isomeric nitropropenes are interesting [22] (Fig. V-10).

$$CH_2{=}CH{-}CH_2{-}NO_2$$
$$BC \quad A \quad D_2$$

XXII

$$\begin{array}{c} H_3C \\ X \end{array}{\diagdown}{}\begin{array}{c} H_B \\ \diagup \end{array}$$
$$C{=}C$$
$$\begin{array}{c} \diagup \\ H_A \end{array}\begin{array}{c} \diagdown \\ NO_2 \end{array}$$

XXIII

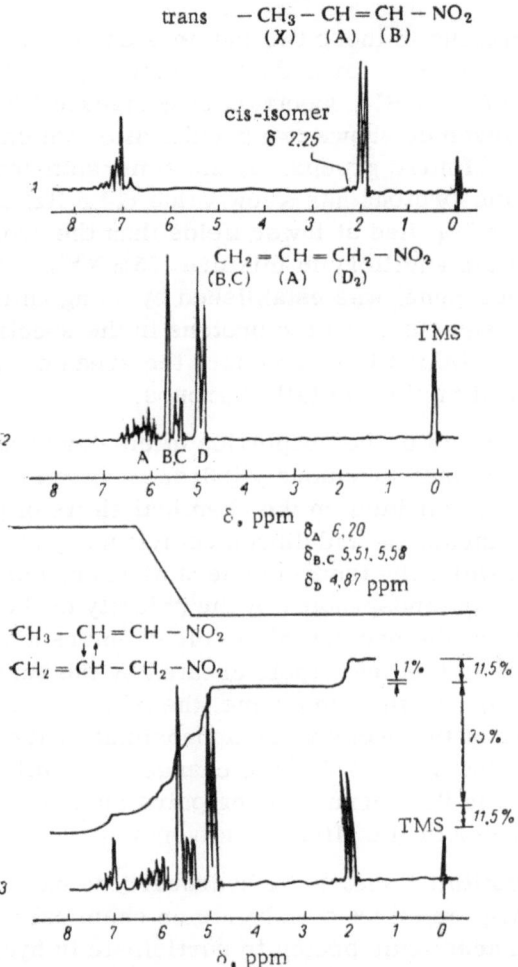

Fig. V-10. Spectra of isomeric nitropropenes. Index H.60.JI=M. 1) Trans-1-nitropropene with a small amount of the cis-isomer; 2) 3-nitropropene; 3) equilibrium mixture of isomeric 1- and 3-nitropropenes.

Although the complete analysis of the spectrum of the allyl isomer XXII has not been carried out in this case, the determination of the chemical shifts of the individual groups was found to be sufficient to elucidate the composition of the equilibrium mixture in the isomerization of the compounds XXII and XXIII, since the signals of the protons in these two isomers do not overlap. Hardly any of the cis-isomer is formed as indicated by the result of complete analysis of the ABX_3 spectrum of compound XXIII. The spectrum of this substance shows two peculiarities which are both due to the presence of nitro groups: 1) the spin—spin coupling constant of the trans-protons has a low value (14.5 Hz) and 2) the signal of the proton H_A lies at lower fields than the proton H_B. The composition of the equilibrium mixture (75% XXII, 23% XXIII, and 2% cis-1-nitropropene) was established by using an integrator for the area of the signals of all the protons in the spectrum. The same result is obtained if we compare the areas of only the signals of the methyl and methylene (allyl) groups.

The difference in the properties of cis- and trans-disubstituted olefins was used ingeniously [23] to investigate the effect of the polarity of the medium on the chemical shifts of protons. Both cis- and trans-dichloro- and dibromoethylenes give identical signals; however, while the trans-isomers are nonpolar and their chemical shifts are independent of the polarity of the medium, the chemical shifts of the protons of the cis-isomers are subject to the influence of an electric dipole created by the surrounding medium (see Ch. II). At the same time, the other contributions to the chemical shifts of the protons are approximately the same for the cis- and trans-isomers so that the change in the difference in the chemical shifts in the corresponding pairs made it possible to assess the character of the effect of solvents.

Monosubstituted acetylenic hydrocarbons have often been used for investigating the effect of solvents on chemical shifts. The tendency of an acetylenic proton to participate in hydrogen bonding has been used both for elucidating the properties of solvents and also for studying electronic interactions in conjugated acetylene systems [24].

The spectra of conjugated compounds with two or more multiple bonds in the molecule are of special interest. Although 1,3-diene hydrocarbons were examined in early work on NMR spectroscopy, the complete analysis of the spectrum of 1,3-butadiene and some of its derivatives was carried out only recently [25]. In the spectrum of unsubstituted 1,3-butadiene at 60 MHz it is possible to distinguish up to 90 lines. Preliminary data for analysis of the spectrum were obtained from the spectra of 2,3-dideutero- and

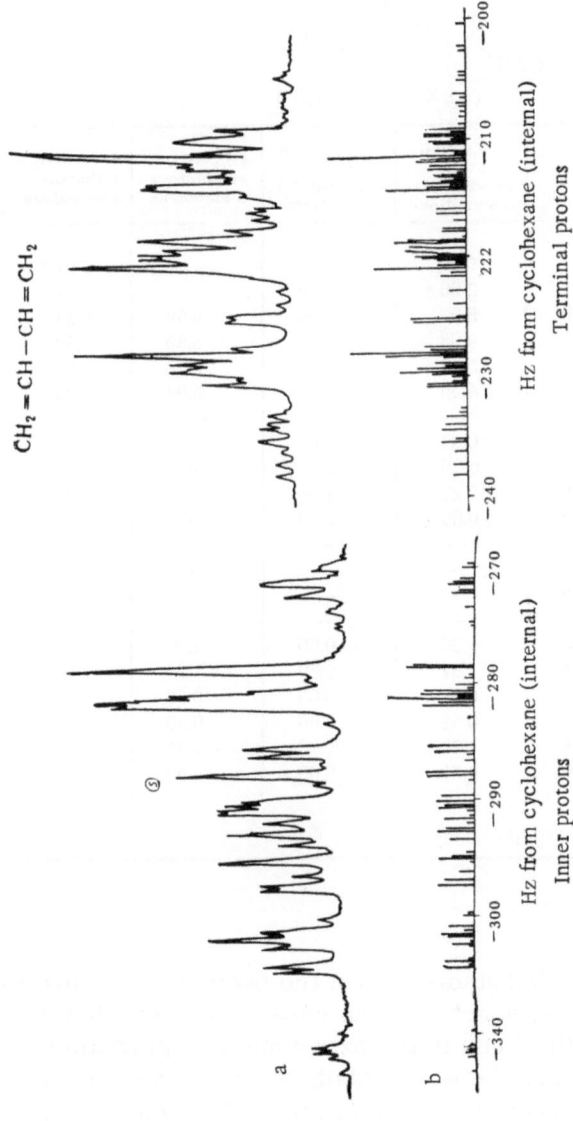

Fig. V-11. Spectrum of 1,3-butadiene (at 60 MHz). Index
H.60.JI=.M.KT. a) Experimental spectrum; b) theoretical spectrum.

TABLE V-4. Spectral Parameters of 1,3-Butadiene and Some of Its Derivatives

$$
\begin{array}{c}
\text{(B)H} \qquad\qquad \text{Y(A)} \\
\diagdown\kern-2pt\diagup \\
\text{C=C} \qquad \text{H(C')} \\
\text{(C)H} \qquad\qquad \text{C=C} \\
\text{(A')X} \qquad\qquad \text{H(B')}
\end{array}
$$

Parameter	X= Y= H	X= OCH₃, Y= H, 2-methoxy-1, 3-butadiene	X = C(CH₃)₃, Y= H, 2-tert-butyl-1, 3-butadiene	X= Y= Cl, 2,3 -dichloro- 1, 3-butadiene	X= Y= CH₃,2, 3-dimethyl-1, 3-butadiene
δ_A	6.27	6 06	6.36	—	1.01 *
$\delta_{A'}$	6.27	3.60 *	1.08 *	—	1.01 *
δ_B	5.08	5.02	4.94	5.59	4.89
$\delta_{B'}$	5.08	4.08	4.99	5.59	4.89
δ_C	5.17	5.49	5.32	5.99	4.99
$\delta_{C'}$	5.17	4.05	4.76	5.99	4.99
$J_{AA'}$	10.41	—	—	—	—
J_{AB}	10.17	10.82	10.80	—	—
$J_{AB'}$	−0.86	0.00	±1.00	—	—
J_{AC}	17.05	17.27	17.00	—	—
$J_{AC'}$	−0.83	0.00	±0.40	—	—
$J_{A'B}$	−0.86	—	—	—	—
$J_{A'B'}$	10.17	—	—	—	—
$J_{A'C}$	−0.83	—	—	—	—
$J_{A'C'}$	17.05	—	—	—	—
$J_{BB'}$	1.30	1.50	0.00	1.90	—
J_{BC}	1.74	1.87	2.30	−1.70	—
$J_{BC'}$	0.60	0.57	0.00	0.55	—
$J_{B'C}$	0.60	0.51	0.00	0.55	—
$J_{B'C'}$	1.74	−1.90	1.70	−1.70	—
$J_{CC'}$	0.69	0.51	0.00	0.59	—

*Chemical shifts for CH₃ groups.

1,1,4,4-tetradeutero-1,3-butadiene. On the basis of these preliminary data, a theoretical spectrum was constructed and this made it possible to assign the lines in the experimental spectrum to the corresponding transitions between steady states. Then the spectral parameters (the chemical shifts of the protons and the spin—spin coupling constants) were varied with an electronic computer until the best fit was obtained between the line frequencies in the experimental and theoretical spectra by the method of least squares.

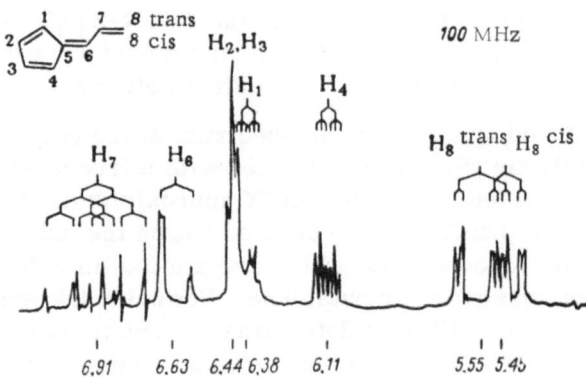

Fig. V-12. Spectrum of 6-vinylfulvene.
Index H.100.Ц + Л=.I + M.

TABLE V-5. Spin–Spin Coupling Constants of Ring Protons
of Fulvene Derivatives (in Dimethylsulfoxide)

Compound	J_{12}	J_{23}	J_{14}	J_{13}
Dimethylfulvene	5.45	2.40	2.40	1.50
Dibenzylfulvene	5.24	1.97	1.94	1.90
Diphenylfulvene	5.17	2.21	1.94	1.38
Spiro(2,4)heptadiene-1,3 (in CCl$_4$)	5.2	2.1	2.0	1.5

In the final variant the mean square deviation was 0.06 Hz with a maximum of 0.14 Hz for only one line. The experimental and theoretical spectra of 1,3-butadiene are shown in Fig. V-11.

As Table V-4 shows, the parameters of the spectra of 1,3-dienes lie within the usual range. It is typical that for 2-methoxybutadiene the signals of the protons $H_{C'}$ and $H_{B'}$ appear at higher fields due to conjugation with the methoxyl group, while the unconjugated protons in it H_B, H_C, and particularly H_A give signals at lower fields. The long-range spin–spin coupling constants in these molecules are interesting. These constants reach considerable values particularly for protons in the trans–trans position at the ends of the molecules. At the same time, all four long-range constants equal zero for 2-tert-butyl-1,3-butadiene. This confirms that the most stable conformation for the 2-tert-butyl derivative, in contrast to other 1,3-dienes, is the cisoid conformation relative to the =C—C= single bond due to steric effects.

A simplified analysis of the spectrum of the polyene hydrocarbon vinylfulvene (Fig. V-12), plotted with a frequency of 100 MHz, may be carried out by the ABXY approximation [26]. The assignment of the bands may be carried out on the basis of the spin—spin coupling constants, which are similar in value to the corresponding constants for butadiene. However, the small values of J_{HH} make it impossible to determine the relative signs of the spin—spin coupling constants of the protons. The position of the ring proton signals in the region of 6.11-6.44 ppm, which is characteristic of the protons at a double bond, indicates that this ring does not have aromatic character [27].

The spin—spin coupling constants of the protons of fulvene rings are similar to the constants for the cyclic diene hydrocarbon spiro(2,4)heptadiene-1,3 (Table V-5) and this also confirms the nonaromatic character of the fulvene ring.

Thus, the characteristics of the NMR spectra of unsaturated compounds are closely connected with their chemical nature. In contrast to saturated compounds, the effect of substituents on the spectral parameters of unsaturated compounds is stronger due to the greater effect of the substituents on the electronic environment of the nuclei investigated. This peculiarity is also characteristic of aromatic compounds.

4. AROMATIC HYDROCARBONS

AND HETEROCYCLES

The main rules for changes in the individual spectral parameters in compounds of aromatic character were examined in detail in Ch. II and III. More attention will be paid here to the analysis of actual spectra of aromatic compounds of different classes. Much experimental material, largely on the chemical shifts of protons of different aromatic rings and particularly polynuclear hydrocarbons and heterocycles, has been examined in a series of reviews [28-33].

It is worthwhile generalizing the main rules which characterize the relation of the spectral parameters of aromatic compounds to the structure.

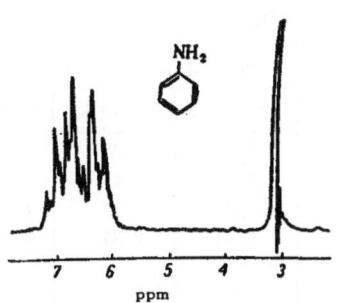

Fig. V-13. Spectrum of aniline. Index H.40.ANH.M.

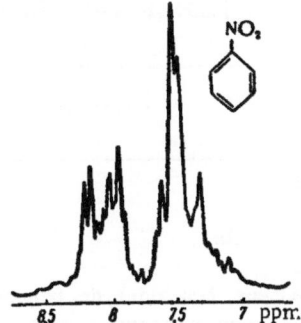

Fig. V-14. Spectrum of nitro-benzene (40 MHz). Index H.40.A.M.

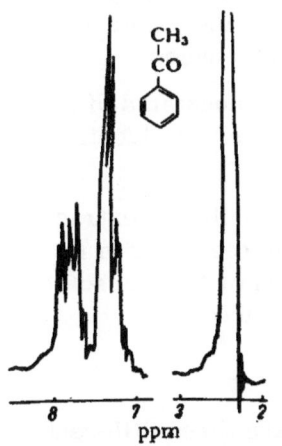

Fig. V-15. Spectrum of aceto-phenone. Index H.40.AAlk.M.

1. The chemical shifts of the signals of protons lying in the plane of an aromatic ring are determined by the aromatic character of this ring so that with an increase in the aromatic character the signal is shifted toward lower fields.

2. The signals of protons attached to condensed rings lie at lower fields as a rule than those from protons of mononuclear compounds. This particularly concerns protons lying close to several rings.

3. The introduction of electronegative substituents or the replacement of a carbon atom of a ring by an electronegative heteroatom produces a shift in the proton signals to lower fields. This has the greatest effect on protons closest to the substituent and also on remote protons conjugated with the substituent or the heteroatom. Electron-donor substituents have little effect on the chemical shifts of ring protons.

4. Dilution by nonpolar substituents usually produces a shift in the resonance of the protons of aromatic compounds toward lower fields.

TABLE V-6. Chemical Shifts of Ring Protons
in Monosubstituted Rings

Compound	Protons of aromatic rings		
	ortho (α)	meta(β)	para(γ)
Benzene	7.27	7.27	7.27
Toluene	7.17	7.17	7.17
Phenol, anisole	6.87	6.87	6.87
Aniline	6.44	6.97	6.70
Nitrobenzene, benzaldehyde, benzoic acid, benzoyl chloride, acetophenone, benzotrichloride	7.97	7.37	7.44
Pyridine	8.50	6.985	7.36

5. The spin—spin coupling constants for groups of similar compounds are characterized by constant values and depend little on the substituents. As a rule, the constants increase with an increase in the bond order and vice versa. There is a tendency for the constants to decrease when electronegative substituents are introduced.

6. In most cases the spin—spin coupling constants of protons of aromatic rings lying in different positions are identical in sign (positive).

The use of these rules often makes it possible to extract valuable information from a spectrum without resorting to its complete analysis. Actual cases of the application of NMR spectroscopy to aromatic compounds are examined below.

Monosubstituted Six-Membered Rings

Despite the fact that in benzene there are three different spin—spin coupling constants between protons (ortho, meta, and para), the protons of the benzene ring give only a single comparatively narrow peak due to the fact that the chemical shifts of its protons are equal. The same may be said of monosubstituted derivatives with donor groups, alkyl, substituted alkyl, and halogens (apart from fluorine). There is usually a slight shift in the signal of the phenyl protons towards higher fields and sometimes broadening or just noticeable splitting. On the other hand, such electron-acceptor substituents as a nitro group, carbonyl, carboxyl, trichloromethyl, etc., lead to a substantial difference in the screening of different ring protons and to the appearance of complex spectra of the AA'BB'C type. The replacement of a carbon atom of the ring by a heteroatom (pyridine) leads to essentially the same result and in this case, as in benzene derivatives with electronegative substituents, the protons in the ortho (α)-position give a signal at lowest fields, while the meta (β)-protons give a signal at highest fields. The amino group also produces splitting of the ring signal, but leads to the reverse distribution (Table V-6).

A complete analysis of the spectrum of pyridine (see Fig. A-31) was carried out by Schneider, Bernstein, and Pople [34, 35] as an example of five strongly coupled spins with definite symmetry. However, the analysis of this spectrum, particularly in the case of compounds with lower chemical shifts between the protons, is very laborious and is rarely used. Replacement of the ring protons by deuterium was used for analysis of the spectrum of nitrobenzene [36, 37]. Examples of spectra of this type are given in Figs. V-13-15 and also A-31, 41-43.

Aromatic Rings with Four

Unsubstituted Protons

The isomerism of disubstituted benzene derivatives (XXIV-XXVI)

XXIV   XXV   XXVI

is established comparatively simply by the NMR method when at least one of the substituents has acceptor properties. In this case ortho-disubstituted derivatives give quite a complex spectrum, which becomes a symmetrical AA'BB' spectrum when the substituents are identical and is readily analyzed. Essentially the spectra of naphthalene, anthracene, and other condensed aromatic hydrocarbons are of the same type. An interesting example of this type is the spectrum of triphenylene [31, 38], whose analysis is possible by the AA'XX' approximation. A comparatively rare example of the spectrum of an unsymmetrical system of four interacting protons which has been analyzed completely is provided by the methyl ester of salicylic acid [39]. All the spectra of aromatic compounds

of this type have a definite similarity because in each case two
spin—spin coupling constants (J_{ortho}) are relatively large and
similar in magnitude, while J_{para} are very small and often do not
appear in the spectra.

Para-disubstituted benzene derivatives of type XXV also give
a symmetrical spectrum of the AA'BB' or AA'XX' type, but they
differ substantially from the spectra of derivatives with four protons in a row since in this case only one of the four constants has
a high value (J_{ortho}), while the other three (J_{meta} and J_{para}) are
small in magnitude and therefore have little effect on the form of
the spectrum or do not appear at all. Spectra of this type are very
similar to simple spectra of the AB type and often they may be
analyzed approximately as AB spectra [40, 41]. In spectra of this
type (Fig. A-29, A-38) the four most intense lines are similar to a
typical AB quartet, but in this case the distance inside each pair
does not characterize J_{AB}, but the sum $J_{AB} + J'_{AB}$, i.e., in this
case $J_{ortho} + J_{para}$. Since the latter is very small, this distance
roughly characterizes the coupling constant of the ortho-protons
and makes it possible to follow the change in this constant with a
change in the substituents. The fact that the constant J_{meta} for
para-disubstituted derivatives is of appreciable magnitude results
in the appearance of additional signals of low intensity in the spectrum and the complete analysis of these makes it possible to determine the two meta constants ($J_{AA'}$ and $J_{BB'}$), which may not equal
to each other, and also the constants J_{ortho} and J_{para} individually.

Spectra of the same type are given by γ-substituted pyridines
(see Fig. A-42), and also five-membered heterocycles such as furan
(see Fig. A-32), thiophene (see Fig. A-33), pyrrole, selenophene,
and others, whose analysis provides classical examples in NMR
spectroscopy. For the five-membered heterocycles the similarity
to para-disubstituted benzenes is due to the fact that the spin—
spin coupling constant of the J_{H_β, H_β} is small in comparison with
J_{H_α, H_β} and similar in magnitude to J_{meta} for benzene derivatives.

Finally, the spectra of meta-disubstituted benzenes (XXVI)
also have a characteristic difference due to the fact that the proton
H_X lying between the two substituents often gives a signal at lower
fields than the other ring protons, while the resonance of the proton H_B is observed at higher fields. Since the spin—spin coupling of
the proton H_X with the three others is small, the protons H_A, H_B, and

TABLE V-7. Spectral Parameters of Six-Membered
Heterocycles with Two Nitrogen Atoms* [42, 43]

Parameter	Pyridazine	Pyrimidine	Pyrazine
δ_2	—	9.26	8.63
δ_3	9.24	—	8.63
δ_4	7.54	8.78	—
δ_5	7.54	7.36	8.63
J_{ortho}	{ 4.9 (3 : 4), 8.4 (4 : 5) }	5.0	1.8
J_{meta}	2.0	{ 0 (2 : 4), 2.5 (4 : 6) }	0.5
J_{para}	3.5	1.5	1.8
$J_{C^{13}-H_\alpha}$	181.5	181.8	183
$J_{C^{13}-H_\beta}$	168.5	168	—

* 3% (weight to volume in $CDCl_3$).

H_C give a spectrum which is close to the ABC type (or A_2B if R and R'
are similar in character or identical), only slightly perturbed by
spin—spin coupling with the nucleus H_X (see Fig. A-43).

The spectra of six-membered rings with heteroatoms are
similar in character to the spectra of the corresponding disubstituted benzenes, but the electronegative heteroatoms manifest themselves both in a decrease in the coupling constants of the protons
and in a low-field shift in their resonance. Table V-7 gives the
spectral parameters of six-membered diazine heterocycles, from
which the degree of change in these parameters can be assessed.

Aromatic Compounds with a Smaller
Number of Interacting Spins

An increase in the number of substituents in aromatic rings
simplifies the analysis of their spectra, reducing them to the ABC,
ABX, or AB type, but at the same time the correlation of the properties of the substituents and the spectral parameters becomes
more difficult. The picture is simpler in the case of condensed
rings, in which the decrease in the number of spins is not connected

TABLE V-8. Spin−Spin Coupling Constants in Nitromethyl
Derivatives of Naphthalene

Compound	$J_{\alpha\beta}$	$J_{\beta\beta'}$	$J_{\alpha\beta'}$	$J_{\alpha\alpha'}$
Dimethylnaphthalenes . . .	8.5	7.1	1.5	~0.4
Dinitronaphthalenes . . .	8.7	7.5	1.4	$\lesssim 1$
Methylnitronaphthalenes . .	8.6	7.7	1.3	$\lesssim 1$

with the introduction of a large number of substituents. The results of investigating nitro derivatives of methylnaphthalenes are
interesting [44]. It was found that the introduction of a nitro group
has little effect on the spin−spin coupling constants of the protons
(Table V-8).

At the same time, the chemical shifts of the protons in an
ortho position relative to the nitro group depend on whether or not
there are other substituents adjacent to the nitro group. The introduction of substituents adjacent to the nitro group produces disruption of the coplanarity of this group with the naphthalene ring
with the result that there is a change in the contribution from the
anisotropy of the nitro group to the chemical shifts of adjacent protons. At the same time the contribution of the nitro group to the
chemical shift of the proton in position 4 in derivatives of 1-nitro-
naphthalene remains unchanged.

In many cases the spectra of tetrasubstituted benzenes make
it possible to establish the structure of the compounds simply. In
contrast to compounds containing two protons in the para position,
which give only singlet peaks as a rule, molecules with two ortho-
protons usually give a spectrum of the AB type. It is interesting
that when the amino group in the methyl ester of 2-amino-3-
hydroxy-4-chlorobenzoic acid (Fig. V-16) is replaced by a nitro
group. there is not only a shift in the signals of the ring protons to
lower fields, but also a decrease in the difference of the chemical
shifts of the two protons. In the case of free 2-nitro-3-hydroxy-

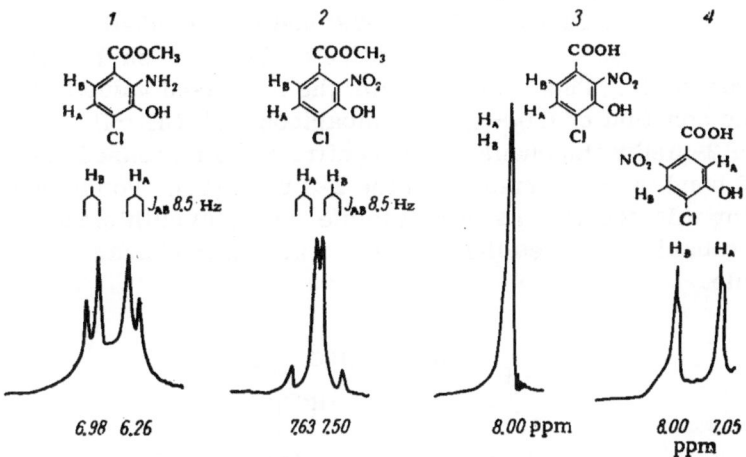

Fig. V-16. Spectra of the aromatic protons of tetrasubstituted benzene derivatives: 1) methyl ester of 2-amino-3-hydroxy-4-chlorobenzoic acid; 2) methyl ester of 2-nitro-3-hydroxy-4-chlorobenzoic acid; 3) 2-nitro-3-hydroxy-4-chlorobenzoic acid; 4) 2-nitro-4-chloro-5-hydroxybenzoic acid.

4-chlorobenzoic acid the effects of the four substituents on the two protons are equivalent so that their chemical shifts become identical and the signal consists of a single line. Comparison of the spectra makes it possible to give the assignment of the signals shown in the figure.

It is important to note that in aromatic compounds the spin—spin coupling of protons of the ring and side chains is usually very slight and as a rule it does not affect the form of the spectrum, producing only broadening of the lines or slight splitting in individual cases. The latter often helps in the assignment of the bands in the spectrum (see Fig. A-29) and in individual cases it makes it possible to assess the nature of the electron distribution in the molecule. Thus, for example, the increased value of the coupling constant of CH_3-H_2 in comparison with CH_3-H_4 in derivatives of 2-methylthiophene XXVII confirms the increased order of the 1-2 bond in comparison with the 2-3 bond [45]. Too high a value for this constant as, for example, in methyldithiolone XXVIII, most probably indicates the absence of aromatic character in this molecule.

XXVII

$J_{CH_3,H_2}0.9 - 1.25\ Hz$

$J_{CH_3,H_4}0.4 - 0.5\ Hz$

XXVIII

$J_{CH_3,H}2.9\ Hz$

Nonbenzoid Aromatic Hydrocarbons

The presence in azulene of a closed system of π-electrons and the ring currents associated with it leads to the appearance of resonance of the azulene ring protons at low fields (7-8 ppm) like other aromatic hydrocarbons. In exactly the same way methyl groups attached to an azulene ring give a signal in a region which is typical of aromatic compounds (2.6-3 ppm). On the other hand, this compound is characterized by a considerable difference in electron density at different carbon atoms of the ring and this leads to a substantial difference in the chemical shifts of protons attached to them. This situation and also a certain symmetry of the molecule and the absence of appreciable spin-spin coupling between protons belonging to different rings of the molecule results in azulene and its derivatives giving spectra which can be analyzed comparatively simply.

The analysis of the spectrum of 6-methylazulene XXIX is simplest (Fig. V-17). Since the spin-spin splitting of remote protons is slight in this case, the spectrum may be broken down essentially into two separate spectra, one of which is of the A_2B type (protons of the five-membered rings), while the other may be inter-

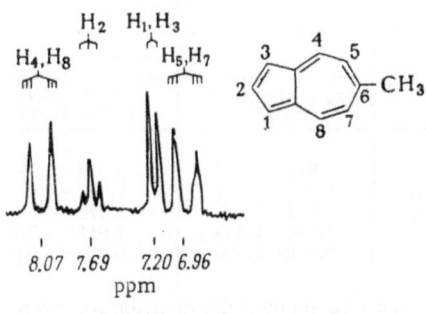

$J_{1,2} \cdot J_{2,3}$ 4.0; $J_{4,5} \cdot J_{7,8}$ 10.0; $J_{5,7} \sim 1.5$ Hz

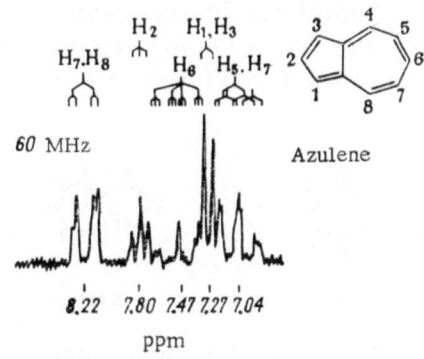

$J_{1,2:2,3}$ 4.0; $J_{4,5:7,8}$ 9.5; $J_{5,6;6,7}$ 10.30; $J_{4,6;6,8}$ 1.5 Hz

Fig. V-17. Spectra of 6-methylazulene and azulene.

TABLE V-9. π-Electron Densities
in Azulene

Position	Values of ρ	
	from chemical shifts*	from MO calculation
2	0.977—0.982	0.979—0.997
1,3	1.055—1.061	1.049—1.096
4,8	0.967—0.968	0.879—0.954
5,7	1.040—1.045	1.011—1.049
6	0.962—0.993	0.938—0.969

*Taking into account the contribution from
ring currents.

preted as a simple AB spectrum [in this case KX, or more accurately, (KX)₂ spectrum] which is only slightly perturbed by spin—
spin coupling with remote protons (protons of the five-membered
ring).

XXIX   XXX

Analysis of the spectrum of azulene itself XXX is more complex, but it is simplified by the fact that the resonance of the protons H_4 and H_8 differs substantially in chemical shift from the other
protons of the seven-membered ring, particularly at 100 MHz, so
that this part of the spectrum may be analyzed by the ABB'XX'
approximation.

It is interesting that the chemical shifts of the protons and
the π-electron densities (ρ) at the carbon atoms of azulene calculated from them (starting from the value of 10.6 ppm per electron,
see Ch. II) correlates well with the values obtained by theoretical
MO calculation (Table V-9) [46].

Interesting information is provided by the spin—spin coupling
constants of the protons of azulene. The equality of the constants
1-2 and 2-3, 4-5 and 7-8, and 5-6 and 6-7 confirms the well-known

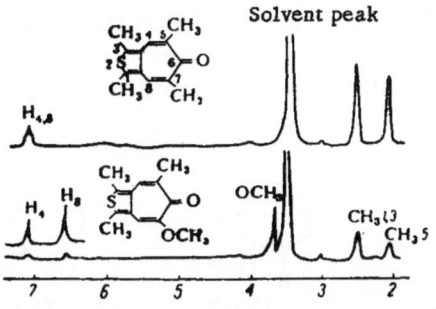

60 MHz, 30% (weight to volume in dioxane)

Fig. V-18. Spectra of aromatic condensed
heterocycles: (the spectra were provided by
A. V. El'tsov and I. V. Tseeteli). 1) 1,3,5,7-
Thienotropone (index H.60.AAlk.I.0); 2)
1,3,5-trimethyl-7-methoxythienotropone
(index H.60.AAlk.I. 0).

fact that the azulene molecule is symmetrical about a line connecting nuclei 2 and 6 so that the given distribution of double and single
bonds is purely arbitrary. On the other hand, the somewhat higher
values of the constants 5-6 (6-7) in comparison with 4-5 (7-8) obviously indicates that in the seven-membered ring of azulene there
is not a completely uniform distribution of the π-electron cloud as
in benzene so that it is very probable that the degree of double-
bond character between the carbon atoms 5-6 and 7-8 is low in
comparison with the other positions in the molecule.

In tropolone derivatives the spin—spin coupling constants of
the ring protons make it possible to establish analogously the character of the tautomeric equilibrium [47]. Tropolone itself is sym-
metical about a line connecting carbon atom 5 and the center of the
bonds between atoms 1 and 2 so that the spin — spin coupling constants
between the protons in the pairs 3-4 and 6-7 and also 4-5 and
5-6 are equal. In the case of the unsymmetrically substituted
tropolones structures XXXIIa and XXXIIb are nonequivalent and a
tautomeric equilibrium exists between them. It was found that the
constants 4-5 (10.2 Hz) and 6-7 (10.6 Hz) are somewhat higher

than for the protons in positions 5–6 (9.6 Hz), indicating an increase
in the multiplicity of the bonds between carbon atoms 4–5 and 6–7,
and, consequently, a shift in the tautomeric equilibrium toward
compound XXXIIb.

XXXI   XXXIIa   XXXIIb

Comparatively simple spectra are given by the methyl derivatives of the peculiar compound with aromatic character thienotropone (Fig. V-18). The presence of spin–spin coupling of the
protons of the seven-membered ring (4 and 8) with methyl groups
in positions 5 and 7 is indicated by broadening of the signal of the
methyl groups in comparison with the methyl groups attached to
the five-membered ring (1 and 3). The appearance of the signal
of these two methyl groups at lower fields evidently indicates that
the thiophene ring has more aromatic character than the tropone
ring. The assignment of the lines is readily established by comparison with the spectrum of the compound in which one of the
methyl groups is replaced by a methoxyl group. It is characteristic that in this case the signal of the proton H_8, which is conjugated with the methoxyl group, is shifted to higher fields due to
the electron–donor effect of the methoxyl.

The examination of this spectrum concludes the short review
of the spectra of compounds in which the protons are attached directly to the carbon skeleton of the molecule. For protons of
this type the different classes of compounds (saturated, unsaturated,
and aromatic) are characterized by quite definite, comparatively
narrow resonance regions and values of the spin–spin coupling constants between protons, which are typical of each molecular structure and change over only comparatively narrow ranges for actual
compounds. These characteristics often make it possible, even
with a cursory examination of the spectrum of the compound, to
estimate its structure and determine the character of the proton-
containing groups and their relative arrangement. In many cases
the analysis of the spectrum and comparison of its parameters
with the parameters of the spectra of related compounds makes it

possible to assess the character of the electron distribution in the organic molecule.

In contrast to these protons, the region of resonance of protons in the functional groups of organic molecules, particularly those which are attached to heteroatoms, is usually quite wide, while the spin—spin coupling of these protons with protons of the skeleton of the molecule is slight or completely absent. Nonetheless, the resonance of the protons of functional groups may provide valuable information both on the structure of the molecule and on the fine electron distribution. Certain special procedures are often used in this case and these are examined in subsequent sections.

5. NMR SPECTRA OF COMPOUNDS WITH PROTON-CONTAINING FUNCTIONAL GROUPS

In this section we examine the characteristics of the NMR spectra of alcohols, phenols, and polyols, organic amines (including compounds with alkylated amino groups) and other nitrogen-containing substances, aldehydes, carboxylic acids and their derivatives, sulfur-containing substances and some other compounds, and also compounds with mixed functions. In contrast to the protons of the hydrocarbon skeleton, the chemical shifts and splitting of the signals attached to heteroatoms depend strongly on the structure of the molecule and also on external factors, i.e., the polarity, acidity, and other properties of the solvent, the concentration, and the sample temperature. These peculiarities are determined by the characteristic properties of the functional groups:

1. the capacity to associate and form hydrogen bonds;
2. the capacity for fast exchange of protons between functional groups and with the solvent;
3. the capacity for addition of a proton or dissociation.

Since these factors have a strong effect on the electron cloud directly at the proton of the functional group, even a small change in them leads to a strong change in the chemical shift and the form of the signal. This is particularly marked in the spectra of hydroxyl and amino compounds.

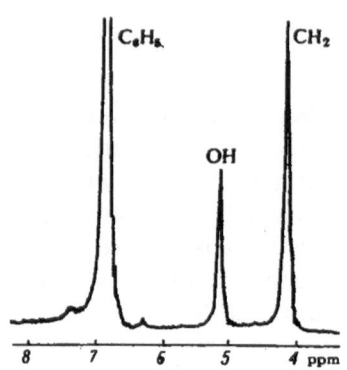

Fig. V-19. Spectrum of benzyl alcohol without special purification. Index H.40.A + ЛОН.I.

Hydroxyl Compounds

When there is no association or exchange of protons, the proton of the hydroxyl group of alcohols is characterized by a signal at high fields (0.5 ppm for ethanol with extrapolation to infinite dilution) and a splitting of 4-5 Hz due to spin—spin coupling with protons at the α-carbon atom. However, under normal conditions the spectra of alcohols have a different form: the hydroxyl group gives a narrow single peak in the region of 4-6 ppm, which is broadened or split due to spin—spin coupling only in special cases. In some cases when the proton exchanges between two positions which differ markedly in chemical shifts, its signal may be completely absent from the spectrum. Therefore, running the spectrum in the usual way gives little information of the structure of a hydroxyl compound as a rule. To investigate substances with hydroxyl groups, several spectra are usually run in different solvents in the presence of certain additives; comparison of these spectra makes it possible to build up a definite picture. In this it is useful to bear in mind the following rules.

1. The addition to a hydroxyl compound of a small amount of acid or base promotes rapid exchange of the hydroxyl protons and narrows the signal of this proton and the protons at the carbon atom attached to the hydroxyl. To narrow the signal it is usual to add strong acids such as sulfuric or trifluoroacetic; bases are used rarely for this purpose. On the other hand, careful removal of traces of acid substances and the use of sample tubes made from alkali-free borosilicate glass helps in the observation of spin—spin coupling of hydroxyl and α-protons.

2. As a rule, the use of the usual solvents such as carbon tetrachloride and chloroform (or deuterochloroform) leads to a high-field shift in the resonance of the hydroxyl proton. The signal of the hydroxyl then appears as a narrow, unsplit line, evidently

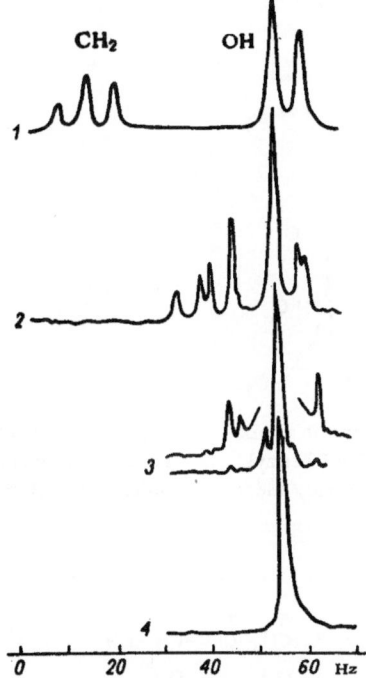

Fig. V-20. Spectra of the protons of methylene and hydroxyl groups in solutions of purified benzyl alcohol in acetone:

Spectrum No.	$C_6H_5CH_2OH$ wt.%	J/ν_{AB}
1	78	0·14
2	55	0·37
3	43	1·3
4	40	—

due to the presence of acidic substances in these solvents. Specially purified carbon tetrachloride does not eliminate the splitting of hydroxyl and α-protons [48].

3. The use of neutral, electron-donor solvents hampers proton exchange and promotes spin—spin splitting of hydroxyl and α-carbon protons.

The specific solvents used to reveal spin—spin splitting are dimethyl sulfoxide and acetone. Both substances promote the

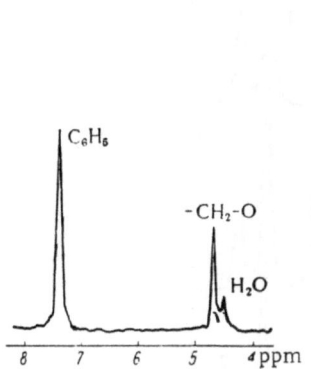

Fig. V-21. Spectrum of a mixture containing 0.2 ml benzyl
alcohol, 0.3 ml acetone, and 7
drops D_2O.

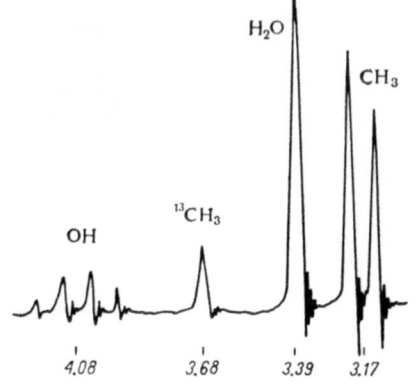

Fig. V-22. Spectrum of a solution of
aqueous methanol in dimethyl sulfoxide.
Index H.60.ЛOH.I.Д,В.

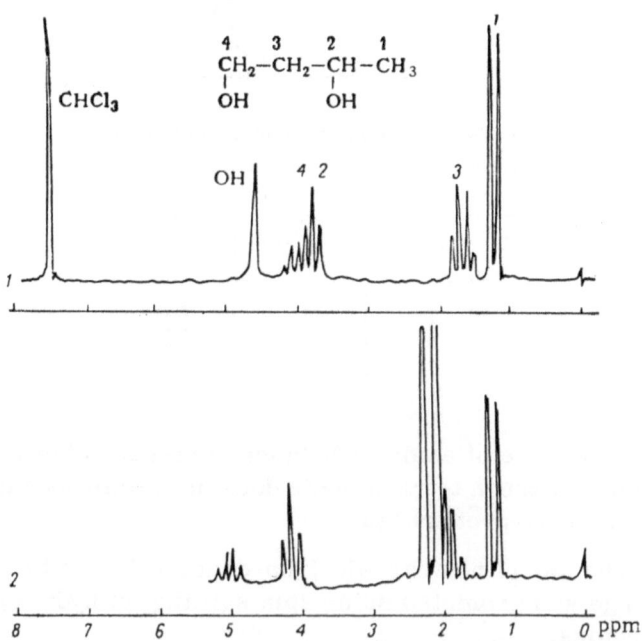

Fig. V-23. Spectrum of 1,3-butanediol: 1) before addition of
acetic anhydride; 2) after treatment with acetic anhydride.

formation of a strong hydrogen bond of the hydroxyl proton with the solvent. The exchange is often slowed down so much that protons of water present in solution in an amount equal to the alcohol appear in the spectrum in the form of a separate signal [49].

4. The addition to a solution of an alcohol of a small amount (a few drops into the tube with the sample) of deuterium oxide leads to replacement of the hydroxyl protons by deuterium and the disappearance of its signal from the spectrum. This property is also characteristic of other functional groups with protons which exchange rapidly.

The spectra of benzyl alcohol (Fig. V-19-21) are a clear example which illustrates the application of these rules. In the spectrum of a solution of purified benzyl alcohol in acetone the signals of the hydroxyl and benzyl protons are split due to spin–spin coupling in an A_2B system, whose parameters vary markedly, depending on the solution concentration. On the other hand, in a solution in carbon tetrachloride or chloroform, particularly with the addition of traces of acid, both of these groups give single peaks. In this case it is not difficult to assign the signals if their intensities are compared with the number of protons in the corresponding groups. In other cases it is useful to add heavy water to determine the signal of the hydroxyl. Figure V-21 shows how the form of the spectrum changes a few minutes after the addition of D_2O to an acetone solution of benzyl alcohol.

Different alcohols are subject to the effect of solvents to a different extent, depending on the character of the substituents. In this sense they may be classified in three groups with respect to dimethyl sulfoxide:

1. Alcohols which give the expected splitting of the signal in dimethyl sulfoxide solution without special preliminary treatment. This includes alcohols with a proton which exchanges weakly such as allyl alcohol and 2-ethoxyethanol and also alcohols with a sterically hindered hydroxyl group such as benzhydrol.

2. Alcohols which give the expected splitting after preliminary treatment with sodium or potassium carbonate to remove traces of acid. These include, for example, 2-chloro- and 2-bromoethanol; in the latter the triplet splitting of the hydroxyl signal is observed only immediately after treatment, and after the sample has stood for 2 h, only a broad single signal of the hydroxyl proton appears in the spectrum again.

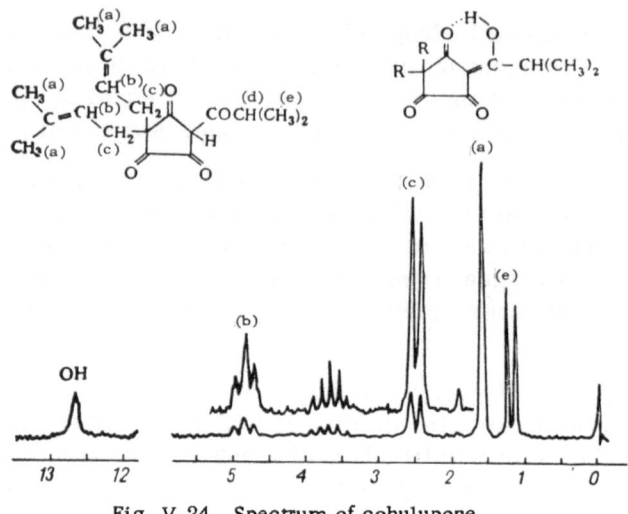

Fig. V-24. Spectrum of cohulupone.
Index H.60.Ц+Л=,Alk,OH.I.

3. Alcohols which give only a narrow or slightly broadened singlet even after treatment with carbonate. These are alcohols with quite an acidic hydroxyl group and also possibly those which readily eliminate acid such as 2–cyanoethanol, trans–2–bromo–cyclooctanol, and 2,2,2–trichloroethanol [50].

Methanol obviously may be assigned to the second group of alcohols. For a long time it was impossible to observe the splitting of the signals of the two proton groups of methanol. When dimethyl sulfoxide was used a spectrum was obtained with the expected quartet and doublet splitting of the signals of the hydroxyl and methyl protons. It is characteristic that water present in the solution in an amount equal to the alcohol gives a separate signal (Fig. V–22) [51].

In practical work in the investigation of compounds containing a hydroxyl group it is often inconvenient to carry out a special purification or to use such a specific solvent as dimethyl sulfoxide; to elucidate the structure of an alcohol it is also possible to use the signal of the protons at the carbon atom. The chemical shift of these protons is usually reduced under the influence of the electronegative oxygen atom with the result that the signal is clearly distinguished from signal of other protons in the molecule and is available for analysis. In the case of tertiary alcohols and phenols, this signal is completely absent. Another approach to the analysis of hydroxyl compounds is the preliminary preparation of derivatives at the hydroxyl group, i.e., acylation, treatment with isocyanates to prepare urethans, etc. Figure V-23 gives the spectrum of 1,3-butanediol before and after treatment with acetic anhydride [52]. The reaction mixture after the treatment gives two signals of protons of the acetyl group — the starting acetic anhydride and butanediol acetate (2-2.3 ppm); at the same time, the signal of the hydroxyl proton at 4.7 ppm disappears. Moreover, there is a change in the chemical shifts of the protons at the carbon atoms bearing the hydroxyl groups and as a result it is quite clear that one of the hydroxyl groups is attached to a primary and the other to a secondary carbon atom.

Alcohols and phenols with an intramolecular hydrogen bond have a specific peculiarity. The spectra of such compounds usually change little with dilution of the solution. In the spectra of enols with a hydrogen bond the signal of the hydroxyl proton appears at very low fields (15 ppm in the case of acetylacetone and down to 19 ppm in β-tricarbonyl compounds) [53] so that the appearance of a low-field signal from ketones may indicate enolization. A clear spectrum of this type is given by the triketone cohulupone (Fig. V-24), one of whose possible enol structures is shown on the spectrum. The signal of the hydroxyl group of this enol lies at comparatively high fields, indicating the formation of only a weak hydrogen bond.

The characteristics of hydroxyl compounds appear to a considerable extent in the spectra of other substances, whose hydrogen atoms are capable of exchanging between different positions in the molecule or between different molecules of the sample.

Amino Compounds

The spectra of organic amines are determined to a still greater extent than for hydroxyl compounds by external factors such as the character of the solvent, the solution concentration, and the presence of impurities. This is connected with the comparatively high basicity of amines and their capacity for salt formation and also the capacity of amines, like hydroxyl compounds, to exchange protons attached to the nitrogen atoms. Another peculiarity of the spectra of nitrogen-containing substances is connected with the fact that the most abundant isotope of nitrogen N^{14} has a spin $I = 1$ and an electric quadrupole moment, whose effect on the form of the spectrum depends on the properties of the amine and is determined to a considerable extent by external factors. The other stable isotope of nitrogen, N^{15}, also has a nuclear magnetic moment and due to the fact that its nuclear spin equals $\frac{1}{2}$, compounds with the nitrogen isotope N^{15} are more convenient for investigation by the NMR method both with excitation of the resonance of the protons and with resonance directly at N^{15} nuclei. However, since the content of this isotope in the natural mixture is only 0.365%, these investigations are a rather special field.

The chemical shift of protons attached directly to the nitrogen atom varies over the wide range of 0.5-5 ppm, depending on the properties of the amine and the acidity of the medium. Therefore, in the identification of amino compounds it is not usual to use the chemical shift of the protons, but to determine the character of the change in the chemical shift and the change in the form of the signal of the amino protons with a change in acidity and also the signals of protons at adjacent and more remote carbon atoms. From the point of view of nuclear magnetic resonance, it is possible to divide amines with respect to the capacity of the amino proton to exchange into those with fast and those with slow exchange. If the exchange is sufficiently rapid, the proton of the NH group appears in the form of a narrow singlet and the signals of the protons at the neighboring carbon atom are not split by the NH proton. This group includes most primary and secondary aliphatic amines. In the other extreme case when there is slow exchange of protons (or no exchange), the signal of the NH protons is strongly broadened due to spin−spin coupling with the N^{14} nucleus and the effect of quadrupole relaxation of this nucleus. In this case there is the possibility of spin−spin splitting of the signal of the protons at the neighboring carbon atom under the influence of the amino nitrogen. Aromatic amines which are not activated by polar groups in the ring usually belong to this group of amines which exchange slowly. In intermediate cases there is more or less considerable broadening of the signal of the NH protons but no spin−spin coupling of these protons with protons at the α-carbon atom.

During the salt formation of amines a positive charge appears at the central nitrogen atom and this leads to deshielding of protons both directly attached to the nitrogen and attached to neigh-

TABLE V-10. Chemical Shifts and Changes in Them in Relation to the Solvent for Methyl Groups in Various Compounds

Type of methyl group	Chemical shifts, ppm			Change in chemical shift, ppm		
	CDCl₃	CD₃COOD	CF₃COOH	CD₃COOD − CDCl₃	CF₃COOH − CD₃COOD	CF₃COOH − CDCl₃
Methyl groups attached to a nitrogen atom						
Tertiary aliphatic amines	1.88-2.56	2.63-3.14	2.76-3.26	0.40-0.81	0.05-0.25	0.53-0.92
Secondary aliphatic amines	2.34-2.44	2.67-2.88	2.78-3.03	0.26-0.54	0.03-0.24	0.43-0.67
Aromatic and heteroaromatic amines	2.66-3.99	2.78-4.00	3.17-4.33	0.01-0.54	0.06-0.54	0.34-0.82
Amides and imides	2.71-3.89	2.72-3.88	2.90-4.20	from −0.09 to +0.07	0.08-0.34	0.08-0.34
Other compounds with neutral nitrogen	3.02-3.68	3.04-3.72	3.57-4.34	from −0.04 to +0.09	0.48-0.62	0.48-0.66
Quaternary ammonium compounds	−	3.30-4.77	3.02-4.79	−	0.01-0.05	−
Methyl groups not attached to a nitrogen atom						
Aromatic compounds with basic nitrogen	2.12-2.82	2.23-3.19	2.45-3.39	0.05-0.37	0.07-0.42	0.21-0.57
Nonbasic aromatic compounds	2.32-2.40	2.20-2.39	2.20-2.46	from −0.06 to +0.04	from −0.02 to +0.07	from −0.08 to +0.06
Acetyl group	1.99-2.59	2.00-2.57	2.15-2.73	from −0.02 to +0.09	from +0.15 to +0.37	from +0.14 to +0.46
Methoxyl group	3.44-3.96	3.49-4.02	3.62-4.17	from −0.03 to +0.11	from +0.09 to +0.55	from +0.12 to +0.62

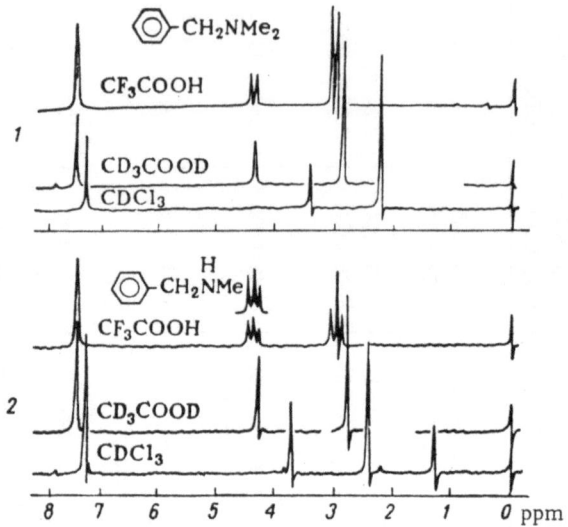

Fig. V-25. Identification of amines by NMR: 1) dimethylbenzylamine; 2) methylbenzylamine. Index for both spectra:
H.60.A+ЛNH.I.Xл.0.

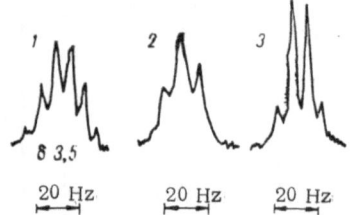

Fig. V-26. Spectra of protons in the
α-position relative to the amino
group in compounds of the type

$RCH_2CH_2\overset{+}{N}H_3$: 1) normal spectrum;
2) with spin−spin decoupling of the

$\overset{+}{N}H_3$; 3) with spin−spin decoupling
of the β-CH_2 protons.

boring carbon atoms. This shifts
the signal of amine protons from
the region of comparatively high
fields, where it usually overlaps the signals of other protons
in the molecule, to a low field
region and this often makes it
possible to identify an amine by
determining the number of protons at the nitrogen by direct integration of the spectrum. Tertiary amines may also be identified by this method. Moreover,
with salt formation it is often
possible to observe splitting of
the signal of adjacent methyl and
methylene protons due to spin−spin coupling with protons of the
ammonium group. The number of peaks in the multiplet unequivocally indicates whether the protonized amine is primary, secondary, or tertiary. Finally, with salt formation by amines in

which the unshared pair of electrons at the nitrogen participate, the electron cloud of the nitrogen approaches spherical symmetry and this leads to a change in the character of the quadrupole relaxation of nitrogen. As a result, in individual cases it is possible to observe triplet splitting of the signals of the protons due to spin−spin coupling with the N^{14} nucleus. In addition to the protons attached directly to the nitrogen, in individual cases protons at the β-carbon atom may also show this splitting (Ch. III), but not the α-carbon protons.

Ma and Warnhoff [54] developed a method of identifying amines containing at least one methyl group at the nitrogen by comparison of spectra of 10-20% solutions of them in $CDCl_3$, CD_3COOD, and CF_3COOH. In some cases instead of completely deuterated acetic acid it is possible to use the normal acid, but then the signal of the methyl groups may be masked by the intense peak of the acetyl protons. With a change from solutions of deuterochloroform to acetic and trifluoroacetic acids the signals of the N-methyl groups are shifted to lower fields. The magnitude of this shift is different for different amines and differs substantially from the shift in the signals of methyl groups that are not attached to a nitrogen atom (Table V-10).

In the case of tertiary aliphatic amines and strongly basic amines of a different type there is the greatest shift in the resonance even with a change to a solution in deuteroacetic acid. On the other hand, for amines of low basicity a substantial shift is observed only in such a strong acid as trifluoroacetic. It is possible to distinguish between aliphatic and aromatic amines by using the fact that the resonance of the methyl groups in aromatic N-methyl compounds appears at higher fields.

Amines may be further identified through the splitting of the resonance signal in trifluoroacetic acid since for most aliphatic and aromatic amines the proton remains at the nitrogen long enough for the observation of spin−spin splitting at 38°C.

A characteristic example of the use of this method is provided by the analysis of the spectra of dimethylbenzylamine and methylbenzylamine in the three given solvents (Fig. V-25) [55]. A large shift in the resonance signal even with the use of deuteroacetic acid indicates that in the first case we have a tertiary aliphatic amine. This is confirmed by the doublet splitting of the

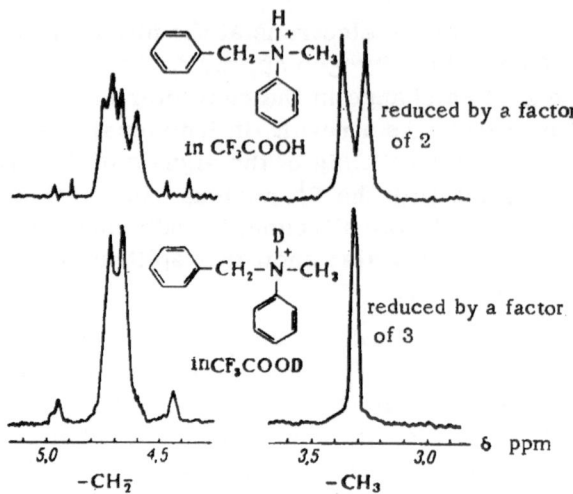

Fig. V-27. Spectra of methyl and methylene protons in
methylbenzylphenylammonium salts.

signals of the methyl (and also methylene) group in trifluoroacetic
acid solution. In the case of the secondary amine, methylbenzylamine, the magnitude of the shift is somewhat less. In the spectrum of this amine in deuterochloroform there is a signal of the
nitrogen of the NH group at about 1.3 ppm, which disappears when
the amine is dissolved in acids. The spectrum of a solution of this
amine in trifluoroacetic acid contains a triplet of the methyl group,
which is produced by spin—spin coupling with two protons at the
nitrogen in the amine salt formed. The methylene benzyl protons
give an analogous signal.

The same procedure is also suitable for investigating amines
containing other groups at the nitrogen instead of a methyl group,
but the interpretation of the spectrum is complicated in this case.
The splitting of the signal of a methylene group at the nitrogen in
ammonium salts is a function of the number of protons both at the
nitrogen and at the β-carbon atom. However, since the spin—spin
coupling constants of H—C—N—H and H—C—C—H are similar in
magnitude, the number of lines in the spectrum may often be determined by using the rules for first-order splitting (the number
of peaks is n + 1, where n is the total number of protons at the nitrogen and at the carbon atoms; see Ch. I) [55]. It is then useful
to use double proton—proton nuclear magnetic resonance. Figure

V-26 shows the spectra of a β-substituted primary amine — the normal spectrum and also spectra with irradiation at the resonance frequency of the NH protons and the β-CH$_2$ protons. Comparison of these spectra makes it possible to determine the number of protons both at the nitrogen atom and at the β-carbon atom.

The spectra of some benzylammonium salts show an interesting peculiarity. In compounds of this type the two benzyl protons may be arranged unsymmetrically relative to the plane of the benzene ring, particularly when bulky substituents are present, and this makes them nonequivalent and leads to complication of the spectrum. In the spectrum of a solution of benzylphenyl-methylamine in trifluoroacetic acid this results in the benzyl protons appearing as the AB part of an ABX spectrum. The use of deuterotrifluoroacetic acid simplifies the spectrum since the nucleus X disappears, but the nonequivalence of the methylene protons is preserved and they appear in the spectrum as a typical AB quartet (Fig. V-27). To improve salt formation a small amount of perchloric acid HClO$_4$ was added to the solution [56].

The spectra of compounds containing a hydrazo group are similar to the spectra of amino compounds. In contrast to them, compounds with a C = N bond are closer in character to aldehydes and the signals of neighboring protons appear at comparatively low fields. The NH protons in amides of acids usually give very broad signals in the low-field part of the spectrum (see Fig. A-25, 34, and 38).

Aldehydes, Carboxylic Acids, and Their Derivatives

Aldehydes are identified comparatively readily. The signals of aldehyde protons lie in a narrow, low-field region of the spectrum (9.3-10.1 ppm) [57]. With the use of high-resolution instruments it is usually possible to detect spin—spin coupling of the aldehyde protons with protons at the α-carbon atom. In unsaturated aldehydes the spin—spin splitting of the aldehyde protons is somewhat greater than in saturated aldehydes and there is often long-range spin—spin coupling with protons, evidently transmitted through the system of π-electrons of the unsaturated or hetero-aromatic system.

For the identification of aldehydes it is very profitable to examine the C^{13} signals, i.e., the satellites of the aldehyde proton, which lie in a region remote from other signals and therefore are detected comparatively readily. As has already been pointed out (Ch. III), the $J_{HC^{13}}$ constants of aldehydes show an additive relation of high accuracy to empirical increments obtained both from analysis of the spectra of aldehydes and various methane derivatives and also correlate with the electronegativity of substituents at the aldehyde group.

The replacement of hydrogen atoms at the α-carbon atom by alkyl groups usually leads to a high-field shift in the signal of the aldehyde proton. It is surprising that the introduction of a double bond also produces a high-field shift in contrast to the other classes of compounds examined previously. In this case there is evidently the competition of the electronegativity of the sp^2-hybridized carbon atom at the double bond, leading to the attraction of electrons from the aldehyde proton, and the positive conjugation effect in the system $C = C - C = O$ with the latter predominating. When the aldehyde molecule contains phenyl and acetylene groups adjacent to the aldehyde group, there is a contribution from the anisotropy leading in the first case to descreening and in the second, to additional screening of the aldehyde proton. The presence of ring currents and the heteroatom in the furan ring evidently leads to the low-field resonance of the aldehyde proton in furfural.

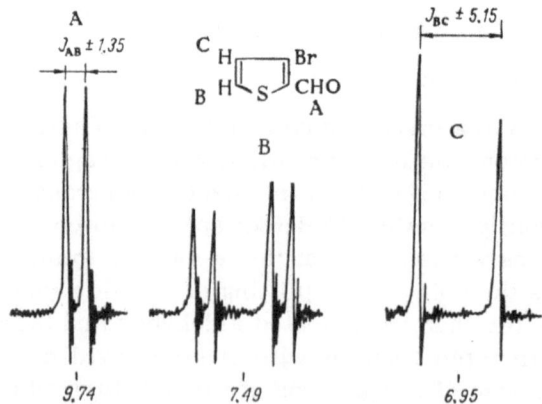

Fig. V-28. Spectrum of 3-bromothiophene-2-aldehyde.
Index H.60H.ACOH.I.

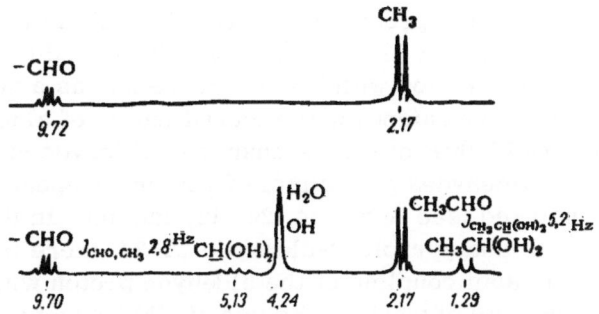

Fig. V-29. Spectra of acetaldehyde (index H.60.ЛСOH.I.)
and a solution of acetaldehyde in water (42 mole %).

It is more difficult to find a rule for the change in the spin—
spin coupling constants of aldehyde and α-carbon protons. It is
possible that in addition to electronic factors inside the aldehyde
molecule, the magnitude of these constants is also affected by association and the formation of a hydrogen bond involving the aldehyde proton. For a more detailed study of the mechanism of spin—
spin coupling in aldehydes, the relative signs of the constants of
short- and long-range spin coupling of the aldehyde proton with
other protons in the molecule were determined. However, results
were obtained here for which it is difficult to find a simple explanation. Thus, for example, while for 2-aldehydes of furan XXXIII
and 3-bromothiophene XXXIV (Fig. V-28) [58] the spin—spin coupling constants of the aldehyde proton with ring protons in positions
4 (and in the first case 5 also) have the same sign as the coupling
constant of the ring protons (J_{45}), in furan-3-aldehyde XXXV
$J_{CHO, 4}$ has the same sign as J_{45}, while $J_{CHO, 5}$ has the opposite
sign to these constants.

XXXIII   XXXIV   XXXV

If we assume that the spin—spin coupling constant of the ring
protons (like J_{ortho} in benzene and J_{cis} in ethylene) is positive,
then the coupling constant of the aldehyde proton with ring protons
is positive in all cases except for furan-3-aldehyde (additional investigations are required to explain this).

The signals of the protons at the carbon atom in the α-position relative to the aldehyde group are usually shifted towards lower fields in comparison with hydrocarbons and as a result of this they are also convenient for the identification of aldehydes. Due to the low-field shift of the resonance of aldehyde and α-carbon protons, aldehydes give simple first-order spectra which are readily analyzed (see A-23, 24, 26, 32, and 34). In the spectra of furfural (A-39) and pyrrole-2-aldehyde (A-34) there is clear long-range spin−spin coupling of the aldehyde proton with ring protons. The results of work on the use of NMR spectroscopy for studying the hydration of aldehydes are interesting (Fig. V-29) [59]. The pure aldehyde gives a spectrum consisting of two typical multiplets, namely, a doublet of the methyl group and a quartet of the aldehyde proton with a constant of 2.8 Hz, which is characteristic of aldehydes. In addition to these signals and the signal of water, an aqueous solution of acetaldehyde gives two additional multiplets, namely, a quartet and a doublet with a constant of 5.2 Hz, which may be assigned only to the hydrated form of the aldehyde $CH_3CH(OH)_2$. Comparison of the areas of the signals at different temperatures makes it possible to determine simply the equilibrium constant and the heat of hydration.

The NMR spectra of carboxylic acids are very similar to the spectra of aldehydes in that the signal of the carboxyl proton lies in the low-field region (9-11 ppm) and the resonance of the protons at the α-carbon atom is also shifted toward lower fields. However, in contrast to aldehydes, in carboxylic acids spin−spin coupling of the carboxyl proton with other protons of the molecule never appears. The position of the signal of the carboxyl proton changes, depending on the solvent and the solution concentration and in the case of a mixture of acids or on addition of water, alcohols, or other hydroxyl-containing substances, the signals of the carboxyl and hydroxyl protons merge into a common bond. Carboxylic acids are usually identified through other protons in the molecule (see A-16-19, 21, 35, 36, and 41).

6. COMPOUNDS CONTAINING ATOMS OF FLUORINE, PHOSPHORUS, AND OTHER MAGNETIC NUCLEI

The introduction of magnetic nuclei other than protons into the molecule of an organic compounds makes it possible to obtain

a large amount of additional information both from the proton spectra by analysis of spin—spin coupling of the protons with these nuclei and through resonance directly at these nuclei. However, up
to now the latter route has presented a technically difficult problem for all nuclei (apart from fluorine and to some extent phosphorus) and essentially it is not a general method of investigating
such compounds. The main approach to the study of organic compounds with different magnetic nuclei is the analysis of spin—spin
splitting in the spectra of protons and fluorine.

The replacement of a group of equivalent protons by a per-
fluoroalkyl group usually leads to simplification of the spectra due
to the conversion of the whole system from a strongly coupled system to one with weak spin—spin coupling. A simple example is
provided by comparing the spectra of ethanol and 2,2,2-trifluoro-
ethanol (Fig. A-10 and 11). In the case of ethanol the difference in
the chemical shifts of the methyl and methylene groups is comparatively small, leading, for example, to perturbation of the ratio
of intensities $1:2:1$ in the triplet of the methyl group, and in spectra run at lower frequencies, to the appearance of additional lines
[60] due to spin—spin coupling of the methyl and methylene groups
in the A_3B_2 system. The proton spectrum of 2,2,2-trifluoroethanol
contains in addition to the single signal of the hydroxyl proton,
which does not participate in spin—spin coupling, only a quartet
of the methylene protons with a clear ratio of intensities of
$1:3:3:1$ due to spin—spin coupling with the trifluoromethyl group
in the A_3X_2 system.

An analogous picture is observed in the spectra of the tri-
fluorohalopropanes, 1,1,1-trifluoro-3-chloropropane (see A-8), and
1,1,1-trifluoro-2,3-dibromopropane XXXVI (see Fig. A-7). The
spectra of compounds with a trifluoromethyl group are interpreted
more simply than their analogs containing protons instead of fluorine. The signal of the protons adjacent to the trifluoromethyl group
in the compound used for Fig. A-8 consists of 12 lines: three
quartets produced by spin—spin coupling with the three magnetically equivalent fluorine nuclei and two protons of the methylene
group bearing the chlorine. The assignment of the lines is confirmed by the spectrum of F^{19}, which consists of a triplet $1:2:1$.
This form of the spectra indicates free rotation about the single
$C-C$ bond. In contrast to this compound, in the dibromide XXXVI
rotation about the $CF_3CHBr-CH_2Br$ bond is hindered so that the
spectrum appears as the superposition of the spectra of at least

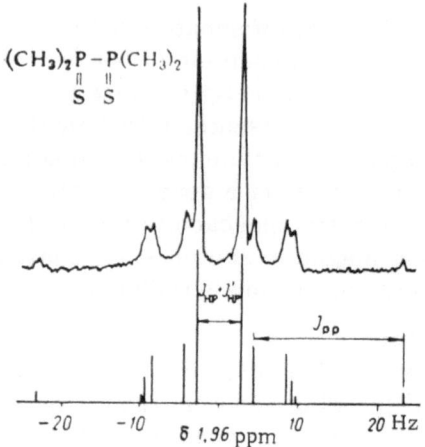

Fig. V-30. Proton spectrum of tetramethyldiphosphine disulfide. Index HP. 60. Л Z.M.Хл.

three conformers, of which one is present in a greater amount than
the other two. This is particularly clear from the F^{19} spectrum.
The F^{19} spectrum cannot be interpreted as three doublets since the
ratio of intensities does not obey the rule $1:2:1$.

However, the assignment of the chemical shifts of fluorine
of the trifluoromethyl group to definite conformers is difficult.

The protons adjacent to the bromine atoms in the spectrum
of the proton analog of compound XXXVI, namely, 1,2-dibromopro-
pane, also lead to the appearance of a difficultly analyzable multiplet, evidently indicating hindrance to rotation about the carbon–
carbon bond in this compound (see Fig. A-6).

The replacement of one of the equivalent protons of the molecule by another magnetic nucleus usually leads to complication of
the spectrum with a considerable increase in the amount of information obtainable. Thus, the replacement of some of the protons in an

equivalent group by deuterium is often used to determine the values of the spin−spin coupling constants between protons in this group. This method was used, for example, to determine the spin−spin coupling constants of the protons in formaldehyde, ammonia, hydrogen sulfide, and also various organic compounds.

The spectrum is still more complicated when the molecule contains several nonequivalent magnetic nuclei of different isotopes. In this case it is often possible to determine the magnitudes and even the relative signs of the spin−spin coupling constants of the nuclei of one isotope by analysis of the spectrum obtained with resonance at nuclei of another isotope present in the same molecule. Thus, for example, the spin−spin coupling constants of $P^{31}-P^{31}$ of a series of diphosphines [61] and bis(dimethylphosphono)ethylene (Ch. IV) were obtained by analysis of the proton spectra of these compounds. The proton spectrum of tetramethyldiphosphine disulfide (Fig. V-30) provides a typical example of such an analysis. Since there is no spin−spin coupling between the protons in the methyl groups through the five bonds, the constant J_{PP} may be obtained directly from the spectrum by measurement. It is interesting that due to the chemical equivalence of the nuclei of each of the magnetic isotopes, these spectra are independent of the magnetic field strength (and, correspondingly the frequency) of the instrument.

The same peculiarity is also shown by some acid fluorides of organic acids of phosphorus, in which the three magnetic isotopes form three groups of equivalent magnetic nuclei. The proton and F^{19} spectra of the diacid fluoride of methylphosphinic acid (see Fig. I-10) makes it possible to reproduce completely the form of the spectra which would be obtained by resonance at the P^{31} nuclei of these compounds (Ch. I). The form of the spectra is determined solely by the values of the constants J_{HP}, J_{HF}, and J_{FP} and therefore it does not change when plotted at different frequencies.

NMR spectroscopy of fluorophosphorus compounds is a convenient method of establishing the chemical structure, stereochemistry, and electron distribution in the molecules of these quite widespread substances. It was shown previously (Ch. III) that there is a relation between the chemical shifts δ_F and the J_{FP} constants in acid fluorides of acids of phosphorus and the character of the substituting groups at the phosphorus atom. Comparison of the

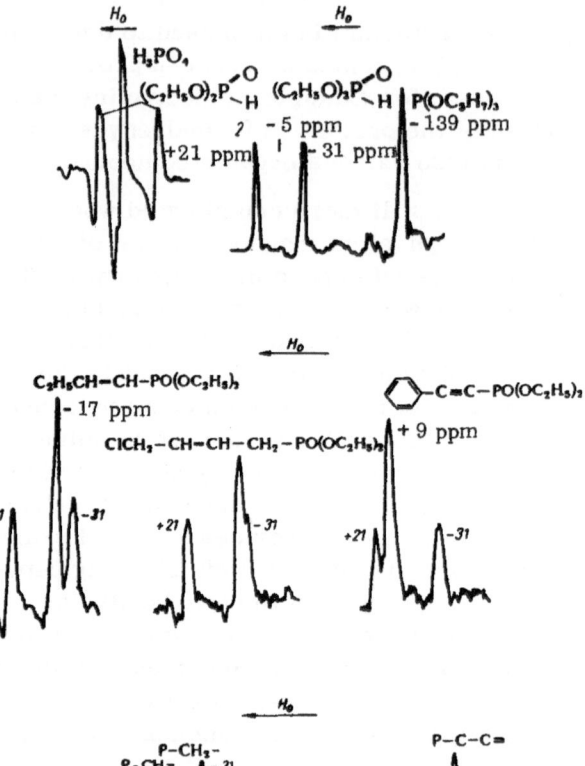

Fig. V-31. P[31] resonance spectra of esters of phosphorus acids:
the tube with the substance investigated (diameter 5 mm) was
placed in a wider tube (diameter ~7 mm) with the reference,
diethyl phosphite.

spin−spin coupling constants J_{FP} of the diacid fluorides of saturated and unsaturated phosphinic acids shows a correlation of
this parameter with the hybridization state of the first carbon atom
of the substituents, possibly due to a change in the contribution of
the d-orbitals of phosphorus to the P−F chemical bond [62]. The
results of Nixon and Schmutzler's studies of the spectra of fluorophosphorus compounds are interesting. Thus, the spectra of hal-

ides with pentacoordinate phosphorus indicate the nonequivalence
of the five positions of the substituents and this may be explained
by a bipyramidal structure for these molecules [63–65]. Nonetheless, it has not been possible up to now to observe a strict general relation between the chemical shifts, spin−spin coupling constants J_{FP}, and the structure of organosphosphorus compounds [66],
indicating the need for refinements of the theory of the origin of
these parameters.

Resonance directly at P^{31} nuclei is important for studying
organophosphorus compounds. Since the appearance of spectrometers with a high magnetic field strength of about 24 kG (100 MHz
for resonance at protons), the plotting of P^{31} spectra has become
a normal method on a par with H^1 and F^{19} spectroscopy (on such
instruments the resonance of P^{31} is observed at a frequency of 40
MHz). However, even with instruments with a weaker magnetic
field it is possible and fruitful to achieve resonance at phosphorus
nuclei. B. I. Ionin together with V. B. Lebedev and A. A.Petrov
used an autodyne oscillator to obtain P^{31} spectra at a frequency of
~13.1 MHz. It was found that a convenient reference (external)
was diethyl phosphite, whose P^{31} spectrum consists of two lines,
produced by splitting due to spin−spin coupling with the directly
attached proton (Fig. V-31). The magnitude of the splitting (~687
Hz) may change very slightly, depending on the temperature and
the nature of impurities, but it is readily determined from the proton spectrum of the substance. The use of such a reference facilitates the construction of a scale for the spectrum. The figure
shows several P^{31} NMR spectra. These spectra make it possible
to distinguish compounds in which the diethylphosphono group is attached to carbon with sp^3, sp^2, and sp hybridization as a result of
which it is possible to estimate the composition of α, β-acetylene,
allene, and β, γ-acetylene phosphonates [67].

At the present time the investigation of resonance at F^{19} nuclei plays a more important part and is used equally with proton
resonance for fluorine-containing compounds. In addition to the
main problem, i.e., establishing the structure of organofluorine
compounds, the resonance of F^{19} in fluoroaromatic compounds is
used for correlation of the chemical shifts of fluorine with the reactivity parameters of other substituents at the aromatic ring
(see Ch. II). An interesting method of determining the acidity of
weak bases was developed from this method [68]. For the rapid

reversible equilibrium

$$FC_6H_4X + H^+ \rightleftarrows FC_6H_4XH^+$$

the chemical shifts of F^{19} observed are the weighted mean values for the protonized and unprotonized forms. If the chemical shifts δ_F for such a system are measured in a series of solvents such as mixtures of acetic and sulfuric acids in different ratios, for which the acidity function H_0 is known, then the curve of δ_F against H_0 has a sigmoid form in the general case. The point of inflection corresponds to equal concentrations of protonized and unprotonized forms and at this point $pK_a = H_0$. The function H_0 for mixtures of acetic and sulfuric acids had been investigated previously. This method is essentially analogous to that used for determining acidity by means of ultraviolet spectroscopy.

Literature Cited

1.   K. Nukada, O. Yamamoto, T. Suzuki, M. Takeuchi, and M. Ohnisi, Anal. Chem., 35:1892 (1963).
2.   R.C. Ferguson and D. W. Marquart, J. Chem. Phys., 41:2087 (1964).
3.   D. Doskǒcilová, and B.Schneider, Coll. Czech. Chem. Comm., 29:2290 (1964).
4.   J. J. Burke and P. C. Lauterbur, J. Am. Chem. Soc., 86:1870 (1964).
5.   F. C. Mortimer, J. Mol. Spectr., 5:199 (1960).
6.   K. L. Williamson, C. A. Lanford, and C. R. Nicholson, J. Am. Chem. Soc., 86:762 (1964).
7.   T. Shaefer, F. Hruska, and G. Kotowycz, Can. J. Chem., 43:75 (1965).
8.   C. A. Reilly and J. D. Swalen, J. Chem. Phys., 32:1378 (1960).
9.   N. Schamp, Mededel Vlaam. Chem. Ver., 26:87 (1964); Chem. Abs., 61:14551c (1964).
10.   K. B. Wiberg, G. M. Lampman, R. P. Ciula, D. S. Connor, P. Shertler, and J. Lavanisch, Tetrahedron, 21:2749 (1965).
11.   S. Winstein, P. Carter, F. A. L. Anet, and A. J. R. Bowin, J. Am. Chem. Soc., 87:5247, 5249 (1965).
12.   L. M. Jackman, Application of NMR Spectroscopy in Organic Chemistry, Pergamon, London (1959), p. 117.
13.   K. M. Wellman and F. G. Bordwell, Tetrahedron Letters, 1963:1703.
14.   A. C. Huitric, J. B. Carr, W. F. Trager, and B. J. Nist, Tetrahedron, 19:2145 (1963).
15.   H. Booth, Tetrahedron, 20:2211 (1964).
16.   H. Booth, Tetrahedron Letters, 1965:411.
17.   D. B. Roll and A. C. Huitric, Tetrahedron, 20:2851 (1964).
18.   B. Coxon, Tetrahedron, 21:3481 (1965).
19.   A. A. Petrov, I. A. Maretina, and N. A. Pogorzhel'skaya, Zh. Obshch. Khim., 36 (1966).

20. E. I. Snyder, L. J. Altman, and J. D. Roberts, J. Am. Chem. Soc., 84:2004 (1962).

21. R. C. Hirst and D. M. Grant, J. Am. Chem. Soc., 84:2009 (1962).

22. Yu. V. Baskov, T. Urbanski, M. Witanowski, and L. Stefaniak, Tetrahedron, 20:1519 (1964).

23. F. Hruska, E. Bock, and T. Schaefer, Can. J. Chem., 41:3034 (1963).

24. A. A. Petrov and N. V. Elsakov, in: The Hydrogen Bond [in Russian], Izd. Nauka (1964).

25. R. T. Hobgood, Jr., and J. H. Goldstein, J. Mol. Spectr., 12:76 (1964).

26. M. Neuenschwander and H. Schaltegger, Helv. Chim. Acta, 47:1022 (1964).

27. W. Smith and B. A. Shoulders, J. Am. Soc., 86:3118 (1964).

28. G. Martin, Chim. et Ind., 89:168 (1963).

29. R. H. Martin, N. Defay, F. Geerts-Evrard, and S. Delaverne, Tetrahedron, 20:1073 (1964).

30. R. H. Martin, Tetrahedron, 20:897 (1964).

31. R. H. Martin, N. Defay, and F. Geerts-Evrard, Tetrahedron, 20:1091, 1495, 1505 (1964); 21:2421, 2435 (1965).

32. R. H. Martin, N. Defay, F. Geerts-Evrard, and H. Figeys, Bull. Soc. Chim. Belge, 73:189, 199 (1964).

33. R. H. Martin, N. Defay, F. Geerts-Evrard, R. H. Given, J. R. Jones, and R. W. Wedel, Tetrahedron, 21:1833 (1965).

34. W. G. Schneider, H. J. Bernstein, and J. A. Pople, Ann. New York Acad. Sci., 70:86 (1958).

35. W. G. Schneider, H. J. Bernstein, and J. A. Pople, Can. J. Chem., 35:1487 (1957).

36. P. L. Corio and B. P. Dailey, J. Am. Chem. Soc., 78:3043 (1956).

37. A. A. Bothner-By and R. E. Glick, J. Chem. Phys., 26:1651 (1957).

38. N. Jonathan, S. Gordon, and B. P. Dailey, J. Chem. Phys., 36:2443 (1962).

39. R. Freeman, N. S. Bhacca, and C. A. Reilly, J. Chem. Phys., 38:293 (1963).

40. B. Gestblom, R. A. Hoffman, and S. Rodmar, Mol. Phys., 8:425 (1964).

41. B. Gestblom, R. A. Hoffman, and S. Rodmar, Acta Chem. Scand., 18:1222 (1964).

42. K. Tori and M. Ogata, Chem. and Pharm. Bull. (Tokyo), 12:272 (1964).

43. G. S. Reddy, R. T. Hobgood, Jr., and J. H. Goldstein, J. Am. Chem. Soc., 84:336 (1962).

44. P. R. Wells, Austral. J. Chem., 17:967 (1964).

45. R. A. Hoffman, Arkiv. Kemi, 17:1 (1961).

46. D. Meuche, B. B. Molloy, D. H. Reid, and E. Heilbronner, Helv. Chim. Acta, 46:2483 (1963).

47. H. Sugiyama, S. Ito, and T. Nozoe, Tetrahedron Letters, 1965:179.

48. S. Forsen, Acta Chem. Scand., 14:231 (1960).

49. D. E. McGeer and M. M. Mocek, J. Chem. Education, 40:358 (1963).

50. J. G. Traynham and G. A. Knesel, J. Am. Chem. Soc., 87:4220 (1965).

51. O. L. Chapman and R. W. King, J. Am. Chem. Soc., 86:1256 (1964).

52. A. Mathias, Anal. Chim. Acta, 31:598 (1964).

53. S. Forsen, M. Nilson, J. A. Elvidge, J. S. Burton, and R. Steven, Acta Chem. Scand., 18:513 (1964).

54.   J. C. Ma and E. W. Warnhoff, Can. J. Chem., 43:1849 (1965).
55.   W. R. Anderson, Jr., and R. M. Silverstein, Anal. Chem., 37:1417 (1965).
56.   C. N. Reilley, Anal. Chem., 37:1298 (1965).
57.   R. E. Klinck and J. B. Stothers, Can. J. Chem., 44:45 (1966).
58.   S. Forsen, B. Gestblom, S. Gronowitz, and R. A. Hoffman, Acta Chem. Scand., 18:313 (1964).
59.   Y. Fujiwara and S. Fujiwara, Bull. Chem. Soc. Japan, 36:574 (1963).
60.   J. T. Arnold, Phys. Rev., 102:136 (1956).
61.   R. K. Harris and R. G. Hayter, Can. J. Chem., 42:2282 (1964).
62.   L. N. Mashlyakovskii, B. I. Ionin, I. S. Okhrimenko, and A. A. Petrov, Zh. Obshch. Khim., 37:(6) (1967).
63.   J. F. Nixon, Chem. and Ind., 1963:1955.
64.   R. Schmutzler, Angew. Chem., 77:530 (1965); International Edition, 4:496 (1965).
65.   J. F. Nixon and R. Schmutzler, Spectrochim. Acta, 20:1835 (1964).
66.   J. F. Nixon and R. Schmutzler, Spectrochim. Acta, 22:565 (1966).
67.   B. I. Ionin and A. A. Petrov, Zh. Obshch. Khim., 34:1174 (1964).
68.   R. W. Taft and P. L. Lewis, Anal. Chem., 34:436 (1962).

Chapter VI

Application of NMR Spectroscopy
in Various Fields of Organic Chemistry

1. INTERMEDIATE REACTION PRODUCTS
AND COMPLEXES

NMR Spectroscopy differs favorably from other physical methods of investigation (IR and UV spectroscopy) in that in some cases it makes it possible to detect directly the formation of intermediate products of chemical reactions (ions, reaction complexes, solvates, etc.). However, it should be noted that the conditions under which the NMR spectra of intermediate products are recorded often differ from the reaction conditions. In these cases some care must be exercised in comparing spectroscopic and chemical data. Despite these stipulations, the use of the NMR method often makes it possible to obtain quite unique data on the structure of intermediate products, making it possible to avoid the need of resorting to various hypotheses, often of a very speculative nature, to explain the mechanism of a process.

There has been a considerable number of studies of p r o - t o n a t i o n of organic compounds. Exact information on the site of addition of the proton is important to interpret the mechanisms of many chemical reactions which proceed in an acid medium. Thus, for example, in some textbooks on organic chemistry two protonation schemes are considered in the description of the mechanism of the hydrolysis of esters in an acid medium (at the oxygen atoms or the carbonyl and ether groups). At the same time, a study of

Fig. VI-1. NMR spectrum of the cation of indeno[2,1-a]phenalene in CF$_3$COOH—
H$_2$SO$_4$ (4:1 by volume), (34°C, 60 MHz).

the NMR spectra of protonated esters showed that the addition of
the proton occurs exclusively at the carbonyl oxygen atom [1].

In actual fact, in the spectrum of a solution of ethyl acetate
in a mixture of SbF$_3$—FSO$_3$H at —80°C at low fields at 10.75 ppm
there appears a single narrow signal of the bound proton. Had protonation occurred at the ethyl oxygen, one would have expected a
change in the multiplicity of the quartet of the methylene group
due to spin—spin interaction with the bound proton. However, this
change is not observed in practice even at very low temperatures
when the rate of proton exchange is markedly reduced. Hence it
follows that the protonation of ethyl acetate occurs at the oxygen
atom of the carbonyl group. These data were confirmed by a study
of the C^{13} NMR spectra [2].

An analogous investigation of N-methylformamides showed
that they are also protonated primarily at the oxygen atom [3].

Studies of the protonation of aromatic compounds are of great
value. The carbocations formed in this way are essentially models
of σ–complexes, i.e., intermediate products in electrophilic substitution reactions.

The protonation of benzene in HF—SbF$_3$ solution occurs only
at quite low temperatures (of the order of —80°C). Then in the
NMR spectrum in the absorption region of aliphatic methylene
protons there is only one signal at 3.60 ppm, whose area is somewhat less than the theoretically expected area for two protons. The
other ring protons absorb in the region of 7.2-8.8 ppm, i.e., at
lower fields than usual [4].

In cases where there is a choice between different nonequivalent positions in which the proton may be, determination of its

site of addition provides valuable information on the fine electronic structure of the substance.

Figuve VI-1 gives the spectrum of the cation formed by dissolving indeno[2,1-a]phenalene in a mixture of trifluoroacetic and sulfuric acids [5]. In contrast to the starting material, whose spectrum consists of a complex group of signals in the region between 7.25 and 7.90 ppm, the spectrum of the catioh may be partly interpreted. In this spectrum there is a singlet in a higher field region (4.51 ppm), which corresponds to the methylene group in position 12; its area corresponds to two protons. The poorly resolved signals between 7.25 and 7.85 ppm with a total area corresponding to approximately four protons, may be assigned to the protons in positions 8-11. The next group with a total area corresponding to 1.8 protons consists of two partly overlapping triplets with chemical shifts of 8.41 and 8.51 ppm; their appearance is most probably connected with the resonance of protons in positions 2 and 5 with the lines split due to spin—spin coupling (J 7.6 Hz) with adjacent protons (1,3 and 4,6 respectively). Finally, the four AB doublets and the singlet line correspond to protons in positions 1, 3, 4, 6, and 7. The position of the signals of protons 1-7 at lower fields in comparison with the other protons indicates localization of the positive charge in the phenalene part of the molecule. The result obtained agrees completely with previous ideas on the structure of this cation, which are confirmed by the results of MO calculation. In particular, the latter indicates concentration of the highest electron density in the molecule of indeno[2,1-a]phenalene in position 12.

An investigation of the protonation of phenol, anisole, and other phenols and aromatic ethers in fluorosulfuric and 70% perchloric acids showed that in all cases the proton adds at a carbon atom of the ring [6]. Thus, for example, anisole is protonated at the para-carbon atom of the ring. At the same time, according to recent investigations of the protonation of methoxybenzenes, in some cases (depending on the structure of the starting material) the proton may add at the oxygen atom [7].

In the NMR spectrum of methylcyclooctatetraene in concentrated sulfuric acid there are signals at 8.60, 6.70, 5.30, and 0.30 ppm with a ratio of areas of $5:1:1:1$, corresponding to the hydrogen atoms in positions 2-6, 7, H_A, and H_B.

The methyl group absorbs at 3.20 ppm. The great difference in the chemical shifts of the protons H_A and H_B is explained
by the presence of a ring current in the cation. The spectrum
demonstrates that protonation occurs exclusively in the α-posi-
tion relative to the methyl group. The position of the signals in a
spectrum determined in concentrated D_2SO_4 is analogous, but their
areas are in the ratio of 5:1:0.75:0.25. (The area of the signal
of the methyl group at 3.20 ppm does not change.) This indicates
that the proton in the methylhomotropylium cation occupies the
endo(H_A)-position more readily than the exo(H_B)-position [8].

In the protonation of indolizines the proton adds to the carbon
atom in position 3 if this atom bears no substituent; otherwise a
mixture of 1- and 3-protonated indolizines is formed [9]. The spectrum of a solution of 3-methylindolizine in trifluoroacetic acid

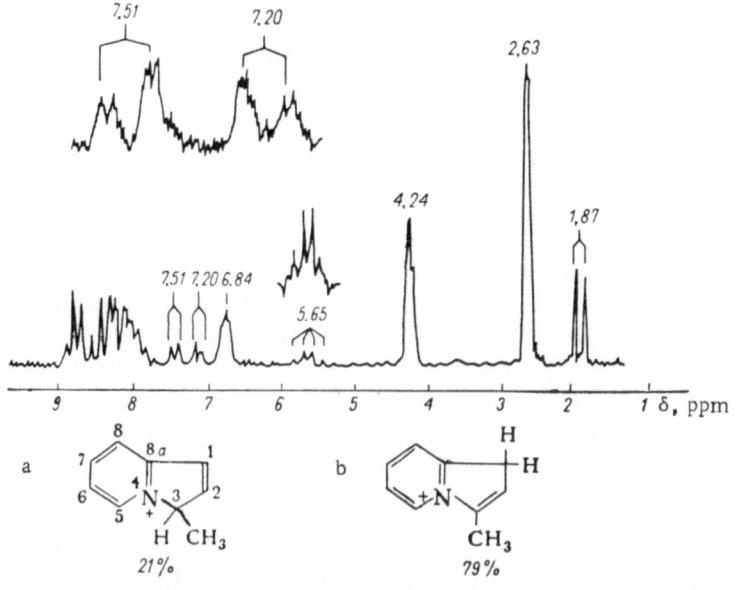

Fig. VI-2. NMR spectrum of a 0.5 M solution of 3-methylindolizine in trifluoroacetic acid (34°C, 60 MHz): a) 3-H cation; b) 1-H cation.

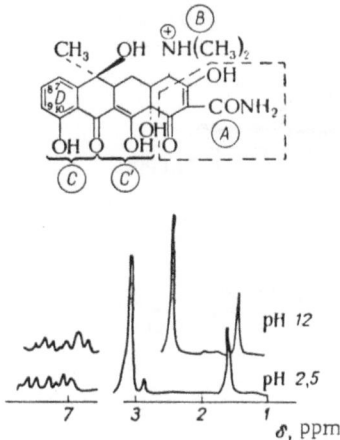

Fig. VI-3. Part of the NMR spectrum of tetracycline hydrochloride
in aqueous methanol (1 : 1 by weight),
(30°C, 60 MHz).

(Fig. VI-2) contains bands corresponding to both structures of the cation (3-H and 1-H) in a ratio of 21 to 79%. The spectrum of the 3-H cation (Fig. VI-2a) contains a quartet at 5.65 ppm and a doublet at 1.87 ppm, corresponding to the proton and the methyl group in position 3; the spin—spin coupling constant between them is 7.2 Hz. The two AB-doublets at low fields (7.51 and 7.20 ppm), which are connected with the resonance of protons in positions 2 and 1 (J_{AB} 6.3 Hz), correspond to the same structure. The spectrum of the 1-H cation (Fig. VI-2b), which is present in the greater amount, has a weakly split signal of the two protons in position 1 at 4.24 ppm and an incompletely resolved doublet of the methyl group in the position 3 at 2.63 ppm, J = 1.7 Hz. (In this structure the latter is connected to a double bond and therefore it gives a signal at lower fields.) The signal of the proton in position 2 of the cation has the form of a broad single peak with a chemical shift of 6.84 ppm. The protons of the six-membered ring in the two cations give difficultly analyzable signals in the region of 7.7-9.1 ppm.

By means of NMR spectroscopy it was possible to determine the protonation scheme of even such a complex compound as the antibiotic tetracycline.

Part of the spectrum of tetracycline is given in Fig. VI-3. The intense singlet at 3.17 ppm (pH 2.5) corresponds to the protons of the dimethylamino group, while the multiplet in the region of 6.6-7.6 ppm corresponds to the protons in positions 7, 8, and 9 of ring D. The chemical shifts of these signals depend on the pH of the medium, while the chemical shifts of the protons of the methyl group in position 6 are independent of it. With an increase in the pH of the medium from 2.5 to 12 the total change in the chemical shift of the dimethylamino group is 0.65 ppm, while that of the protons of ring D is 0.28 ppm. If we assume that the change in the chemi-

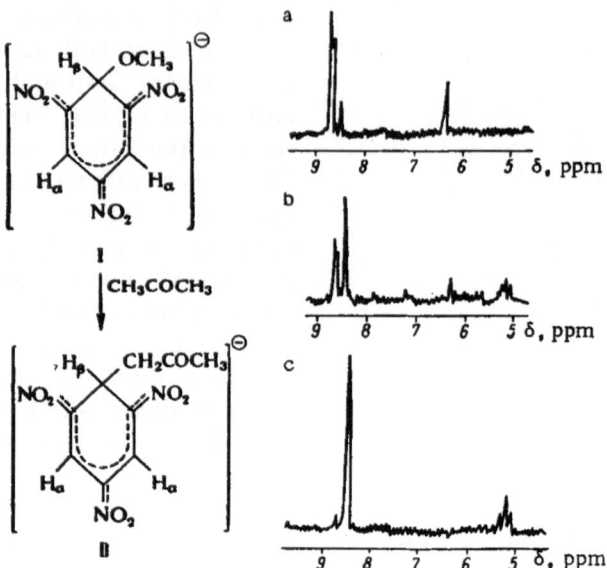

Fig. VI-4. Change in the NMR spectrum of a solution of the Meisenheimer complex in acetone (33°C, 60 MHz) with time: a) after 0 min; b) after 2 min; c) after 90 min.

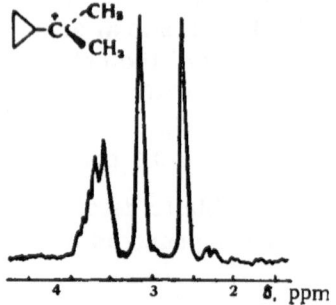

Fig. VI-5. NMR spectrum of di-methylcyclopropylcarbinol in SO_2—SOClF—SbF_5 at −75°C.

cal shift of the protons of the di-methylamino group is produced solely by protonation of position B, while the change in the shift of the protons of ring D is produced exclusively by protonation of position C, then the fraction of the change in the chemical shift on addition of each equivalent of acid equals the concentration of protons in each of the exchanging states. The dissociation in position C' is insignificantly small and is ignored so that the degree of protonation of position A is determined by difference. It was found that on addition of one equivalent of acid to a solution of tetracycline hydrochloride position C' was 100% protonated, while the other positions were not protonated. On addition of two equivalents of acid position A was 33% protonated, position B 16%, position C 51%, and position

C' 100%. Knowing the percentage protonation of each of the main positions and the macroscopic dissociation constants, the authors determined the microscopic dissociation constants in the individual positions [10].

A series of investigations has been devoted directly to the study of c o m p l e x e s formed during electrophilic and nucleophilic substitution reactions in the aromatic series. Thus, Christer and his co-workers used NMR spectroscopy to show [11] that bromination of 2-naphthol-6,8-disulfonic acid is accompanied by the intermediate formation of a σ-complex, which is then converted into the reaction product. At the same time, no iodination occurs under analogous reaction conditions. Judging by the NMR spectrum, in the latter case an intermediate charge-transfer complex (π-complex) is formed and this explains the difference in the direction of the reaction.

Figure VI-4a shows the spectrum of the protons of the aromatic ring in the Meisenheimer complex I, obtained by the reaction of 1,3,5-trinitrobenzene with sodium methylate. The doublet at 8.66 ppm (J < 1 Hz) corresponds to two H_α protons, while the triplet at 6.33 ppm (J < 1 Hz) corresponds to the proton H_β in structure I. When the complex I is dissolved in acetone the methoxyl group is replaced by an acetonyl group. Figure VI-4, b and c, shows the change in the spectrum of a solution of complex I in acetone with time. The intensity of the doublet at 8.66 ppm falls while there is an increase in the intensity of a new doublet at 8.47 ppm (J < 1 Hz), corresponding to the protons H_α in structure II. Simultaneously with the disappearance of the triplet at 6.33 ppm, in the spectrum there gradually appears a new triplet at 5.18 ppm (J 9 Hz) corresponding to the proton H_β in structure II. The sum of the intensities of these peaks remains constant during the whole of the exchange process. Hence it follows that there is the same number of ring protons in the reacting substance I and the product II. If the exchange is carried out in hexadeuteroacetone, then instead of the triplet at 5.18 ppm there appears in the spectrum a broad signal due to the constants $J_{H_\alpha H_\beta}$ and $J_{H_\beta D}$ being small in magnitude [12].

The NMR method has been used widely for the direct observation of stable classical carbocations [13]. The latter are usually obtained at quite a high concentration during the dissolution of ap-

propriate alcohols, alkyl halides, and ethers in mixtures of the
type SO_2-SbF_5, $HF-SbF_5$, $FSO_3H-SbF_5-SO_2$. Thus, for example,
when tert-butyl fluoride is treated with SbF_5 a tert-butyl cation is
formed and this leads to a shift in the signal of the methyl groups
by 3.05 ppm to lower fields and the disappearance of the doublet
splitting of the starting material, produced by the presence of the
fluorine atom. In SbF_5 solution this cation is quite stable at room
temperature. In contrast to tert-butyl fluoride, a solution of tert-
amyl fluoride in antimony pentafluoride at room or at elevated
temperature gives an unresolved spectrum with a broad band in
the region of 4.6 ppm. At 0-2°C, there appear in the spectrum
separate, weakly resolved signals of methyl and methylene groups,
indicating rapid exchange of the fluorine atom in the equilibrium
system:

$$CH_3-\underset{\underset{CH_3}{|}}{\overset{\overset{F}{|}}{C}}-CH_2-CH_3+SbF_5 \rightleftarrows CH_3-\underset{\underset{CH_3}{|}}{\overset{+}{C}}-CH_2-CH_3+SbF_6^-$$

At −30°C the exchange rate is markedly reduced so that there
appear in the spectrum sharp multiplets of three nonequivalent proton groups with a ratio of intensities of 2.96 : 5.97 : 2.00, which is
close to the theoretical ratio (3 : 6 : 2) [14]. It is interesting that in
this ion there is considerable spin−spin coupling (~7 Hz) of the
methyl and methylene protons through the sp^2-hybridized carbon
atom which bears the positive charge. When the ion is formed
there is a sharp increase in the constant $J_{HC^{13}}$ of the proton with
the positive carbon atom, namely, from 151 in the starting compound to 382 Hz in the tert-amyl cation, while $J_{HC^{13}}$ for the methyl
groups changes insignificantly, namely, from 127 to 130 Hz. This
change is difficult to explain within the framework of the usual concepts of the contact mechanism of $H-C^{13}$ spin−spin coupling (see
Ch. III).

Interesting information is provided by the spectrum of the
dimethylcyclopropylium cation formed by the addition of dimethylcyclopropylcarbinol to the system $SO_2-SOClF-SbF_5$ at −75°C. The
spectrum reveals the presence of two nonequivalent methyl groups,
which differ by 0.54 ppm in chemical shift (Fig. VI-5). This may
be explained by assuming that the plane of the cyclopropane ring

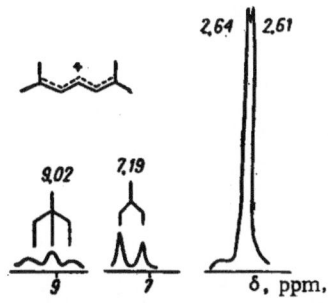

Fig. VI-6. NMR spectrum of 2,6-dimethylheptadien-2-ol-4 in 96% H_2SO_4 (60 MHz).

is perpendicular to the plane

$$\overset{+}{C}\overset{CH_3}{\underset{CH_3}{\diagdown}}$$ of the system and parallel to the axis of the vacant orbital. In this case the two methyl groups are in nonequivalent positions relative to the ring and due to the magnetic anisotropy of the latter the protons of the methyl group in the cis-position relative to the ring give a signal at higher fields [15].

When dimethylcyclopropylcarbinol is dissolved in FSO_3H at $-50°C$ the spectrum has a different form and the signals close to 7.74 ppm indicate isomerization of the cation and the appearance of olefinic protons. In this case the 2-methylpentenyl cation is evidently formed [16].

It is well known that in carbonium ions containing multiple bonds, the positive charge is delocalized due to conjugation in the system. NMR spectra provide clear confirmation of this fact.

Thus, the spectra of allyl and 2-methallyl cations (important intermediates in many chemical reactions) indicate the complete symmetry of the cation [17]. The four methylene protons of the allyl cation formed by dissolving allyl fluoride in the mixture SO_2-SbF_5 (at $-60°C$) give quite a broad doublet at 8.97 ppm, while the proton of the methyne group gives a broad unresolved signal at 9.64 ppm. The spectrum contains a whole series of broad peaks due to secondary products, namely, cycloalkenyl cations. The four methylene protons of the 2-methallyl cation under analogous conditions also give a broad signal at 8.95 ppm; the signal of the protons of the methyl group lies at 3.85 ppm, i.e., in approximately the same region as the signals of the protons of the tert-butyl cations. In the case of the 2-methallyl cation much less of the sec-

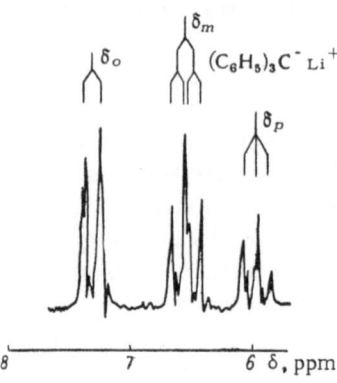

Fig. VI-7. NMR spectrum of triphenylmethyllithium in tetrahydrofuran (60 MHz).

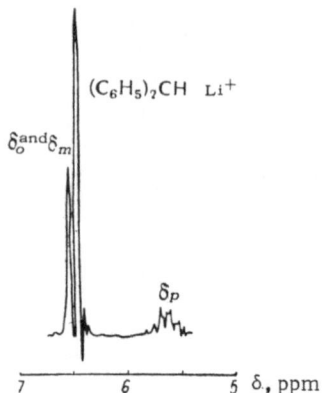

Fig. VI-8. NMR spectrum of
diphenylmethyllithium in
tetrahydrofuran (60 MHz).

ondary cyclic products are formed and this indicates its greater stability.

The F^{19} NMR spectrum of a solution of allyl fluoride in SO_2-SbF_5 shows the absence of aliphatic fluorine and the presence of the anion SbF_6^-, in which the fluorine atoms are equivalent. The considerable paramagnetic shift of the signal of the methyne proton in the allyl cation and the methyl protons in the methallyl cation indicates strong 1,3-π-interaction.

Similarly the spectrum of a solution of 2,6-dimethylheptadien-2-ol-4 in 96% sulfuric acid also indicates the complete symmetry of the cations formed (Fig. VI-6) [18]. A peculiarity of the spectrum of the cation, in contrast to the starting diene, is the substantial increase in the spin—spin coupling of the equivalent protons at the multiple bonds (in positions 3 and 5) with the proton in position 4, which is then 13.2 Hz. This undoubtedly indicates an increase in the double bonding of the carbons C_3-C_4 and C_4-C_5. The signal of the proton in position 4 is at lower fields (9.02 ppm) than for the protons in positions 3 and 5 (7.19 ppm) and this agrees with a MO calculation, which indicates a higher residual positive charge at C_4. It is interesting that this spectrum shows the non-equivalence of the methyl groups (2.61 and 2.64 ppm); this indicates the absence of free rotation about the carbon—carbon bonds in the cation.

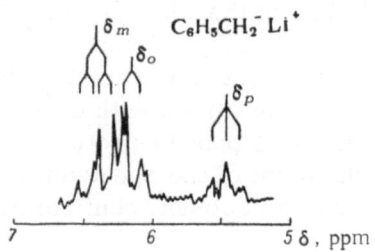

Fig. VI-9. NMR spectrum of benzyllithium in tetrahydrofuran (60 MHz).

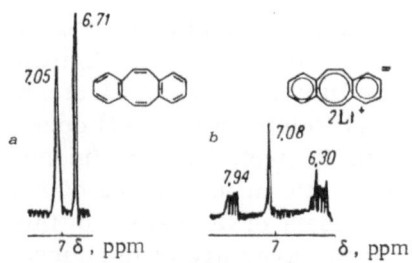

Fig. VI-10. NMR spectrum of sym-dibenzocyclooctatetraene (a) and its dianion (b) in tetrahydrofuran (0.5 M solution, 60 MHz).

The NMR spectrum of the tricyclopropylmethyl cation in sulfuric acid contains only one singlet signal at 2.26 ppm due to the identical values of the chemical shifts of the α- and β-protons of the cyclopropane ring [13]. This coincidence of the chemical shifts may be explained by the presence of conjugation, which produces a considerable shift in the signals of the more remote β-methylene protons into the region of the signals of the α-protons. The direct observation of a large number of cyclopropyl substituted carbocations is a good illustration of the capacity of the cyclopropane ring for conjugation and stabilization of alkenyl cations [13].

A considerable achievement of NMR spectroscopy is the direct recording of nonclassical carbocations, particularly since some authors have doubted the possibility of their existence until recently. Eberson and Winstein observed the nonclassical anthrylethyl cation [19]. Olah and his co-workers recorded the formation of bridge p-anisoyl and 2,4,6-trimethylphenonium ions from the corresponding β-ethyl chlorides by dissolving them in a mixture of SbF_5-SO_2 at low temperatures [20].

Naturally, the results obtained cannot be applied mechanically to the conditions under which solvolytic processes are studied since the strongly acidic and strongly ionizing medium (SbF_5-SO_2) favors a change from equilibrium systems of classical ions, which are stable in more basic solvents, to static bridge (nonclassical) ions.

The NMR method has been applied with as much success to the direct observation of carbanions.

Thus, Sandel and Freedman investigated the NMR spectra of solutions of triphenylmethyllithium, diphenylmethyllithium, and benzyllithium in tetrahydrofuran [21]. The spectrum of triphenyl-methyllithium (Fig. VI-7) consists of three multiplets with centers at 5.96 (p-protons), 6.52 (m-protons), and 7.31 ppm (o-protons) with a ratio of intensities of $2:2:1$. The form of the spectrum is independent of the nature of the cation and the solvent, confirming the ionic structure of the substance.

The spectrum of the phenyl protons of diphenylmethyllithium (Fig. VI-8) consists of a multiplet at 5.65 ppm and two peaks at 6.52 ppm with a ratio of intensities of $1:4$. At the same time in the spectrum of p,p'-dideuterodiphenylmethyllithium in the same region there is only one singlet signal at 6.52 ppm. Hence it follows that the multiplet in the high-field region belongs to the p-protons, while the signals at lower fields belong to the o- and m-protons (A_2B_2C system).

In the spectrum of the phenyl protons of benzyllithium (Fig. VI-9) the multiplet at high fields corresponds to the p-protons and those at low fields correspond to the o- and m-protons. It is possible to determine directly from the spectrum the chemical shift of the p-proton (5.50 ppm) and also J_{o-m} and J_{m-p} (8 and 6.2 Hz, respectively). When the hydrogen in the p-position is replaced by deuterium the spectrum is converted into a typical A_2B_2 quartet (J_{o-m} 8 Hz), confirming the position of the o- and m-protons and making it possible to determine more accurately the magnitudes of their chemical shifts (δ_m 6.30 and δ_o 6.09 ppm).

The electron densities at different positions in the phenyl rings were calculated from NMR data. The values obtained agree well with data calculated by the self-consistent field method. The charge distribution in the triphenylmethyl anion is particularly interesting since it is found that the m-carbon atoms of the ring bear more negative charge than the o-atoms (an analogous distribution of positive charge is observed in the triphenylmethyl cation [22]). This is evidently explained by the redistribution of the charge due to the mutual repulsion of the electronic clouds inside the ring. The effect becomes less noticeable in diphenyl and benzyl carbanions.

In the NMR spectrum of a solution of sym-dibenzocyclo-octatetraene in tetrahydrofuran (Fig. VI-10a) there are two bands at 6.71 and 7.05 ppm, which belong to olefinic and aromatic protons, respectively. When the solution is treated with metallic lithium the dianion is formed (Fig. VI-10b) and its spectrum contains three groups of lines of equal intensity, namely, a singlet at 7.08 ppm and A_2X_2 multiplet at 7.94 and 6.30 ppm [23].

Thus, NMR spectra provide a simple and reliable method of recording organic ions. Another method of using NMR spectroscopy consists of determining the formation constant of an ion by determining the amounts of ionized and unionized form when the substance investigated is dissolved in solvents of different acidities. This method, which was described above for the case of fluorobenzene derivatives, is essentially the same in principle as other spectroscopic methods.

The NMR method is used widely to study c o m p l e x f o r m a - t i o n by organic compounds. Thus, a study of the H^1, B^{11}, and F^{19} NMR spectra of complexes of organic compounds with boron halides made it possible to find a correlation between the chemical shifts of these nuclei and the s t r e n g t h of the Lewis acids, and also to compare the donor power of different organic ligands.

As a rule, in the NMR spectra of complexes of N,N–dimethylformamide with a series of Lewis acids the methyl groups are nonequivalent, indicating coordination of the acid to the oxygen atom. In actual fact, had coordination occurred at the nitrogen atom this would have been accompanied by disruption of the double–bond character of the C — N bond and would have led to the equivalence of the methyl groups. There are no peaks of free dimethylformamide in the spectra of the stoichiometric adducts. When excess dimethylformamide is present the spectrum contains signals of two pairs of methyl protons, one of which corresponds to the complex and the other to free dimethylformamide. These data show that the equilibrium is shifted toward the complex. By comparing the chemical shifts of the methyl protons in different complexes it was possible to compare the relative strengths of Lewis acids [24]. The series obtained agrees with data from IR spectroscopy.

Foster and Fyce showed that the NMR method may be used successfully to determine the association constants K of donor-acceptor complexes. For this purpose the chemical shift δ of the nucleus of an acceptor (A) is determined in a series of solutions containing different concentrations of the donor (D) (in all solutions

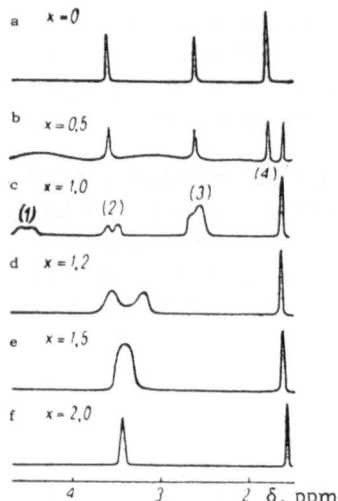

Fig. VI-11. NMR spectra of meth-allylpalladium chloride with x moles of triphenylphosphine added in chloroform (33.5°C, 60 MHz).

[D] ≫[A]). It may be shown that

$$\frac{\delta}{[D]} + \delta K = \delta_0 K$$

where δ_0 is the chemical shift of the given nucleus of the acceptor in the pure complex. A graph of $\delta/[D]$ against K should consist of a straight line with a slope equal to K. The method has been applied successfully to the calculation of the association constants of the complex of 1,3-difluoro-2,4,6-trinitrobenzene with hexamethylbenzene simultaneously from the chemical shifts of H^1 and F^{19}. Good agreement of the results was obtained [25].

The study of the NMR spectra of organometallic compounds [26] and complexes which are intermediate products or catalysts of many chemical processes deserves particular attention. There has been a considerable number of studies of the structure and stereochemistry of Grignard reagents and allyl complexes of transition metals.

Powell and his co-workers studied the intermediate products of the conversion of π-methallylpalladium chloride into the corresponding σ-complex in the presence of triphenylphosphine [27]. Figure VI-11a shows the spectrum of the starting π-complex (x = 0), while Fig. VI-11b shows the change in its spectrum when 0.5 mole of triphenylphosphine is added (x = 0.5). When x = 2 the spectrum (Fig. VI-11) shows two comparatively close signals with a ratio of intensities of 4 : 3 and this, together with the absence of spin–spin interaction between the allyl proton and the phosphorus atom, indicates the formation of a σ-allyl ligand in which there is rapid exchange:

$$H_2C=C-CH_2PdCl \rightleftarrows ClPdH_2C-C=CH_2$$
$$\underset{CH_3}{|} \underset{CH_3}{|}$$

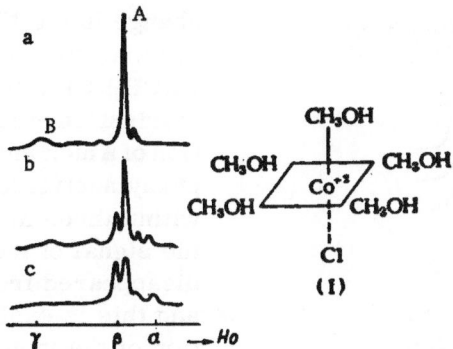

Fig. VI-12. NMR spectra of 0.9 solution of $Co(ClO_4)_2$ in methanol (−80℃, 60 MHz). (The spectra a, b, and c correspond to solutions containing 0, 1.8, and 3.5 Cl⁻ ion.)

A similar exchange was observed previously in allyl Grignard reagents. The spectrum of the stable intermediate complex of methallylpalladium chloride and triphenylphosphine (1 : 1) is shown in Fig. VI-11b. It consists of three broad doublets and one narrow signal, whose intensity ratio is 1 : 1 : 2 : 3. The authors assigned to this intermediate compound the structure shown.

$$
\begin{array}{c}
\text{(3)} \\
\text{CH}_2 \\
\diagup \quad \diagdown \\
\text{CH}_3\!-\!\underset{\text{(4)}}{\text{C}}\cdots\text{Pd}\!-\!\text{P(C}_6\text{H}_5)_3 \\
\diagdown\!\!\diagdown \quad | \\
\text{C} \quad \text{Cl} \\
\diagup \quad \diagdown \\
\text{H} \quad \text{H} \\
\text{(1)} \quad \text{(2)}
\end{array}
$$

(The numbering of the resonance signals of the complex (Fig. VI-11c) corresponds to the numbering of the protons.) The signals of protons 1 and 2 are split due to spin−spin interaction with the phosphorus atom, while the signal of the proton 3 is split either for the same reason or due to hindrance to rotation. The width of the signals indicates quite rapid exchange of protons in positions 1 and 2 and slower exchange of the protons in these positions with proton 3. When triphenylphosphine is added the exchange rates increase markedly (Fig. VI-11d-f).

Bystrov and other investigators studied the nature of the

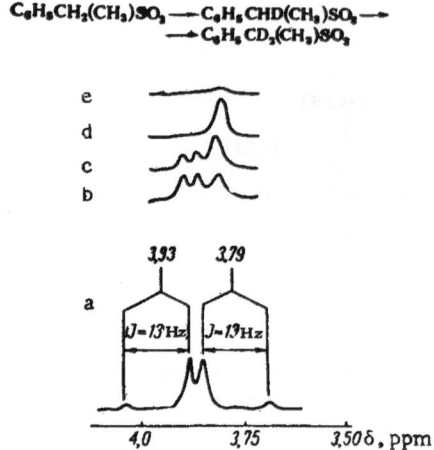

$C_6H_5CH_2(CH_3)SO_2 \longrightarrow C_6H_5CHD(CH_3)SO_2 \longrightarrow$
$\longrightarrow C_6H_5CD_2(CH_3)SO_2$

Fig. VI-13. Change in the NMR spectrum of the methylene protons of benzyl methyl sulfone (solution in D_2O) during deuterium exchange: a) in the absence of alkali; b) in the presence of 1 M NaOD (25% exchange occurred); c, d, e) 40, 50, and 95% exchange, respectively (the spectra in Figs. b, c, d, and e are on a different scale).

change in the NMR spectrum of the system $(C_5H_5)_2TiCl_2 \cdot Al(CH_3)_2Cl$, which initiates polymerization processes, on addition of a monomer, namely, phenylacetylene. It was found that within three minutes of mixing the signal of the $Ti-CH_3$ group disappeared from the spectrum and this is explained by insertion of the monomer at the $Ti-C$ bond [28].

The NMR method is used widely to study the solvation of ions in solution and interionic interactions, particularly of paramagnetic complexes of transition metals. Thus, Luz and Meiboom found that for the ions Co^{2+}, Ni^{2+}, and Mn^{2+} at low temperature the rate of exchange between f r e e solvent molecules (methanol) and those bound in a complex is so low that it is possible to observe separate signals for the free and bound molecules [29]. Figure VI-12 gives the spectra of solutions of Co^{2+} in anhydrous methanol at $-80°C$. Signals A and B are assigned to the methyl and hydroxyl protons of the complex $[Co(CH_3OH)_6]^{2+}$ (Fig. VI-12a). When chlorine ion is added to the solution there appear in the spectrum three new peaks, α, β, and γ (Fig. VI-12b). With an increase in the concentration of Cl^- the intensity of these peaks also increases while the intensity of the peaks of the complex $[Co(CH_3OH)_6]^{2+}$ gradually decreases (Fig. VI-12c). The ratio of the intensities of the peaks β and α remains approximately constant (4:1). The authors assigned the new peaks α, β, and γ to protons of the complex $[Co(CH_3OH)_5Cl]^+$, which has a tetragonally symmetrical structure. The peak β belongs to protons of methyl groups occupying an equatorial position, the peak α to protons of the axial methyl group, and the peak γ to hydroxyl protons of the new complex.

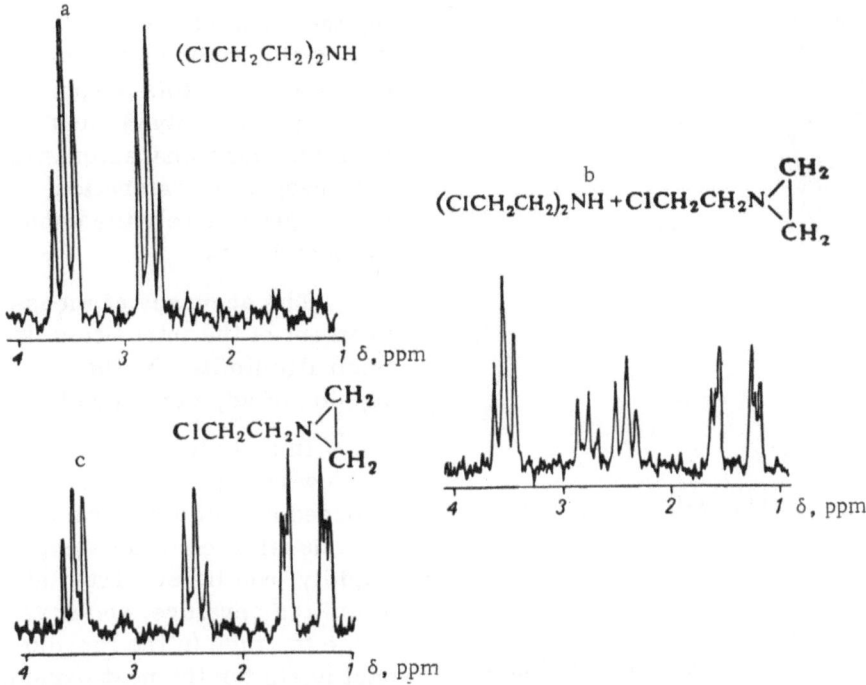

Fig. VI-14. Change in the NMR spectrum during the solvolysis of bis(2-chloroethyl)-amine in D_2O (37 °C, 60 MHz): a) 5 min after addition of NaOH; b) after 40 min; c) at the end of the reaction.

2. KINETICS AND MECHANISM

OF ORGANIC REACTIONS

An NMR spectrometer can be used as an analytical instrument to record changes in concentrations of substances during a chemical reaction. Since the minimum time required to record a spectrum is ~1 min, the method can only be used to study chemical reactions whose half-period does not exceed this time. A further limitation to the method is the need to work with quite concentrated solutions (not less than 5%) due to the comparatively low sensitivity of contemporary spectrometers. Finally, the method is applicable, strictly speaking, only under conditions which are far from saturation and this cannot always be achieved in practice. Therefore, the intensity of the signals of a reacting sample is often compared with the intensity of other lines which do not change dur-

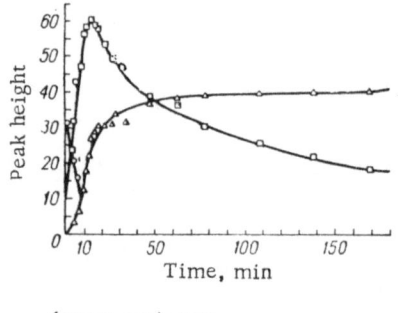

o $(ClCH_2CH_2)_2NCH_3$

□

Δ $2Cl^-$

Fig. VI-15. Change in the concentration of samples during the solvolysis of 2,2'-dichloro-N-methyldiethylamine in D_2O at 37°C.

ing the chemical reaction. Added standards are sometimes used for this purpose. The signals involved must have the same characteristics with respect to saturation, i.e., approximately the same product $T_1 \cdot T_2$.

The accuracy of measurement of the intensities is naturally limited by the overlapping of adjacent signals.

In a study of the kinetics of slow reactions which only proceed at high temperatures it is usual to cool the sample rapidly, run its spectrum at room temperature, and then place the tube in the thermostat again for the next period. A series of identical samples is used most frequently.

The method is used particularly often for investigating deuterium exchange reactions. Figure VI-13a gives the NMR spectrum of the methylene protons of benzyl methyl sulfone in heavy water and this consists of a quartet. As Fig. IV, b and c shows, as deuterium exchange proceeds the quartet gradually disappears and is replaced by a broad signal with a center at 3.79 ppm. Then this peak gradually disappears. The broad signal at 3.79 ppm is converted into a narrow singlet by decoupling of the deuterium; hence it follows that the methylene proton absorbing at 3.93 ppm exchanges more rapidly than the second methylene proton. Integration of the spectra makes it possible to determine the exchange rate of each of the methylene protons separately [30].

By using the NMR method for kinetic investigations it is often possible to determine simultaneously the rate constants of reactions from the disappearance of the signals of the starting materials and the increase of the signals of the final reaction products. Thus, Levins and Papanastassiou studied the kinetics and mechanism of

the solvolysis of 2-haloethylamines in an alkaline medium [31].
Figure VI-14 gives spectrograms of different stages in the solvolysis of bis(2-chloroethyl)amine hydrochloride. The spectrum plotted 5 min after the addition of alkali (Fig. VI-14a) shows two triplets of equal intensity of the α- and β-methylene protons of the
starting material at 2.77 ppm and 3.55 ppm, respectively. The
remaining signals are difficult to distinguish from the background
noise. In the spectrum plotted 40 min after the beginning of the
reaction (Fig. VI-14b) the intensity of the peak at 2.77 ppm is
markedly reduced and at the same time there appear three new
triplets of equal intensity with centers at 1.21, 1.41, and 2.41 ppm,
which belong respectively to the methylene protons of the aziridine ring and the α-methylene protons of the β-chloroethyl group
in the reaction product, namely, N-β-chloroethylaziridine. The
intensity of the triplet at 3.55 ppm does not change since the chemical shifts of the β-methylene protons of bis(2-chloroethyl)amine
and N-β-chloroethylaziridine are identical. Finally, when the reaction is complete there appear in the spectrum only the signals
of the final product, namely, the aziridine (Fig. VI-14c).

The study of the solvolysis of 2,2'-dichloro-N-methyldiethyl-
amine hydrochloride is of particular interest since by applying the
NMR method to this compound it was possible to determine simultaneously the changes in the concentrations of the starting amine,
the intermediate aziridine ion, and the final product, namely, the
piperazinium ion (Fig. VI-15). Measurements were made 1 min
after neutralization of the amine hydrochloride. The amount of
the starting substance was measured from the height of the singlet
peak of N-CH_3 at 2.35, the aziridinium ion from the height of the
singlet of N-CH_3 at 3.17, and the piperazinium ion from the signal
of the equivalent methylene protons of the ring at 4.1 ppm. The
reaction proceeded quite rapidly and this made it difficult to use
an integrator. Therefore the concentrations of the substances were
determined by direct measurement of the peak heights. This example clearly emphasizes the particular advantages of NMR techniques, which make it possible to record the formation and accumulation of intermediate and final reaction products in a number of
cases.

The NMR method has been used successfully for qualitative
observation of the course of chemical reactions. The data obtained are often of great value for determining mechanisms. Thus,

for example, Eisch and Husk studied the allyl rearrangement of organic compounds of aluminum. The authors added diisobutyl-aluminum hydride (R_2AlH) to a series of conjugated alkenes. The adducts obtained were treated with heavy water and the hydrocarbons formed were analyzed by NMR spectroscopy. The position of the $C-D$ bonds in the hydrocarbons should correspond to the position of the $C-Al$ bonds in the adducts if no rearrangement occurs during the hydrolysis. To check the reliability of the labeling the authors studied the NMR spectrum of the adduct of R_2AlH to 1,1-diphenyl-1,3-butadiene directly. The signal of the phenyl protons of the adduct at 7.17 ppm was taken as a standard with an intensity corresponding to 10 protons. Then the number of olefinic protons (H_v) absorbing in the range of 5.7-6.5 ppm was determined relative to this standard. The value of H_v was found to equal 1.50 ± 0.05. After treatment of the adduct with heavy water and plotting of the NMR spectrum of the hydrolysis products it was found that they contained 64% of trans-1-deutero-1,1-diphenyl-2-butene and 36% of 4-deutero-1,1-diphenyl-1-butene. The close agreement between the values of H_v observed experimentally and expected for hydrolysis if there is no rearrangement leads to the conclusion that in the reactions investigated the 1,4-adduct is formed predominantly and that the hydrolysis is not accompanied by rearrangement [32].

3. KINETICS OF EXCHANGE PROCESSES

NMR spectroscopy is used widely in physical organic chemistry to study the rates and activation parameters of so-called exchange processes, i.e., equilibrium chemical conversions during which there is a periodic change in the magnetic environment of the nucleus examined (group of nuclei). They include such unimolecular processes as isomerization, valence tautomerism, conformational conversions (hindered internal rotation, hindered inversion of rings, etc.), and also chemical exchange reactions of a higher order (exchange of atoms, groups of atoms, molecules, or electrons). Exchange processes affect the form of the resonance signals. Therefore, it is possible to relate some spectral parameters (most frequently the signal width and rarely the height or the distance between signals) to the rate of the exchange process.

The method may be used to study rates of fast exchange processes with a mean lifetime of the reagents τ from 10^{-1} to 10^{-3} sec [33-35].

To determine the rate of an exchange process it is first necessary to calculate the value of τ from spectral data.

A universal method of determining it consists of first calculating theoretical spectra for different values of τ, starting from the most general equations, taking into account the effect of the exchange process on the spectrum [36]. Then the parameters of the observed spectrum are compared with the parameters of the calculated spectra and the case where they are identical gives the value of τ directly from the theoretical spectrum. The calculation is carried out either analytically or graphically. The theory of the method was described in detail by Kaplan [37], Alexander [38], and Johnson [39]. The method is suitable for the whole range of rates of exchange processes (10^{-1}-10^{-3} sec), but its application to real systems is usually very laborious, even despite the use of computing techniques.

In the cases when the exact solution of general equations is laborious, equations are used which analyze the system under conditions approaching fast or slow exchange.

It may be shown that in the general case for the n-th line under conditions approaching slow exchange the following relation holds [33]:

$$\frac{1}{\pi T_2^*} = \frac{1}{\pi T_2} + \frac{1}{\tau} \qquad \text{(VI-1)}$$

where $1/\pi T_2^*$ is the width of the line at the half-height (in Hz), $1/\pi T_2$ is the total width of the line at the half-height in the absence of exchange (in Hz), T_2^* is the spin-spin relaxation time, and T_2 is the spin-spin relaxation time in the absence of exchange.

Similarly, also for the case of the n-th line under conditions approaching fast exchange the following relation is valid [33]:

$$\frac{1}{\pi T_2^*} = \frac{1}{\pi T_2} + \tau \sum_{i=1}^{n} p_i \delta_i^2 \qquad \text{(VI-2)}$$

where p_i is the mole fraction and δ_i the chemical shift (or the constant J) of the i-th line.

With small values of $\delta(J)$ it is necessary to use quite small changes in the width of the spectral line and this naturally limits the accuracy of the experiment. In practice the experiment is most often carried out under conditions which satisfy equation VI-1 since in this case there is no need to determine exact values of δ (or J) for determination of the value of τ.

To study hindered rotation and inversion the method used most frequently is based on the determination of the point of merging of peaks, when the following relation holds [33]:

$$\tau\delta \cong 1 \ (\ \text{or} \ \ \tau J \cong 1) \tag{VI-3}$$

This point is determined by gradually changing the observed chemical shifts of resolved peaks in the region of intermediate exchange rates (see Fig. VI-14). Then the value of τ is determined from relation (VI-3). Although this method is used widely, it has substantial drawbacks since the chemical shift is measured over a limited range of temperatures (or concentrations). It is often quite difficult to determine the p o i n t o f m e r g i n g with sufficient accuracy particularly when the populations of the exchanging states are not equal. A frequent example of the use of the method is the case where a change in the form of the signals occurs due to exchange between states, one of which is paramagnetic [40] or contains a nucleus with an electric quadrupole moment [41].

The mean lifetime of a nucleus in one of the states A between two successive exchange acts τ_A is related to the rate constant of the exchange process (k) by the equation [34]

$$\frac{1}{\tau_A} = \frac{1}{[A]} \cdot \frac{d[A]}{dt} = k[A]^{m-1}[B]^n \ldots \tag{VI-4}$$

where m, n, etc., are the orders of the reaction with respect to the concentrations of A, B, etc. For calculation of the values of m and n, the value of τ_A is determined as a function of the concentrations A and B.

If the reaction is unimolecular and first order (hindered rotation and inversion), the the following relation holds:

$$k = \frac{1}{\tau} \tag{VI-5}$$

By measuring the temperature dependence of the quantity τ it is possible to determine the thermodynamic parameters of exchange processes by using Eyring's or Arrhenius' equation.

Thus, in kinetic investigations of this type the time scale of the experiment is determined by the molecular and atomic properties, namely, the magnitudes of the chemical shifts and the spin—spin interaction constants. In this case these frequencies replace the usual time readings which are made in most other kinetic measurements. The exchanging states are labeled by the nuclear spins themselves and this is analogous to the introduction of an isotope in work with tracer atoms. This characteristic of the method makes it possible to study exchange between identical molecules. By using the procedure examined it is possible to determine the rates of exchange between two or several states. However, it is not always possible to draw strict conclusions on the mechanism or intermediate stages since the data obtained often refer to the combination of a series of processes.

Let us examine some examples of the application of high-resolution NMR spectroscopy to the study of the kinetics of exchange processes.

In the NMR spectra of a series of para-substituted acetophenones in FSO_3H, run at low temperatures (down to $-80°C$), there is as a rule a narrow singlet signal of the bound proton of the $C\overset{+}{=}OH$-group in the region of 12.50–15.30 ppm [1]. With an increase in temperature this signal broadens due to a decrease in the mean lifetime of the conjugate acid. Since the signals of the $C\overset{+}{=}OH$-group and the solvent do not overlap, for determining the rate of proton exchange in this case it is possible to use equation (VI-1), in which $1/\pi T_2^*$ is the width of the signal of the bound proton at the given temperature, $1/\pi T_2$ is the width of its signal in the absence of exchange (i.e., at a temperature of the order of $-80°C$, when the exchange rate is negligibly small), and $1/\tau$ is the reciprocal of the lifetime of the conjugate acid. The rate constant of the loss of the proton of the conjugate acid $k = 1/\tau$, while its temperature dependence is expressed by the Arrhenius equation. The activation energies are determined in the usual way from a graph of log k against $1/T$, where T is the absolute temperature.

The most intense signals in the spectrum of an aqueous solution of isobutyraldehyde are the doublets of the methyl groups
of the free aldehyde and its hydrate. With successive additions of
a strong acid these doublets are correspondingly broadened and
finally become indistinguishable, merging into one doublet. To
calculate the rate constants of the second-order hydration reaction catalyzed by hydrogen ions k_H, Hine and Houston used the
equation [42]:

$$k_H = \frac{v^\infty}{\pi T^\infty} \cdot \frac{d(1/v_B)}{d[H^+]} \qquad \text{(VI-6)}$$

where v^∞ is the height of the signal of the methyl protons of isobutyraldehyde under conditions where the hydration rate is negligibly low (without acid added), v_B is the height of the same signal
under the reaction conditions, πT^∞ is the width of the peak at the
half-height (in Hz) in the absence of a chemical reaction, and $[H^+]$
is the hydrogen ion concentration. Since the values of v^∞ are difficult to reproduce, they were determined by comparison with a
reference, namely, tert-butanol of known concentration.

Sheinblatt and Alexander studied the exchange of acetyl groups
between molecules of acetic acid and acetic anhydride catalyzed
by perchloric acid [43]. In the absence of the acid the NMR spectrum of a solution of acetic acid and acetic anhydride in chloroform contains narrow individual signals of the methyl protons of
the two substances. When perchloric acid is added the rate of exchange of the acetyl groups increases with the result that the signals of the methyl groups of the acid and the anhydride gradually
broaden and finally merge at a high rate of exchange. The authors
studied the rate of the process under conditions of fast exchange.
The width of the common peak of the methyl groups was determined from the time for the wiggles to die away with fast passage
(see Ch. I, section 6). Each measurement was repeated 4-5 times
and the mean value taken. The scatter of the experimental points
was ~10%. Under the conditions examined the rate of exchange
between the two states is related to the signal width observed experimentally by the equation

$$\frac{1}{\pi T_2^*} = \frac{1}{\pi T_2} + (\Delta v)^2 p_A^2 (1 - p_A)^2 (\tau_A + \tau_B) \qquad \text{(VI-7)}$$

where $1/\pi T_2^*$ is the observed signal width, $1/\pi T_2$ is the signal width in the absence of exchange, $\Delta\nu$ is the chemical shift between the exchanging states (in Hz), p_A is the fraction of methyl protons of component A [CH_3COOH or $(CH_3COO)_2$]; and $\tau_{A(B)}$ is the mean lifetime of component A(B) between successive exchanges (see also Ch. I, section 4).

We should remember that this equation is applicable only when $\Delta\nu \ll (1/\tau_A + 1/\tau_B)$, i.e., the exchange rate is high in comparison with the difference in the chemical shifts of the exchanging states and provided that condition $p_A\tau_B = p_B\tau_A$ holds, i.e., for equilibrium systems. To determine the value of τ_A the authors calculated the rate constant of the exchange from the equation (VI-4).

Saunders and his co-workers studied the kinetics of rearrangement of the 2-norbornyl cation [44].

The NMR spectrum of 2-exo-norbornyl fluoride is quite complex (Fig. VI-16), but it is very much simplified when the fluoride is dissolved in SbF_5 or SbF_5-SO_2: over the temperature range from −5 to +37°C it contains only one broad signal at 3.75 ppm (in pure SbF_5) or at 3.10 ppm (in SbF_5-SO_2). When the solution is cooled to −60°C there appear in the spectrum three broad signals at 5.35, 3.15, and 2.20 ppm with a ratio of intensities of 4 : 1 : 6 (Fig. VI-17). With an increase in the temperature the peaks broaden and at −23°C they merge into one peak, whose width at the half-height at +3°C is 20 Hz. If the solution is again cooled to −60°C, the spectrum again consists of three signals. The heating−cooling cycle may be repeated many times, indicating the stability of the cation in solution.

The study of the temperature dependence of a spectrum makes it possible to determine the rates and activation parameters of the migration of hydride ions in the norbornyl cation.

It is known that the 2-norbornyl cation may undergo three rearrangements, namely, a Wagner−Meerwein rearrangement (1), a 6,2-hydride shift (2), and a 3,2-hydride shift (3). (The formulas of the classical carbocations are given for simplicity.)

Wagner—Meerwein
rearrangement (1)

6,2-hydride shift (2)

3,2-hydride shift (3)

The NMR spectrum at low temperatures (Fig. VI-17) confirms the hypothesis that processes (1) and (2) proceed rapidly,
while process (3) proceeds slowly. The protons at the carbon atoms in positions 1, 2, and 6 are chemically equivalent due to the
rapid interconversion of these positions; they give one signal A,
which lies at lowest fields; its intensity corresponds to four protons. The protons at carbon atoms 3, 5, and 7 are also chemically
equivalent and give a signal at high fields, whose intensity corresponds to six protons C. Finally, the single proton attached to

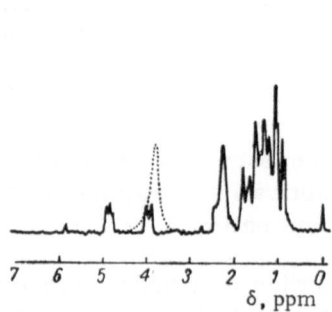

Fig. VI-16. NMR spectra of 2-exo-
norbornyl fluoride in CCl$_4$ and the
2-norbornyl cation in SbF$_5$; 37°C,
60 MHz.

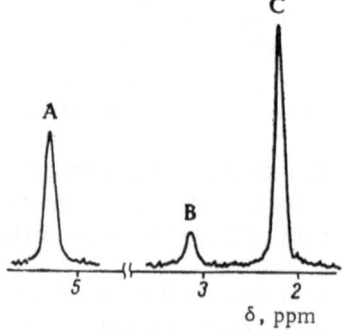

Fig. VI-17. NMR spectrum of 2-
norbornyl cation in SbF$_5$—SO$_2$
(−60°C, 60 MHz).

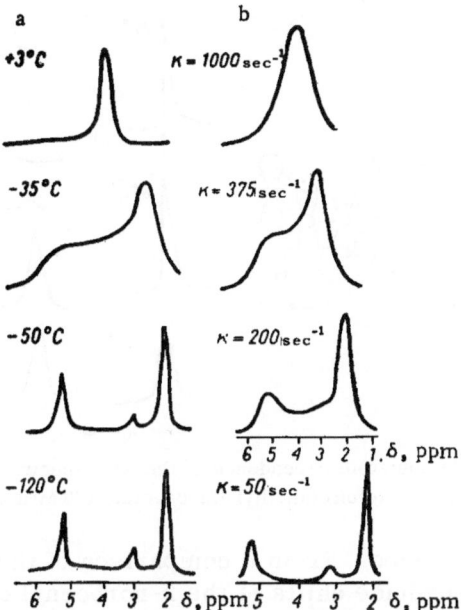

Fig. VI-18. NMR spectra of the 2-norbornyl
cation determined at different temperatures
(a) and calculated spectra of the 2-norbornyl
cation for different rates of the 3,2-hydride
shifts (b).

the carbon atom 4, which is at the bridge head, gives a signal at
intermediate fields B. At higher temperatures all the rearrangements proceed so rapidly that all the protons become chemically
equivalent on the NMR time scale with the result that one peak is
observed in the spectrum.

As a result of one 3,2-hydride shift three protons A move
to C and one proton A to B. The same shift results in the movement of one proton B and three protons C to A:

$$B\,(H) \underset{\longleftarrow}{\overset{1\text{ proton}}{\longrightarrow}} A\,(H) \underset{\longleftarrow}{\overset{3\text{ protons}}{\longrightarrow}} C\,(H)$$

By means of an electronic computer the authors calculated
the theoretical NMR spectra for various values of the rate constants of the 3,2-hydride shift. As Fig. VI-18 shows, the calculated spectra agree well with the experimental spectra plotted at

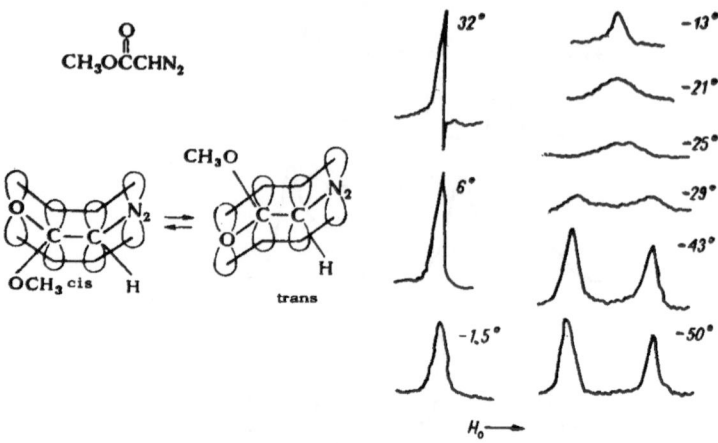

Fig. VI-19. Temperature dependence of the NMR spectrum of the
methine proton of methyl diazoacetate (60 MHz).

different temperatures. From a comparison of these spectra the
rates of the 3,2-hydride shifts of the 2-norbornyl cation were calculated at various temperatures and the enthalpy of activation
determined.

It is interesting to examine data on the rates of the Wagner—
Meerwein rearrangement (1) and the 6,2-hydride shift (2). Had the
rates of these processes been reduced considerably at low temperatures, broadening of the signal A at low fields would have been
observed. However, this does not occur even at −120°C. The authors calculated the form of the signals of peak A, assuming arbitrarily that the rates of processes (1) and (2) are approximately
the same. It was found that appreciable broadening of the low-
frequency peak A should be observed if the rate constant of these
reactions were less than or equal approximately 300,000 sec^{-1}.
Hence it is clear that the rate of processes (1) and (2) must exceed this value even at −120°C. Thus, at −120°C the rate of the
3,2-hydride shift is less by a factor of approximately $10^{8.8}$ than
the rate of the Wagner—Meerwein rearrangement (1) and the 6,2-
hydride shift (2).

NMR spectroscopy is a convenient method of studying the
rates and thermodynamic parameters of various conformational
conversions (hindered internal rotation, inversion of rings, inversion at a nitrogen atom, etc.).

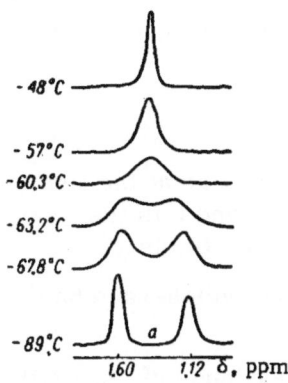

-48°C

-57°C

-60.3°C

-63.2°C

-67.8°C

-89°C

a

1.60   1.12 δ, ppm

Fig. VI-20. NMR spectra of
a 50% solution of C_6HD_{11} in
CS_2 (60 MHz), (all the spectra except a were plotted
with decoupling of the
deuterium).

Kaplan and Meloy investigated
hindered rotation in diazoketones [45].
This work is of particular interest
because the composition of the products of a series of chemical conversions of diazoketones is determined by conformational factors. Figure VI-19 gives the temperature dependence of the spectrum of the methine proton in methyl diazoacetate. The
signal at low fields is ascribed to the
cis-isomer and the one at high fields,
to the trans-isomer. The intensity of
the latter signal is lower. From the
area of the signals the authors determined the populations of each conformation (p_{cis} and p_{trans}), the
equilibrium constant (K = trans/cis),
and the difference in the standard free energies.

To determine the mean lifetime of the individual conformations the authors used a general equation which describes the form
of the signals in exchange processes [46]. In this equation they
substituted the mean values of the chemical shift of the methine
proton and data on the population of the isomeric conformation.
Then by means of an electronic computer they calculated the the-
oretical spectra for different values of $\tau = (\tau_{cis}\tau_{trans})/\tau_{cis} + \tau_{trans})$
and constructed a graph of τ against the signal width at its half-
width ω. This graph was used to calculate τ from the experimental
values of ω. The mean lifetimes of the individual conformations
τ_{cis} and τ_{trans} were determined from the equations: $\tau_{cis} = \tau/p_{trans}$ and $\tau_{trans} = \tau/p_{cis}$ [33]. The rates of interconversion
of the conformations cis → trans and trans → cis equal $1/\tau_{cis}$ and
$1/\tau_{trans}$, respectively. The activation energy ΔE, which is identical to the energy barrier of hindered rotation about the C−C
bond of diazoketone, and the frequency factor A were determined
by the method of least squares from the Arrhenius equation

$$\log\left(\frac{1}{\tau_{cis}}\right) \text{or } \log\left(\frac{1}{\tau_{.trans}}\right) = \log A - E_A/2.303\ RT$$

The free energy of activation was calculated from the equation:

$$\Delta F_T^{\neq} = 2.303 \ RT \log(\tau_x KkT/h)$$

where $\tau_x = \tau_{cis} (\tau_{trans})$ and is the least square value at the temperature T obtained from the Arrhenius equation and k is the translational coefficient, which is taken to be equal to unity.

The rate of conformational conversions of cyclohexane has been studied by many authors.

Figure VI-20 shows the temperature dependence of the NMR spectra of d_{11}-cyclohexane. The spectra were determined with decoupling of the deuterium in order to reduce the width of the signal to a minimum. At $-89°C$ the spectrum (Fig. VI-20a) contains two broad signals at 1.12 ppm (width 7.5 Hz) and 1.60 ppm (width 6 Hz), corresponding to axial and equatorial protons of the two conformations, respectively. The rate constants of the inversion and its thermodynamic parameters were calculated by comparing the experimental spectra with calculated spectra [47].

In some cases the NMR method makes it possible to determine the rates of two exchange processes proceeding simultaneously.

Thus, Anet and his co-workers investigated the temperature dependence of the NMR spectra of cyclooctatetraenyl-2,3,4,-5,6,7-d_6-dimethylcarbinol and determined the rate constants of inversion of the ring k_1 and valence isomerization k_2 separately [48].

At $-35°C$ the ring proton gives two narrow signals of equal intensity separated by 2.6 Hz (at 60 MHz). The signal at higher fields (at 5.76 ppm) is assigned to the proton in structures Ia and Ib and that at low fields (at 5.80 ppm) to the proton in structures Ic and Id (see following page).

Under these conditions the methyl protons are also nonequivalent and give two separate signals at 1.16 and 1.21 ppm, which are separated from each other by 3.3 Hz.

With an increase in temperature the signals of the methyl groups broaden and then (at $-2°C$) merge, subsequently changing to a narrow single peak. The signals of the ring proton behave analogously with the only difference that they merge at $+41°C$.

The distance between the signals of the ring proton is somewhat less than the distance between the signals of the methyl groups at the same temperature. This denotes that the rate of the process which is accompanied by a change in the magnetic environment of the ring proton must be considerably lower than the rate of the change in the magnetic environment of the methyl group. The magnetic environment of the methyl groups is averaged out in the processes Ia → Ib and Ia → Id, while the magnetic environment of the ring proton is averaged out in the processes Ia → Ib, Ia → Ic, and Ia → Id.

The NMR spectroscopic data show that $(k_1 + k_2)/2 \gg k_2$, i.e., $k_1 \gg k_2$. Therefore the rate constant measured from the change in the chemical shifts of the methyl protons actually equals k_1, i.e., the rate constant of inversion of the ring.

To conclude this section we should point out that with the use of the high-resolution NMR method to measure the kinetics of exchange processes the errors in the determination of the values ΔH^* and ΔS^* are considerably greater than in the determination of ΔF^*. This is because the range over which it is possible to measure the rates of exchange by high-resolution methods are considerably limited by the lack of instruments of the required accuracy.

In this method the contribution of the exchange process to the line width to the upper and lower limits of the range approaches the experimental error in the determination of the total width of the signal. These limitations are very substantial. It is sufficient to state that the temperature range over which it is possible to measure the rate of inversion of the ring of d_{11}-cyclohexane is only 28°C at 60 MHz [47].

Another limitation of the method is the impossibility of using it to determine activation barriers less than 5 kcal/mole [49].

In recent years the pulsed spin—echo method has been used particularly successfully to study the kinetics of exchange processes [36]. This method has considerable advantages in comparison with stationary high-resolution methods since it makes it possible to completely eliminate the instrumental contribution to the signal width. This considerably extends the region over which it is possible to measure with sufficient accuracy the contribution of the exchange process to signal width. Thus, for example, the use of the spin—echo method made it possible to measure the rate of inversion of d_{11}-cyclohexane over the range from 0.5 sec^{-1} to $4 \cdot 10^4$ sec^{-1} and this, in its turn, made it possible to calculate the values of ΔH^* and ΔS^* more accurately [50].

4. TAUTOMERISM AND VALENCE ISOMERISM

The NMR method is used successfully to study the tautomerism of organic compounds. Like most physical methods, it does not affect the equilibrium position. At the same time, NMR spectroscopy has indisputable advantages over IR and UV spectroscopy since there is no need to measure the intensities of bands of each

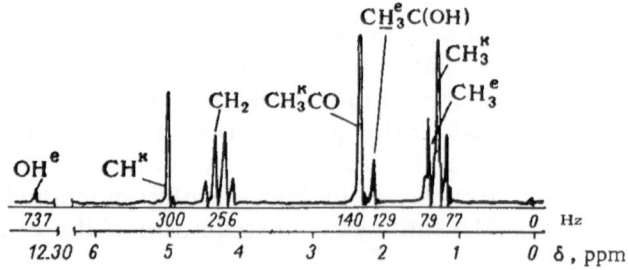

Fig. VI-21. NMR spectrum of pure ethyl ester of α-chloro-acetoacetic acid (33°C, 60 MHz).

TABLE VI-1. Direction of Enolization of Cyclic α-Formyl Ketones*
(10% Solutions in CCl_4, 25°C)

Cyclic ketones	K	Hydromethylene ketone form (IIb), %
α-Formyl		
Cyclopentanone	3.54	78
Cyclohexanone	0.32	24
Cycloheptanone	1.94	66
Cyclooctanone	0.79	44
Cyclononanone	0.45	31
Cyclodecanone	0.22	18
Cyclododecanone	0.39	28
2-Isopropylcyclohexanone	0.39	28
Camphor	—	100

* All the β-keto aldehydes with the exception of α-formylcamphor (94% enol) and
α-formylcyclopentanone (85% enol), are more than 99% enolized.

of the tautomers separately beforehand to calculate the constants
of the tautomeric equilibrium. The equilibrium constant is determined directly from the spectrum by comparing the intensities of
the signals of the tautomeric forms. Thus, for example, in the pure
ester of α-chloroacetoacetic acid at 33°C, judging by the ratio of
the areas of the methine proton of the ketonic form and the methylene protons of the ethoxyl group (Fig. VI-21), the equilibrium constant K = enol/ketone = 0.18, which corresponds to the presence of
15% enol in the tautomeric mixture [51]. From the temperature
dependence of the equilibrium constant it is possible to determine
the thermodynamic characteristics of tautomeric systems.

The use of the NMR method for studying tautomeric processes
has made it possible to obtain a large amount of fundamentally important information. Thus, for example, previously there was a
widely developed idea of the existence of trans-enols of open chain
α-alkyl substituted β-dicarbonyl compounds. In recent years it
has been shown clearly by means of NMR spectra that the trans-
enol form of these substances is completely absent. What was previously taken as the trans-enol was found to be the cis-enol form.
This is indicated by signals of the cis-chelate bound hydroxyl group
in the region of 12.9-17.5 ppm [52].

In the enolization of β-keto aldehydes there is the possibility of the formation of two tautomeric forms, namely, the aldenol form IIa and the hydroxymethylene ketone form IIb:

IIa   IIb

Until recently it was assumed that the hydroxymethylene ketone form (IIb) predominates in the equilibrium mixture in the enolization of β-ketone aldehydes. It was shown by means of NMR spectroscopy that this hypothesis is valid for acyclic compounds [53], but does not correspond to reality for cyclic β-keto aldehydes.

Garbish investigated the direction of enolization of a large number of cyclic β-keto aldehydes [54]. It was found that the equilibrium between the ketone form II and the enol forms IIa and IIb is established quite slowly so that it does not produce averaging of the spectrum, while the equilibrium between the forms IIa and IIb is established rapidly. Therefore the signals of the ketone form II and the averaged signals of the enol forms IIa and IIb are observed in the spectrum. The mole fraction N of the enol IIb in the equilibrium may be determined from the relation

$$N_{\mathrm{IIb}} = \frac{\delta_0 - \delta_A}{\delta_B - \delta_A}$$

where δ_0 is the observed averaged chemical shift of the protons H_A and H_B and δ_A and δ_B are the chemical shifts of the protons H_A and H_B. By determining the value of N_{IIb} it is possible to calculate the value of N_{IIa} since the overall ratio of ketone/enol

is readily determined directly from the spectrum. The author determined the equilibrium constants K for a whole series of cyclic
β-keto aldehydes (Table VI-1).

In all cases it is possible to observe spin — spin interaction
between vinyl and hydroxyl protons at low temperatures. The
value of the constant J corresponds to the combined contribution
of structures (IIa) and (IIb).

The chemical shifts of protons participating in the formation
of strong intramolecular hydrogen bonds depend little on concentration as a rule. However, in some cases on dilution with solvents of low polarity (such as carbon tetrachloride, hexane, and
carbon disulfide) and also with an increase in temperature a slight
shift is still observed in the signal of the OH groups of cis-enol
forms to high fields (by 0.5 ppm) and this indicates partial rupture
of intramolecular hydrogen bonds [51].

With an increase in the conjugation in molecules of enols of
β-diketones, just as with an increase in their acidity, there is a
shift in the signal of the hydroxyl group to low fields and this is
explained by a decrease in the electron density of the hydrogen
atom. Then an almost linear correlation is observed between the
chemical shift of the proton of the enolic OH group and the dissociation constants of the β-diketone [55].

In some cases NMR spectral data make it possible to determine the lifetime of a state with a hydrogen bond or to establish
its limits. As a rule, the lifetime of these states is short in enolic
forms of β-diketones. In particular, this is indicated by the equivalence of the chemical shifts of the methyl groups in the acetylacetone molecule.

The form of the signal of the OH group of the enol form of
β-diketones depends strongly on the substituents R_1 and R_2 [55]:

(A)   (B)

If R_1 and R_2 are electron–acceptor groups or groups conjugated with a quasiaromatic chelate ring (for example, $R_1 = C_6H_5$; $R_2 = CH_3$), then the signal of the OH group in all cases is very narrow (8–10 Hz). On the other hand, groups which do not participate actively in delocalization give rise to broad diffuse signals (up to 50 Hz for $R_1 = (CH_3)_2CHCH_2$; $R_2 = CH_3$), and this may be produced by averaging of the chemical shifts of at least two unsymmetrical tautomeric forms (A and B).

An approximate estimate of the lifetime of a proton in the potential well at the oxygen atom in the enol of hexafluoroacetylacetone showed that this value is $\sim 1.5 \cdot 10^{-4}$ sec [56]. At the same time, the recording of spin−spin interaction of the enol proton with the nearest protons in the enol forms of β-keto aldehydes (J 5–12 Hz) indicates that the lifetime of the enols in this case is comparatively long [53, 54].

NMR spectra give interesting information on the tautomeric equilibrium of sugars in solutions [57]. It has been shown that d-ribose and 2-deoxy-d-ribose in D_2O consist of tautomeric mixtures of pyranose and furanose structures, while with d-xylose, d-lyxose, d-arabinose, etc., no furanose rings are detected.

The NMR method is very convenient for studying valence isomerization [58]. In particular, by this method it was possible to observe directly for the first time the norcaradiene−cycloheptatriene equilibrium using 7–cyano–7–trifluoromethylcycloheptatriene–7–cyano–7–trifluoro-methylnorcaradiene [59].

Marvell and his co–workers studied valence isomerization in the equilibrium system cis–β–ionone (III) − 1–oxa–2,5,5,9–tetramethyl–1,5,6,7,8,10–hexahydronaphthalene (IV) [60].

The α-pyran (IV) predominantes in the equilibrium mixture at room temperature, while with an increase in temperature there is an increase in the content of cis-β-ionone (III). The equilibrium constant was determined from the ratio of the areas of the signals of the olefinic protons at 6.38 ppm (III) and 5.60 ppm (IV). The rate of the valence isomerization was measured during the return of the heated mixture to the equilibrium state. An NMR tube containing 150 mg of pyran IV in 500 μl of tetrachloroethane was heated at 120°C for 10 min and then immersed in a Dewar flask of solid carbon dioxide. The tube was then placed in the spectrometer and allowed to gradually come to room temperature. The rate of return to equilibrium at room temperature was determined from the decrease in the height of the large peak of the doublet at 6.03 ppm of (III). From the equation

$$2.3 \log \frac{x_0 - x_e}{x_t - x_e} = (k_1 + k_2) \, t$$

(where x_0 is the height of the peak at zero time, x_e is the equilibrium position, and x_t is the height at time t) and from the values of the equilibrium constants for different temperatures the values of k_1 and k_2 were determined and the activation energy was found from the temperature dependence.

5. COMPOUNDS CONTAINING AN UNPAIRED ELECTRON

The presence of unpaired electrons in an organic substance leads to a decrease in the relaxation time, i.e., to considerable broadening of the lines; as a result it becomes impossible to observe the high-resolution NMR spectrum. However, if the complex of a transition metal contains a system of conjugated bonds, the spin density due to the presence of the unpaired electron may be delocalized and in individual cases the relaxation time of the unpaired electron is very short and therefore the broadening of the lines in the NMR spectra less than the distance between them. In these cases it is possible to observe clear nuclear resonance spectra for at least part of the molecule.

CH$_3$
|
N
Br—⟨⟩Ni$_{1/2}$
N
|
CH$_3$

A characteristic of the spectra of these substances is the
contact shift due to the contact action of the unpaired electron on
the nuclear spins. In this case the spin density at neighboring
carbon atoms is opposite in sign (Ch. II). This property has been
used for analysis of spin–spin coupling in aromatic molecules
forming ligands in complexes of nickel [61, 64]. As a result of the
increase in the chemical shift between neighboring protons (for
which the coupling constant is particularly great), the analysis of
such spectra may be carried out by first-order rules. A simple
example is the spectrum of the chelate complex N(II)-γ-bromo-
N,N'-dimethylaminotroponiminate in carbon disulfide. In contrast
to the compound without the nickel, which gives a spectrum of the
AA'BB' type, the spectrum of the nickel complex can be analyzed
by the rules for AA'XX' spectra. It is important that the signal
of the α-protons, lying closer to the nickel atom, which is the
source of paramagnetism, be broader than the signal of the β-
protons. The complex of the analogous troponimine which does
not contain bromine in the γ-position gives a somewhat more complex spectrum because there are three coupled protons in this ring.
Nonetheless, the spectrum is analyzed readily because the signal
of the γ-proton is at a distance from the signals of the other two
protons of the ring. The spin density determined from the contact shifts is distributed in this ligand in the following way:

CH$_3$
+0.0386 |
−0.0210   N
⟨⟩Ni$_{1/2}$
+0.0539   N
|
CH$_3$

It is characteristic that the highest spin density (in absolute
value) is concentrated at the γ-carbon atom and this leads to the
greatest broadening of the signal of the γ-proton and, consequently,
to a decrease in the amplitude of this signal.

Another application of paramagnetic substances in NMR spectroscopy is connected with the effect of the dynamic polarization of nuclei. This phenomenon, which was extensively investigated by Abragam [62], appears as an increase in the intensity of the nuclear resonance signal by 2-3 orders on addition of a paramagnetic compound to a substance and with additional irradiation at a frequency close to the electron resonance frequency (with the use of strong magnetic fields). This method was used to observe the resonance signal of C^{13} with the natural content of the isotope. In benzene the signal of C^{13} consisted of a doublet with a splitting of ~159 Hz, corresponding to spin—spin coupling of the C^{13} nuclei of the ring with protons [63].

Finally, in individual cases, when the lines in the nuclear resonance spectrum are far from each other so that considerable broadening of them does not interfere with the analysis of the spectrum, the addition of paramagnetic substances may be used to reduce the spin—lattice relaxation time in order to avoid saturation of the signal when an rf magnetic field of high strength is used. In some cases this makes it possible to increase the amplitude of the nuclear resonance signal.

Literature Cited

1.   T. Birchall and R. J. Gillespie, Can. J. Chem., 43:1045 (1965).
2.   G. E. Maciel and D. D. Traficante, J. Phys. Chem., 69(3):1030 (1965).
3.   S. J. Kuhn and J. S. McIntyre, Can. J. Chem., 43:995 (1965).
4.   G. A. Olah, J. Am. Chem. Soc., 87:1103 (1965).
5.   W. Bonthrone and D. H. Reid, J. Chem. Soc., 1966B:91.
6.   A. N. Bourns, R. J. Gillespie, and R. J. Smith, Can. J. Chem., 42:1433 (1964).
7.   D. M. Brouwer, E. L. Mackor, and C. MacLean, Rec. Trav. Chim., 85(I):109 (1966).
8.   C. E. Keller and R. Pettit, J. Am. Chem. Soc., 88:604 (1966).
9.   M. Fraeser, S. McKenzie, and D. H. Reid, J. Chem. Soc., 1966B:44.
10.   N. E. Rigler, S. P. Bag, D. E. Leyden, J. L. Sudmeier, and C. N. Reilly, Analyt. Chem., 37:872 (1965).
11.   M. Christer, W. Koch, W. Simon, and Hch. Zollinger, Helv. Chim. Acta, 45:2077 (1962).
12.   R. Foster and C. A. Fyfe, Tetrahedron, 21:3363 (1965).
13.   N. C. Deno, in: Progress in Physical Organic Chemistry, Vol. 2, ed. S. G. Cohen, A. Streitweiser, Jr., and R. W. Taft, Wiley, New York (1964), p. 129.
14.   G. A. Olah, E. B. Baker, J. C. Evans, W. S. Tolgyesi, J. McIntyre, and I. J. Bastien, J. Am. Chem. Soc., 86:1360 (1964).
15.   C. U. Pittman, Jr., and G. A. Olah, J. Am. Chem. Soc., 87:2998 (1965).

16. N. C. Deno, J. S. Liu, J. O. Turner, D. N. Lincoln, and R. E. Fruit, Jr., J. Am. Chem. Soc., 87:3000 (1965).

17. G. A. Olah and M. B. Comisarow, J. Am. Chem. Soc., 86:5682 (1964).

18. N. C. Deno and C. U. Pittman, Jr., J. Am. Chem. Soc., 86:1871 (1964).

19. L. Eberson and S. Winstein, J. Am. Chem. Soc., 87:3506 (1965).

20. G. A. Olah, E. Namanroorth, and M. B. Comisarow, J. Am. Chem. Soc., 89: 711 (1967).

21. V. R. Sandel and H. H. Freedman, J. Am. Chem. Soc., 85:2328 (1963).

22. R. S. Berry, R. Dehl, and W. R. Vaughan, J. Chem. Phys., 34:1460 (1961).

23. Th. J. Katz, M. Joshida, and L. C. Siew, J. Am. Chem. Soc., 87:4516 (1965).

24. S. J. Kuhn and J. S. McIntyre, Can. J. Chem., 43:375 (1965).

25. R. Foster and C. A. Fyce, Chem. Comm., 24:642 (1965).

26. M. L. Maddox, S. L. Stafford, and H. D. Kaesz, Adv. Organometal. Chem., 3:1 (1965).

27. J. Powell, S. D. Robinson, and B. L. Shaw, Chem. Comm., 78 (1965).

28. F. S. D'yachkovskii, P. A. Yapovitskii, and V. F. Bystrov, Vysokomolek. soed., 6:659 (1964).

29. Z. Luz and S. Meiboom, J. Chem. Phys., 40:1748 (1964).

30. A. Rauk, E. Buncel, R. J. Moir, and S. Wolfe, J. Am. Chem. Soc., 87:5498 (1965).

31. P. L. Levins and Z. B. Papanastassiou, J. Am. Chem. Soc., 87:826 (1965).

32. J. Eisch and G. R. Husk, J. Organometal. Chem., 4:415 (1965).

33. A. Loewenstein and T. M. Connor, Z. Elektrochem. Ber. Bunsenges. Phys. Chem., 67:280 (1963).

34. J. J. Delpuech, Bull. Soc. Chim. France, 10:2697 (1964).

35. L. W. Reeves, in: Advances of Physical Organic Chemistry, Vol. 3, ed. S. G. Cohen, A. Streitweiser, Jr., and R. W. Taft, Wiley, New York (1965), p. 187.

36. J. Pople, W. Schneider, and N. Bernstein, High-Resolution Nuclear Magnetic Resonance, McGraw-Hill, New York (1959).

37. J. I. Kaplan, J. Chem. Phys., 28:278 (1958); 29:462 (1958).

38. S. Alexander, J. Chem. Phys., 37:967, 974 (1962); 38:1787 (1963).

39. Johnson, Jr., J. Chem. Phys., 41:3277 (1964).

40. Z. Luz and S. Meiboom, J. Chem. Phys., 40:1058 (1964).

41. Z. Luz, G. Gill, and S. Meiboom, J. Chem. Phys., 30:1540 (1959).

42. J. Hine and J. Houston, J. Org. Chem., 30:1328 (1965).

43. M. Sheinblatt and S. Alexander, J. Am. Chem. Soc., 87:3905 (1965).

44. M. Saunders, P. v. R. Schleyer, and G. A. Olah, J. Am. Chem. Soc., 86:5679 (1965); see also F. R. Jensen and B. H. Beck, Tetrahedron Letters, 36:4287 (1966).

45. F. Kaplan and G. K. Meloy, J. Am. Chem. Soc., 88:950 (1966).

46. M. T. Rogers and J. C.Woodbrey, J. Phys. Chem., 66:540 (1962).

47. F. A. Bovey, F. P. Hood, III, E. W. Anderson, and R. L. Kornegay, J. Chem. Phys., 41:2041 (1964).

48. F. A. Anet, A. J. R. Bourn, and J. S. Lin, J. Am. Chem. Soc., 86:3576 (1964).

49. J. E. Anderson, Quart. Rev., 19:426 (1965).

50. A. Allerhand, Fu-ming Chen, and H. S. Gutowsky, J. Chem. Phys., 42:3040 (1965).

51.   J. L. Burdett and M. T. Rogers, J. Am. Chem. Soc., 86:2105 (1964); Can. J. Chem., 43:1516 (1965); J. Phys. Chem., 70:939 (1966).

52.   S. T. Ioffe, E. I. Fedin, P. V. Petrovskii, and M. I. Kabachnik, Tetrahedron Letters, 24:2661 (1966).

53.   S. Forsen and M. Nilson, Arkiv. Kemi, 19:569 (1962).

54.   E. W. Garbish, Jr., J. Am. Chem. Soc., 85:1696 (1963).

55.   R. L. Lintvedt and H. F. Holtzclaw, Jr., Inorg. Chem., 5:239 (1966).

56.   E. G. Popova, D. N. Shigorin, N. N. Shapet'ko, A. P. Skoldinov, and G. A. Gol'der, Zh. Fiz. Khim., 39:2726 (1965).

57.   R. U. Lemieux and J. D. Stevens, Can. J. Chem., 44:249 (1966).

58.   G. Schröder, Angew. Chem., Int. Ed., 4:752 (1965).

59.   E. Ciganek, J. Am. Chem. Soc., 87:1149 (1965).

60.   E. N. Marvell, G. Caple, T. A. Gosink, and G. Zimmer, J. Am. Chem. Soc., 88:619 (1966).

61.   D. R. Eaton, A. D. Josey, W. D. Phillips, and R. E. Benson, J. Chem. Phys., 39:3513 (1963).

62.   A. Abragam, Principles of Nuclear Magnetism, Oxford Univ. Press, New York (1961).

63.   K. H. Hausser, J. Chim. Phys., 61:204 (1964).

64.   J. S. Waugh, ed., Advances in Magnetic Resonance, Academic Press, New York (1965).

Appendix

In this appendix there are collected examples of NMR spectra of compounds belonging to the characteristic classes of organic substances in accordance with the classification in Ch. V. The spectra are arranged in order of increasing complexity. Each spectrum is indexed in accordance with the system examined in Ch. I (section 11). Spectra marked with an asterisk are taken from the catalog of Varian Associates and the others (except as noted) were obtained on a JNM-3 instrument in the Lensovet Leningrad Technological Institute.

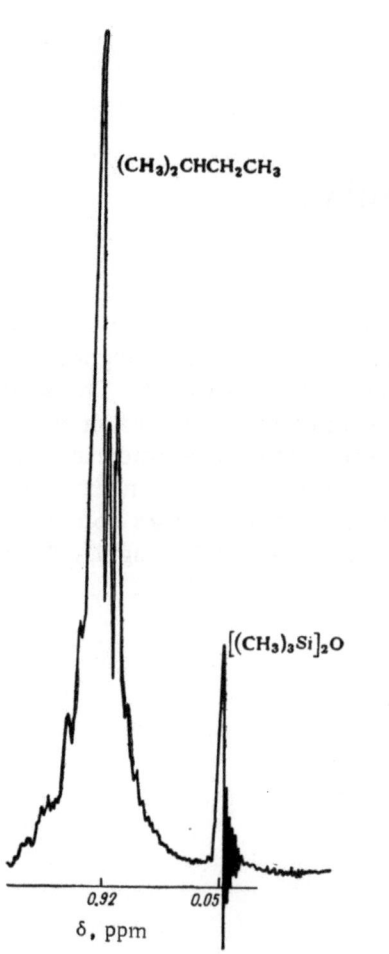

1. Isopentane H.40.ЛAlk.M

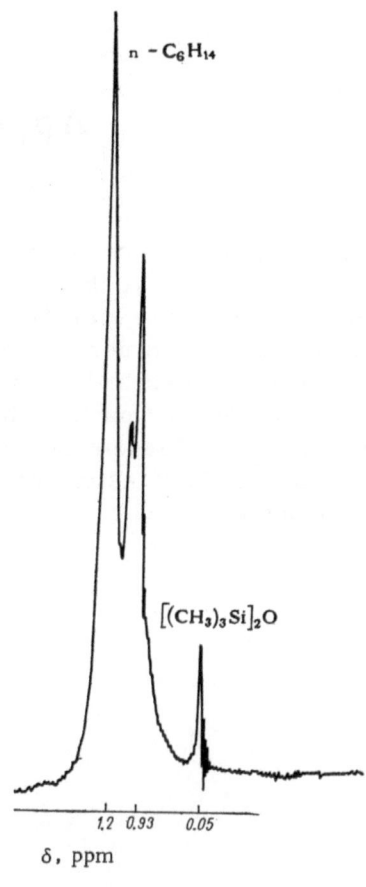

2. n-Hexane H.40.Л.M

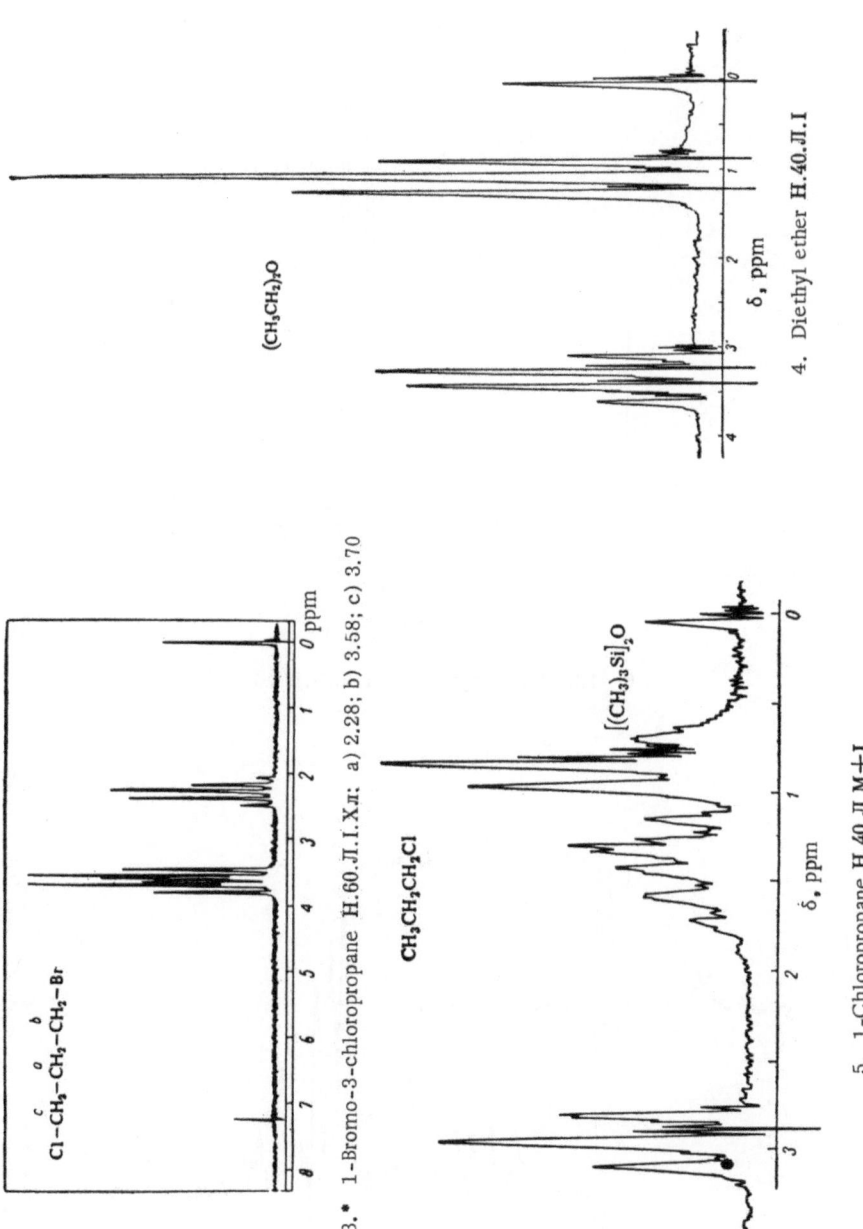

3.* 1-Bromo-3-chloropropane Н.60.Л.I.Хп: a) 2.28; b) 3.58; c) 3.70

$Cl-\overset{c}{C}H_2-\overset{a}{C}H_2-\overset{b}{C}H_2-Br$

$CH_3CH_2CH_2Cl$

$[(CH_3)_3Si]_2O$

5. 1-Chloropropane Н.40.Л.М+I

$(CH_3CH_2)_2O$

4. Diethyl ether Н.40.Л.I

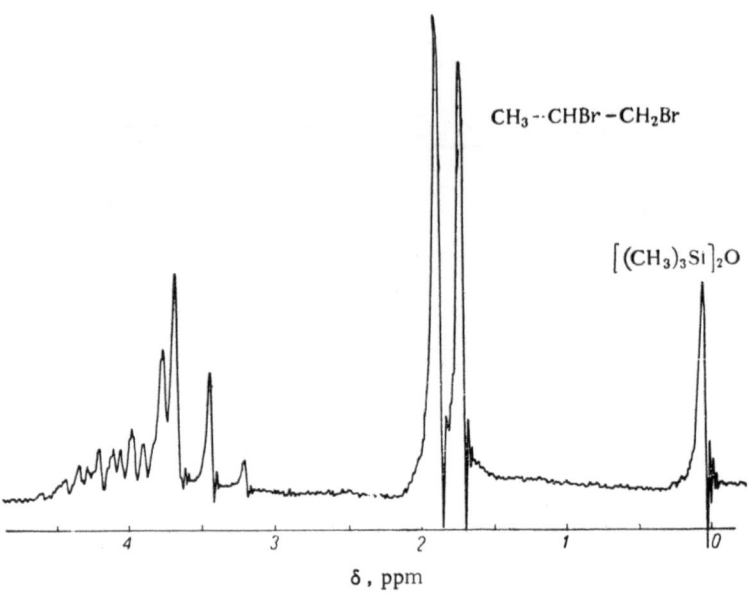

6. 1,2-Dibromopropane H.40.Л.M+I

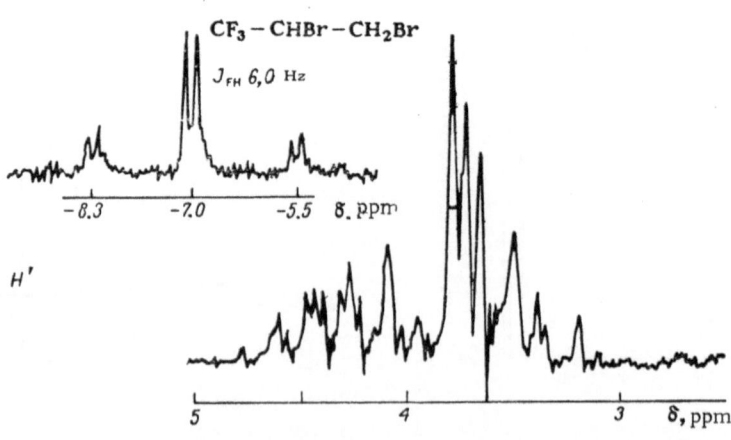

7. 1,1,1-Trifluoro-2,3-dibromopropane HF+FH.40.ЛZ.M+I.+Bш

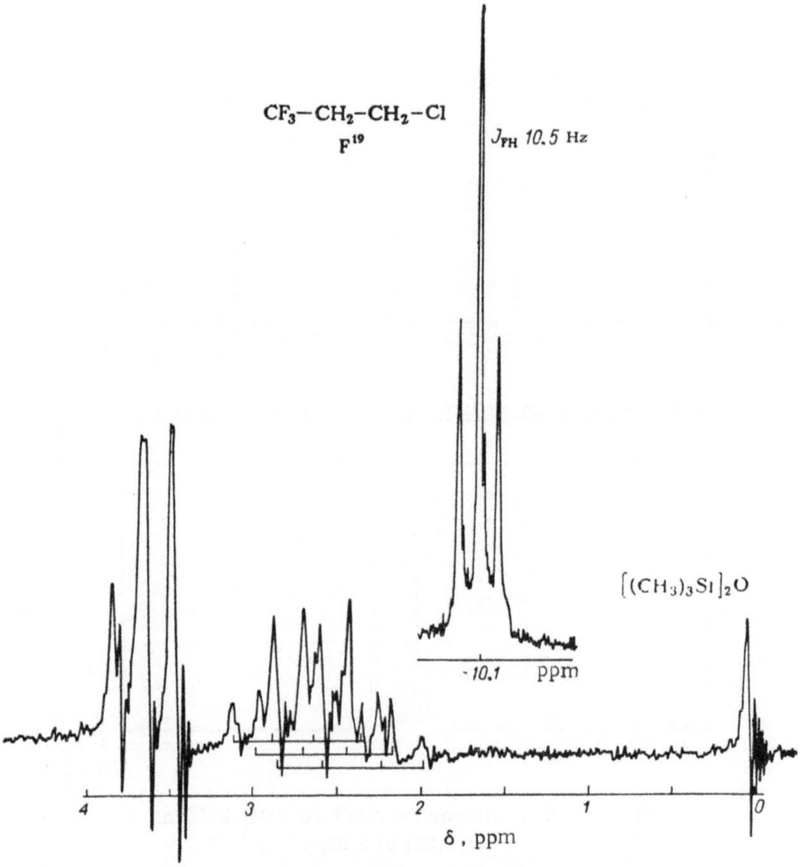

$CF_3-CH_2-CH_2-Cl$
F^{19}
J_{FH} 10.5 Hz
$[(CH_3)_3Si]_2O$
-10.1
ppm
4
3
2
1
0
δ, ppm

8. 1,1,1,-Trifluoro-3-chloropropane HF+FH.40.ЛZ.M+I.+Вш

Cl-C-C-CH3
H H
Cl
b
d
a
c
8
7
6
5
4
3
2
1
0
ppm

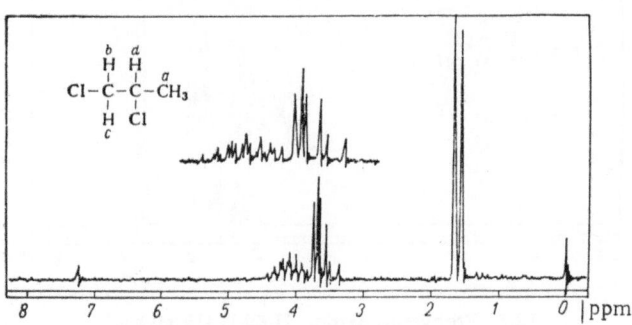

9.* 1,2-Dichloropropane H.60.Л.M.Xл : a) 1.60; b) 3.52; c) 3.78; d) 4.10.

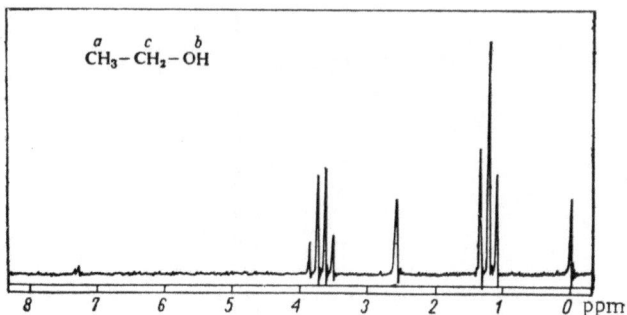

10.* Ethanol **H.60.ЛOH.I.Xл** : a) 1.22; b) 2.58; c) 3.70.

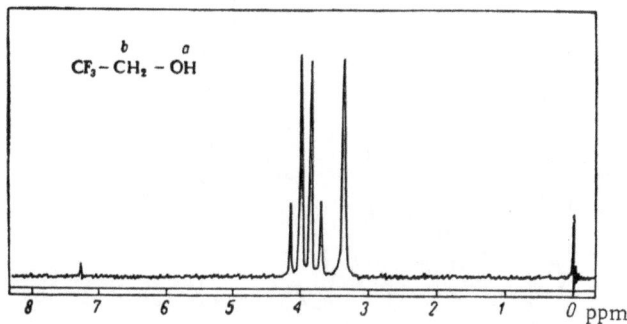

11.* 2,2,2,-Trifluoroethanol **H F.60.ЛOH,Z.I.Xл**:
a) 3.38; b) 3.93.

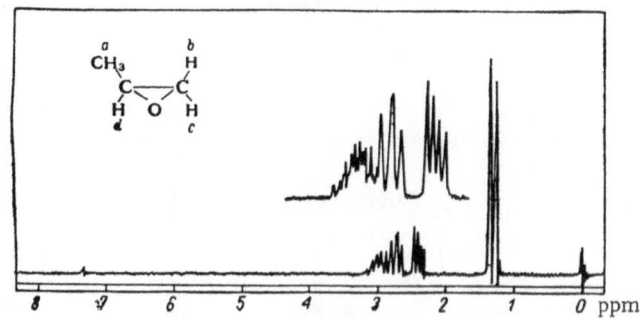

12.* Propylene oxide **H.60.ЦAlk.M.Xл**:
a) 1.32; b) 2.42; c) 2.72; d) ~ 2.98.

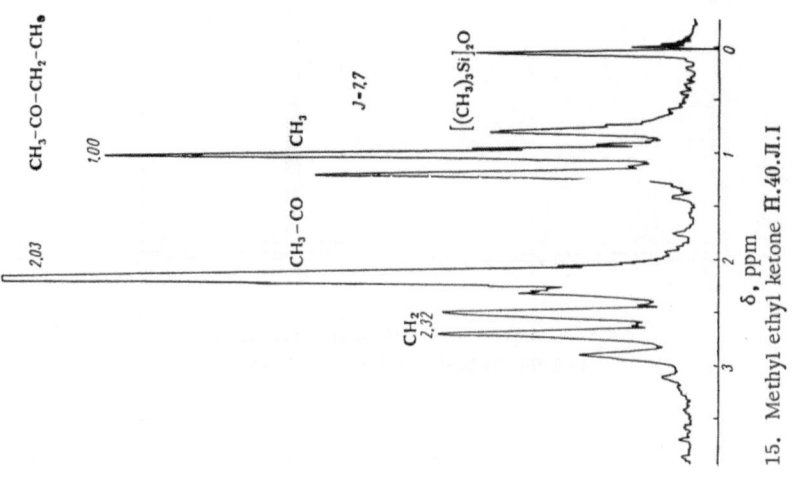

15. Methyl ethyl ketone Н.40.Л.I

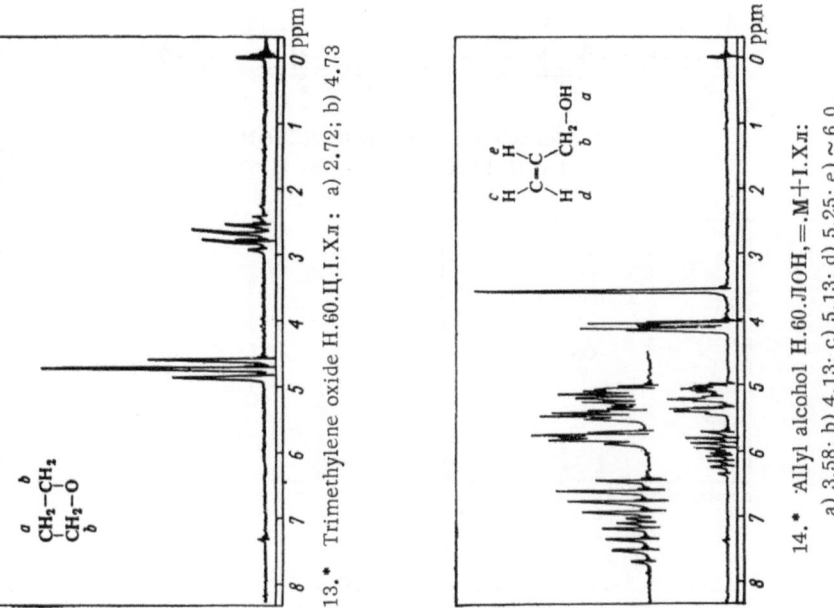

13.* Trimethylene oxide Н.60.Ц.I.Хл: a) 2.72; b) 4.73

14.* Allyl alcohol Н.60.ЛОН,═М+I.Хл:
a) 3.58; b) 4.13; c) 5.13; d) 5.25; e) ∼ 6.0.

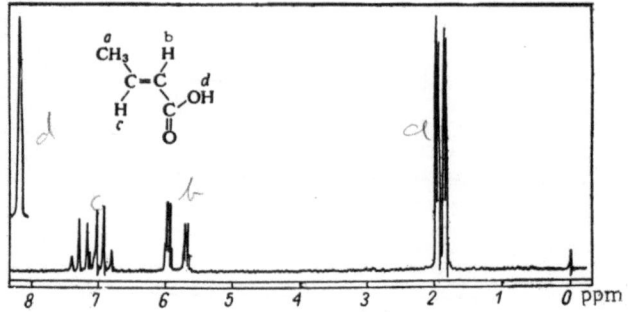

16.* Crotonic acid   **Н.60.ЛСООН.I.Хл;**
a) 1.90; b) 5.83; c) 7.10; d) 12.18.

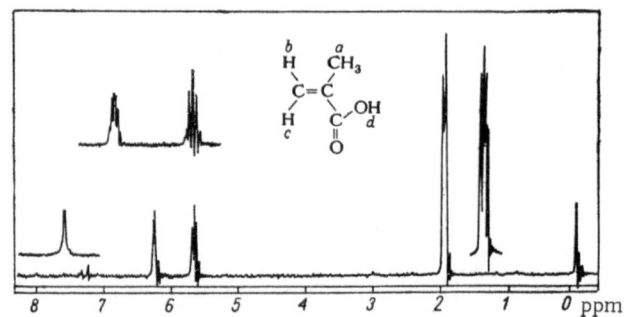

17.* Methacrylic acid   **Н.60.ЛСООН.I.Хл;**
a) 1.97; b) 5.72; c) 6.30; d) 11.57.

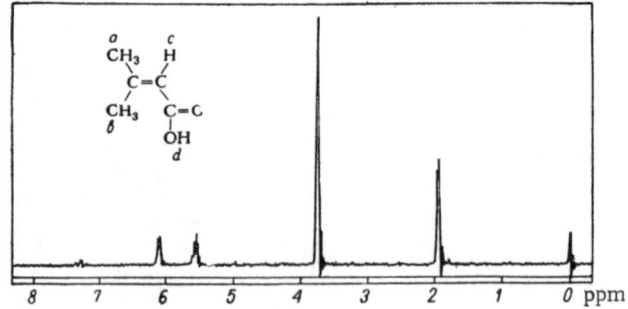

18.* β,β-Dimethylacrylic acid **Н.60.ЛСООН,Alk.I.Хл;**
a) 1.93; b) 2.18; c) 5.72; d) 11.95.

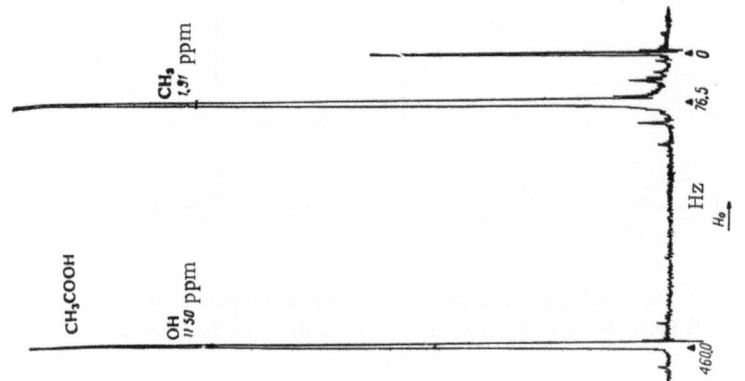

21. Acetic acid H.ЧU.JICOOH.I

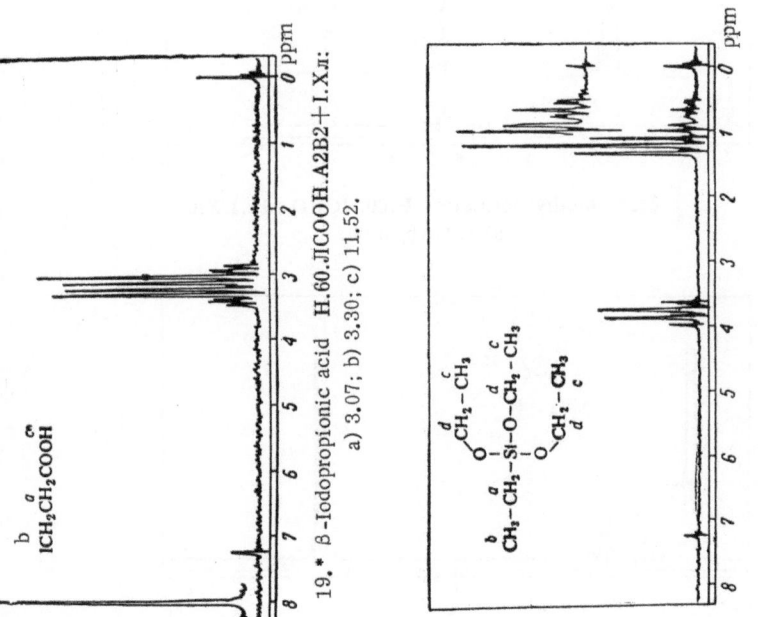

b $\overset{a}{ICH_2}\overset{a}{CH_2}\overset{a}{COOH}$

19.* β-Iodopropionic acid H.6U.JICOOH.A2B2+I.Xи:
a) 3,07; b) 3,30; c) 11,52.

$$\begin{array}{c}\overset{d}{CH_2}-\overset{c}{CH_3}\\ \underset{b}{|}\\ \overset{a}{O}\\ \underset{b}{|}\\ \overset{a}{CH_3}-\overset{a}{CH_2}-Si-O-\overset{d}{CH_2}-\overset{c}{CH_3}\\ |\\ O\\ |\\ \overset{d}{CH_2}-\overset{c}{CH_3}\end{array}$$

20.* Ethyltriethoxysilane H.6U.ЛAlk.A3B2+I.Xи:
a) ∼ 0,62; b) ∼ 1,00; c) 1,23; d) 3,82.

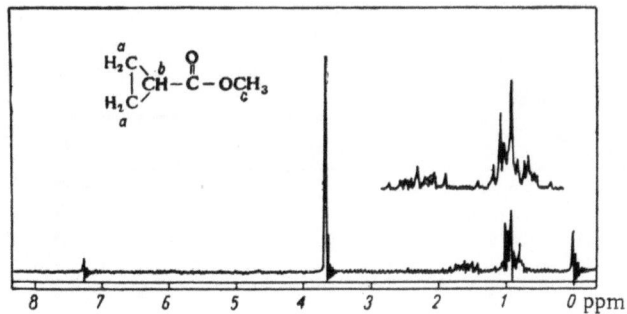

22.* Methyl ester of cyclopropanecarboxylic acid
H.60.ЦAlk,COOH.M+I.Xл : a) ~ 0.93; b) 1.63; c) 3.67.

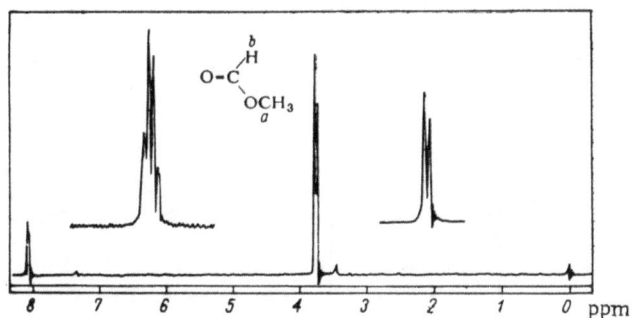

23.* Methyl formate   H.60.ЛCHO,Alk.I.Xл:
a) 3.77; b) 8.08.

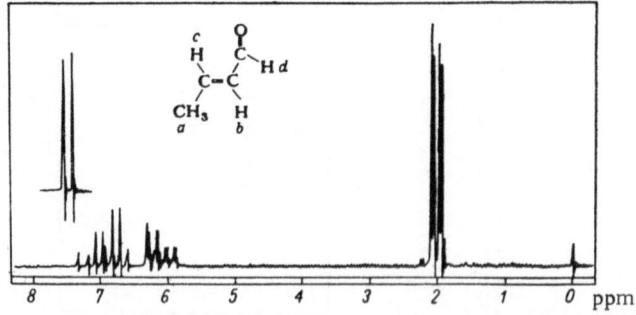

24.* Crotonaldehyde H.60.ЛCHO,=.I.Xл:
a) 2.03; b) 6.13; c) 6.68; d) 9.48.

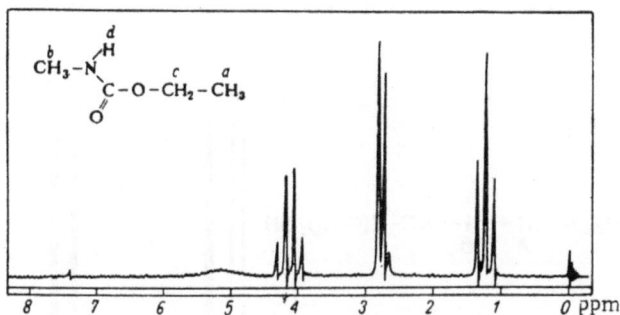

25.* Ethyl ester of N-methylcarbamic acid
H.60.ЛNH.I.Xл; a) 1.23; b) 2.78; c) 4.14; d) 5.16.

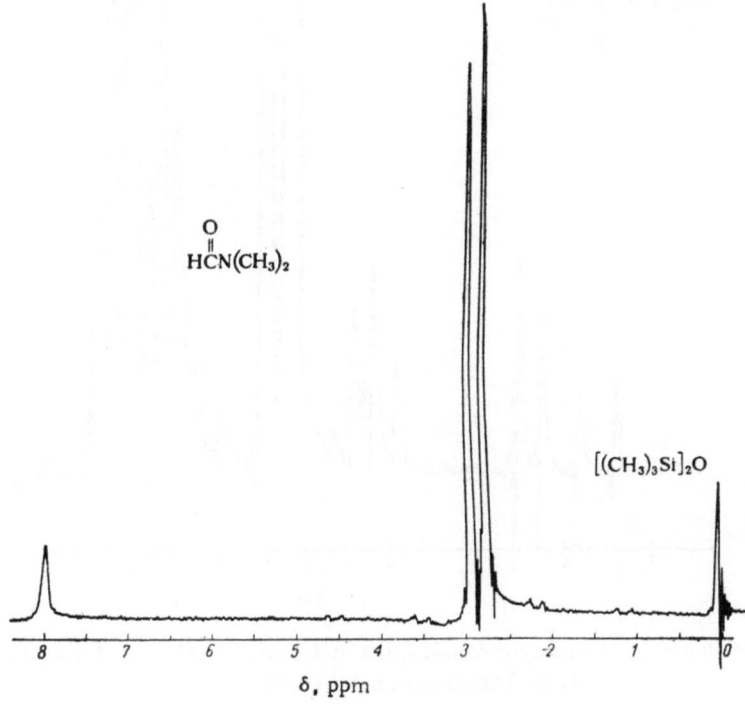

26. Dimethylformamide H.40.ЛCHO.I

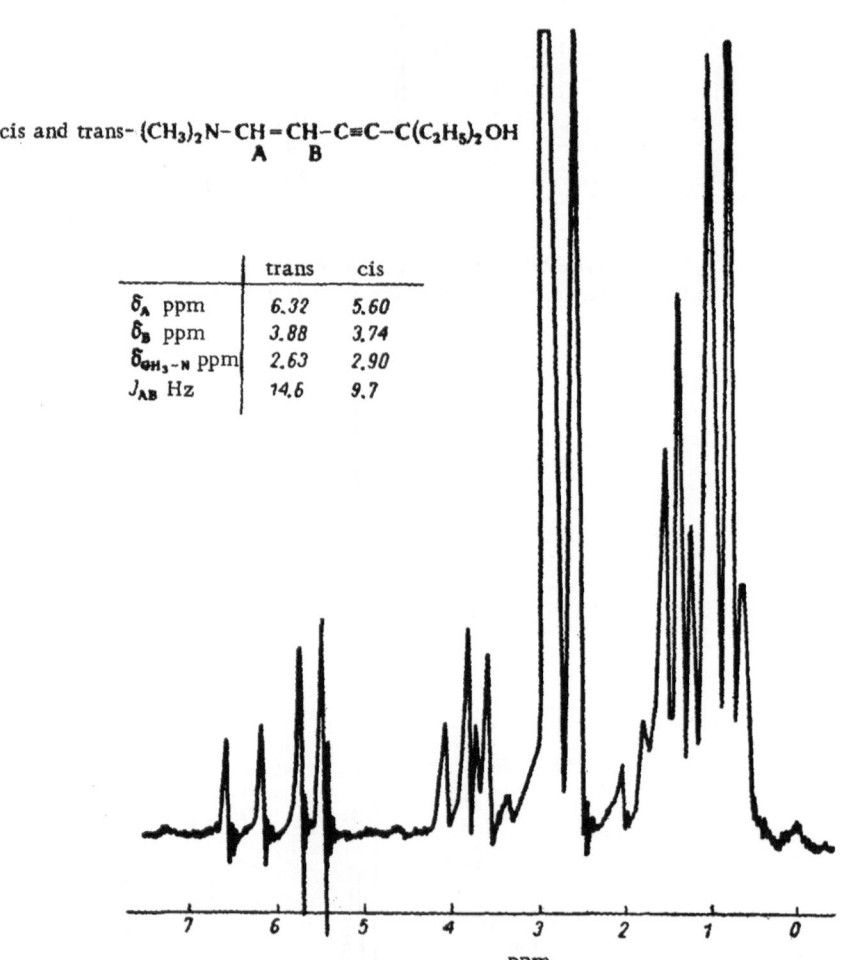

cis and trans- $(CH_3)_2N-CH=CH-C\equiv C-C(C_2H_5)_2OH$
                                    A   B

	trans	cis
δ_A ppm	6.32	5.60
δ_B ppm	3.88	3.74
δ_{CH_3-N} ppm	2.63	2.90
J_{AB} Hz	14.6	9.7

27. 1-Dimethylamino-5-methylhepten-1-yn-3-ol-5 (mixture of cis- and trans-isomers).
H.40.ЛAlk,OH,=,≡.AB+I.Bm

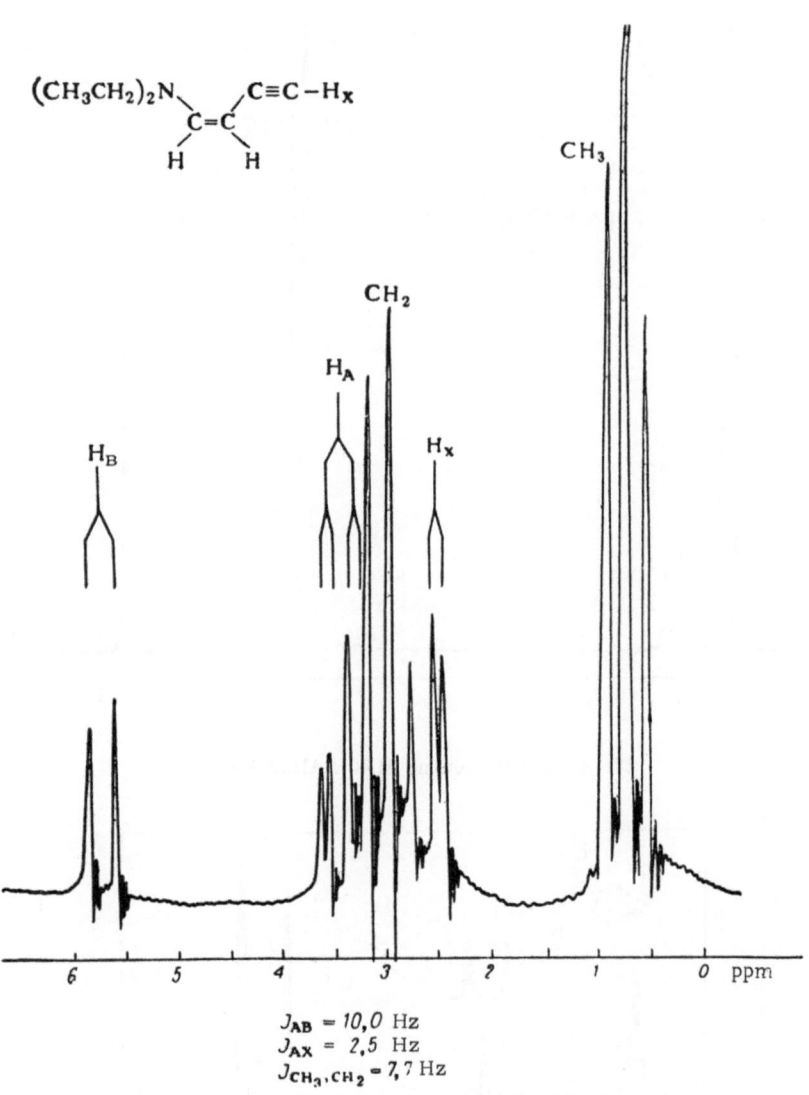

$$J_{AB} = 10,0 \text{ Hz}$$
$$J_{AX} = 2,5 \text{ Hz}$$
$$J_{CH_3, CH_2} = 7,7 \text{ Hz}$$

28. cis-1-Diethylaminobuten-1-yne-3 H.40.ЛAlk,=,≡.ABX+I.Bᴴ

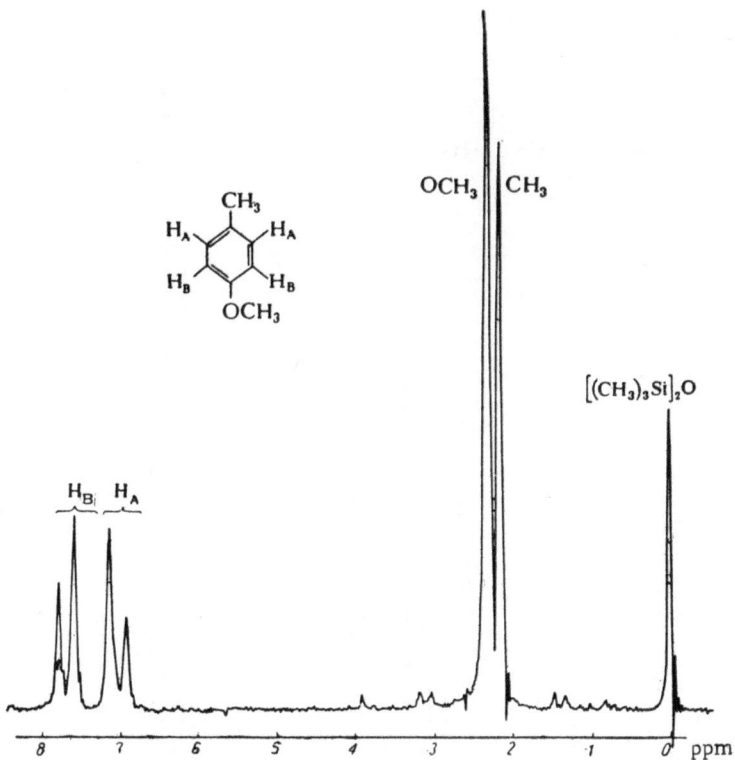

29. 4-Methoxytoluene H.40.AAlk.AABB+I.

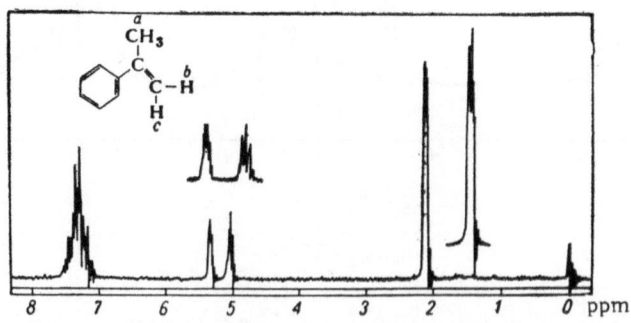

30.* α-Methylstyrene H.60.A=,Алк.М.Хл;
a) 2.12; b) 5.05; c) 5.36.

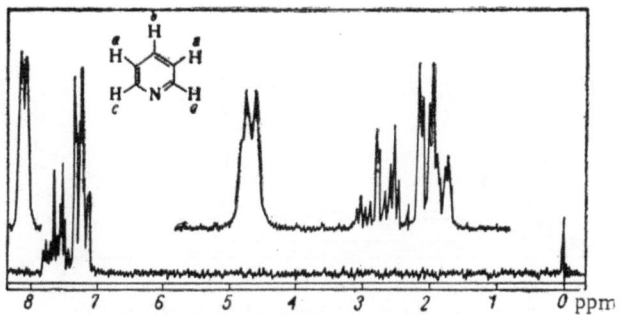

31.* Pyridine **H.60.A.M.Xл** : a) 7.00; b) 7.60; c) 6.60.

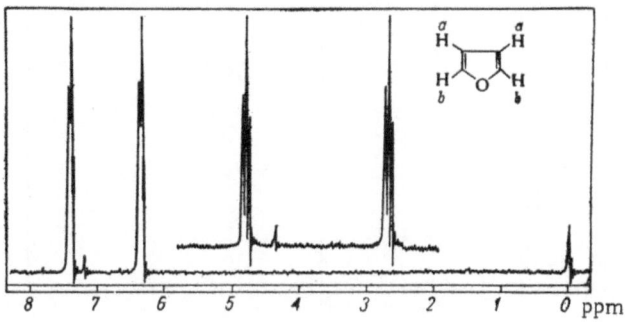

32.* Furan **H.60.A.AABB.Xл** a) 6.37; b) 7.42.

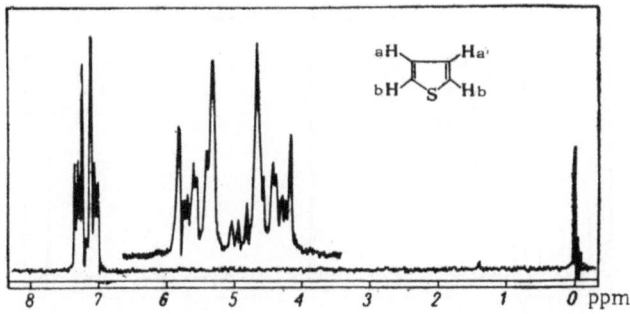

33.* Thiophene **H.60.A.AABB.Xл** : a) 7.10; b) 7.30.

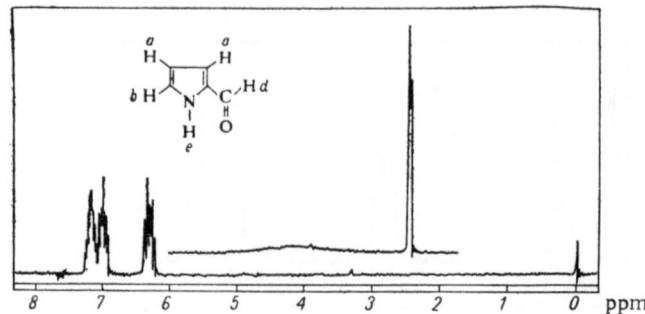

34.* Pyrrole-2-aldehyde **H.60.ACHO.M.Xл:**
a) 6.30; b) 6.98; c) 7.17; d) 9.45; e) ~11.08.

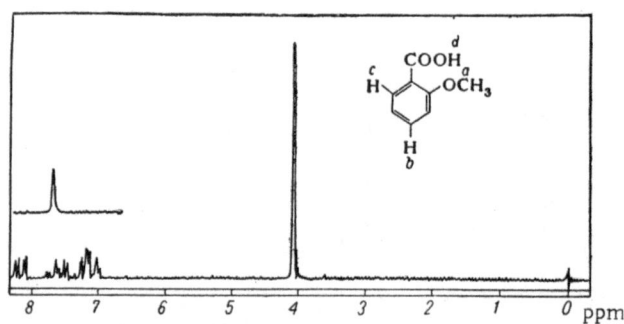

35.* o-Methoxybenzoic acid **H.60.AAlk,COOH.ABC+ I.Xл:**
a) 4.07; b) 7.60; c) 8.17; d) 11.00.

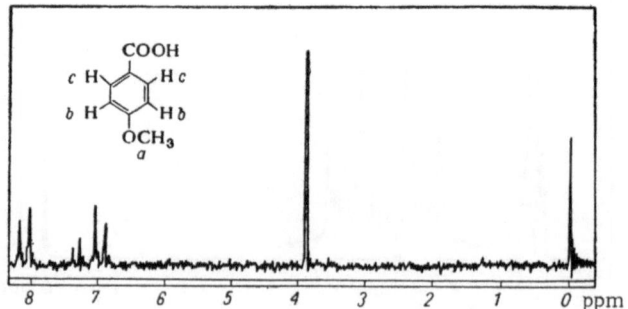

36.* o-Methoxybenzoic acid **H.60.AAlk,COOH.AABB+**
I.Xл; a) 3.88; b) 6.98; c) 8.08.

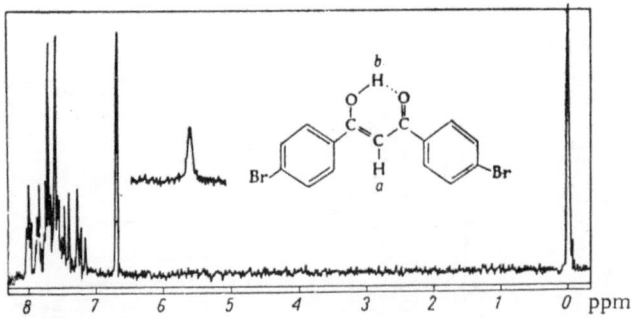

37.* 1,3-Bis(p-bromophenyl)propen-1-ol-1-one-3 [1,3-bis-
(p-bromophenyl)propanedione-1,3] H.60.AAlk,=,OH.AABB + I.Xл:
a) 6.70; b) 16.61.

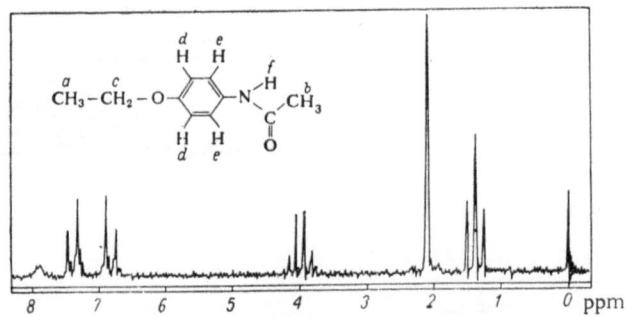

38.* Phenacetin (p-ethoxyacetanilide) H.60.AAlk NH.AABB+
+I.Xл: a) 1.38; b) 2.12; c) 4.00; d) 6.83; e) 7.41; f) 7.91.

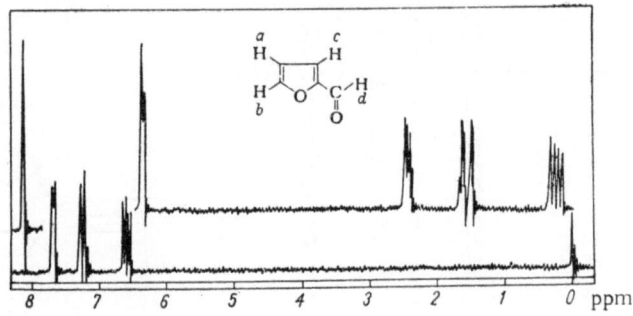

39.* Furfural (furan-2-aldehyde) H.60.ACHO.I.Xл:
a) 6.63; b) 7.28; c) 7.72; d) 9.67.

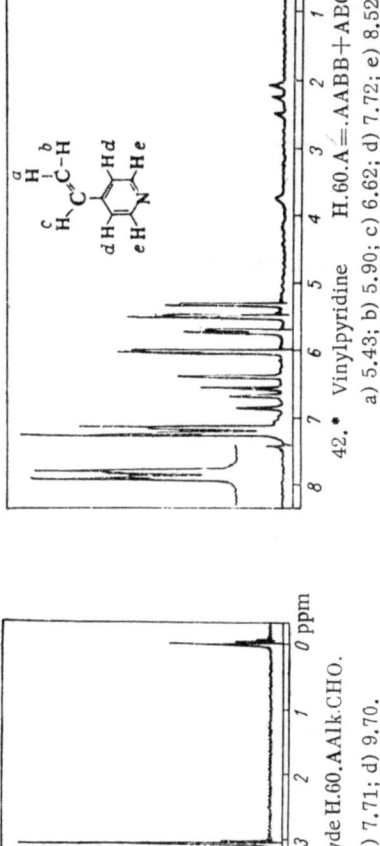

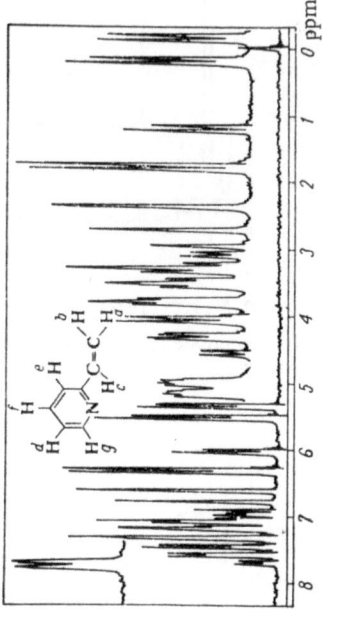

42.* Vinylpyridine   H.60.A=.AABB+ABC.Хл:
a) 5.43; b) 5.90; c) 6.62; d) 7.72; e) 8.52.

43.* 2-Vinylpyridine .H.60.A=.ABC+M.Хл;   a) 5.45;
b) 6.22; c) 6.75; d) 7.08; e) 7.27; f) 7.55; g) 8.52.

40.* p-Dimethylaminobenzaldehyde H.60.AAlk.CHO.
AABB.Хл: a) 3.05; b) 6.69; c) 7.71; d) 9.70.

41. 5-Methylhexen-1-yn-3-ol-5
H.40.ЛAk.=,OH.ABC+I.

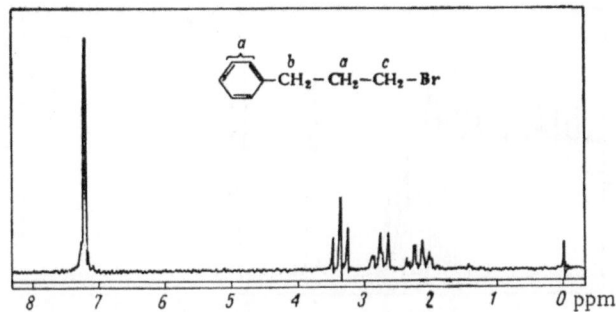

44.* 1-Bromo-3-phenylpropane H.60.AAlk.I.Xл:
a) 2.15; b) 2.75; c) 3.38; d) 7.22.

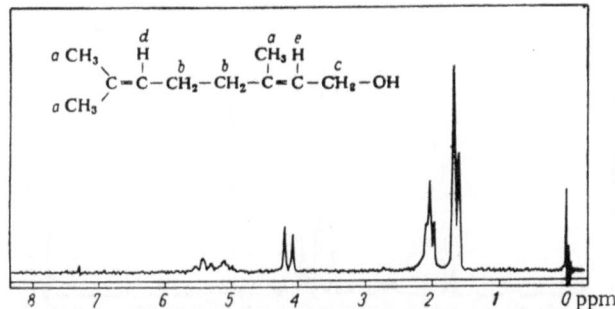

45.* Geraniol (3,7-dimethyloctadiene-2,6-ol-1)
H.60.ЛAlk,OH,=.M+I.Xл : a) 1.62 or 1.68; b) 2.05;
c) 4.15; d) 5.12; e) 5.45.

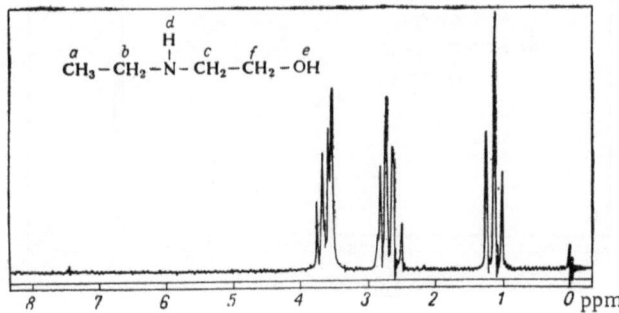

46.* 2-Ethylaminoethanol H.60.ЛOH,NH.I.Xл

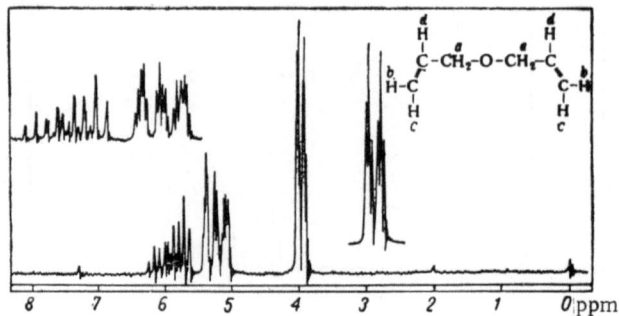

47.* Diallyl ether H.60.Л=.M.Хл;
a) 3.97; b) 5.17; c) 5.25; d) 5.88.

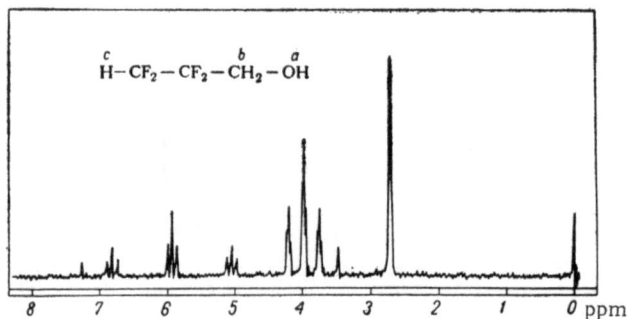

48.* 2,2,3,3-Tetrafluoropropanol HF.60.ЛOH,Z.I.Хл;
a) 2.72; b) 3.97; c) 5.93.

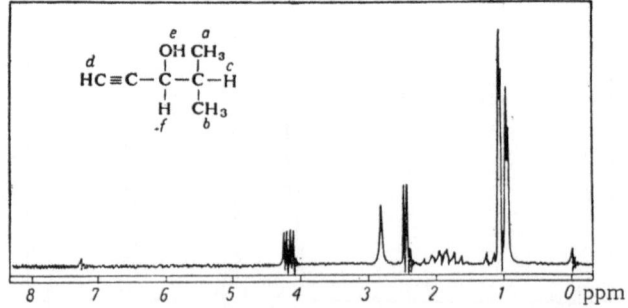

49.* 4-Methylpentyn-1-ol-3 H.60.ЛOH,≡,Alk.I.Хл:
a) 1.00; b) 1.02; c) 1.85; d) 2.47; e) 2.82; f) 4.18.

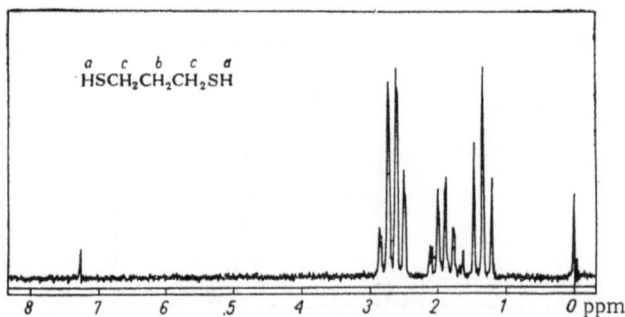

50.* Trimethylene dithioglycol H.60.ЛН.I.Хл:
a) 1.35; b) 1.88; c) 2.68.

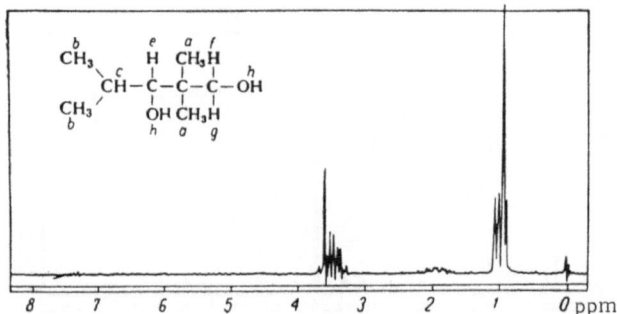

51.* 2,2,4-Trimethylpentanediol-1,3 H.60.ЛОН.I.Хл:
a) 0.92; b, c) 0.95 or 1.00; d) 1.92; e) 3.38; f, g) 3.44
or 3.53; h) 3.58.

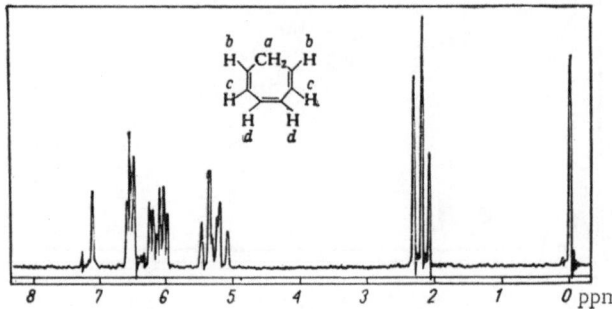

52.* Cycloheptatriene (the sample contained some toluene)
H.60.Ц=.М.Хл: a) 2.20; b) 5.28; c) 6.12; d) 6.55.

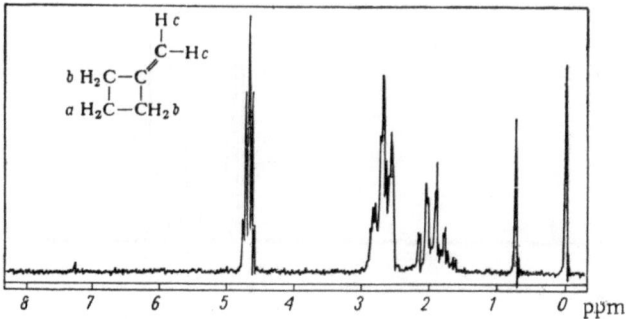

53. * Methylenecyclobutane 60.H.Ц=.AB+M.Xл:
a) 1.92; b) 2.70; c) 4.70.

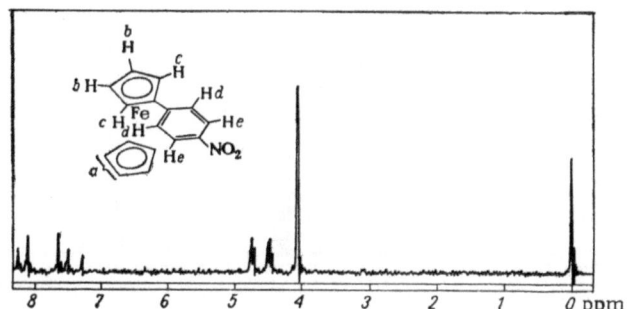

54. * p-Nitrophenylferrocene   H.60.A.AABB+I.Xл:
a) 4.07; b) 4.50; c) 4.77; d) 7.62; e) 8.18.

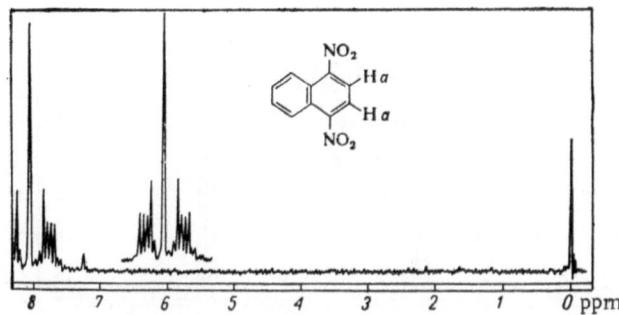

55. * 1,4-Dinitronaphthalene H.60.A.AABB+I.Xл:
a) 8.07.

Index